China's Embedded A

China has been remarkable in achieving extraordinary economic transformation, yet without fundamental political change. To many observers this would seem to imply a weakness in Chinese civil society. However, though the idea of democracy as multitudes of citizens taking to the streets may be attractive, it is simultaneously misleading as it disregards the nature of political change taking place in China today: a gradual shift towards a polity adapted to a pluralist society. At the same time, one may wonder what the limited political space implies for the development of a social movement in China. This book explores this question by focusing on one of the most active areas of Chinese civil society: the environment.

China's Embedded Activism argues that China's semi-authoritarian limitations on the freedom of association and speech, coupled with increased social spaces for civic action, have created a milieu in which social activism occurs in an *embedded* fashion. The semi-authoritarian atmosphere is restrictive of, but paradoxically also conducive to, nationwide collective action with less risk of social instability and repression at the hand of the governing elite.

Rich in case studies about green activism and environmental civic organizations in China, and written by a team of international experts on social movements, NGOs, democratization, and civil society, this book addresses a wide readership of students, scholars and professionals interested in development, geography and environment, political change, and contemporary Chinese society.

Peter Ho is Professor in International Development Studies and Director of the Centre for Development Studies (CDS), University of Groningen. **Richard Louis Edmonds** is Visiting Professor in the Geographical Studies Program and Member, Center for East Asian Studies, University of Chicago.

Routledge studies on China in transition
Series Editor: David S.G. Goodman

1 **The Democratisation of China**
Baogang He

2 **Beyond Beijing**
Dali Yang

3 **China's Enterprise Reform**
Changing state/society relations after Mao
You Ji

4 **Industrial Change in China**
Economic restructuring and conflicting interests
Kate Hannan

5 **The Entrepreneurial State in China**
Real estate and commerce departments in reform era Tianjin
Jane Duckett

6 **Tourism and Modernity in China**
Tim Oakes

7 **Cities in Post Mao China**
Recipes for economic development in the reform era
Jae Ho Chung

8 **China's Spatial Economic Development**
Regional transformation in the Lower Yangzi Delta
Andrew M. Marton

9 **Regional Development in China**
States, globalization and inequality
Yehua Dennis Wei

10 **Grassroots Charisma**
Four local leaders in China
Stephan Feuchtwang and Wang Mingming

11 **The Chinese Legal System**
Globalization and local legal culture
Pitman B. Potter

12 **Transforming Rural China**
How local institutions shape property rights in China
Chi-Jou Jay Chen

13 **Negotiating Ethnicity in China**
Citizenship as a response to the state
Chih-yu Shih

14 **Manager Empowerment in China**
Political implications on rural industrialisation in the reform era
Ray Yep

15 **Cultural Nationalism in Contemporary China**
The search for national identity under reform
Yingjie Guo

16 Elite Dualism and Leadership Selection in China
Xiaowei Zang

17 Chinese Intellectuals Between State and Market
Edward Gu and Merle Goldman

18 China, Sex and Prostitution
Elaine Jeffreys

19 The Development of China's Stockmarket, 1984–2002
Equity politics and market institutions
Stephen Green

20 China's Rational Entrepreneurs
The development of the new private business sector
Barbara Krug

21 China's Scientific Elite
Cong Cao

22 Locating China
Jing Wang

23 State and Laid-Off Workers in Reform China
The silence and collective action of the retrenched
Yongshun Cai

24 Translocal China
Linkages, identities and the reimagining of space
Tim Oakes and Louisa Schein

25 International Aid and China's Environment
Taming the yellow dragon
Katherine Morton

26 Sex and Sexuality in China
Edited by Elaine Jeffreys

27 China's Reforms and International Political Economy
Edited by David Zweig and Chen Zhimin

28 Ethnicity and Urban Life in China
A comparative study of Hui Muslims and Han Chinese
Xiaowei Zang

29 China's Urban Space
Development under market socialism
T.G. McGee, George C.S. Lin, Mark Y.L. Wang, Andrew M. Marton and Jiaping Wu

30 China's Embedded Activism
Opportunities and constraints of a social movement
Edited by Peter Ho and Richard Louis Edmonds

China's Embedded Activism

Opportunities and constraints of a social movement

Edited by Peter Ho and Richard Louis Edmonds

LONDON AND NEW YORK

First published 2008
by Routledge
2 Park Square, Milton Park, Abingdon, Oxon OX14 4RN

Simultaneously published in the USA and Canada
by Routledge
711 Third Avenue, New York, NY 10017.

Routledge is an imprint of the Taylor & Francis Group, an informa business

First issued in paperback 2011

Typeset in Times by Wearset Ltd, Boldon, Tyne and Wear

British Library Cataloguing in Publication Data
A catalogue record for this book is available from the British Library

Library of Congress Cataloging in Publication Data
A catalog record for this book has been requested

ISBN10: 0-415-43374-6 (hbk)
ISBN10: 0-415-66650-3 (pbk)
ISBN10: 0-203-94644-8 (ebk)

ISBN13: 978-0-415-43374-7 (hbk)
ISBN13: 978-0-415-66650-3 (pbk)
ISBN13: 978-0-203-94644-2 (ebk)

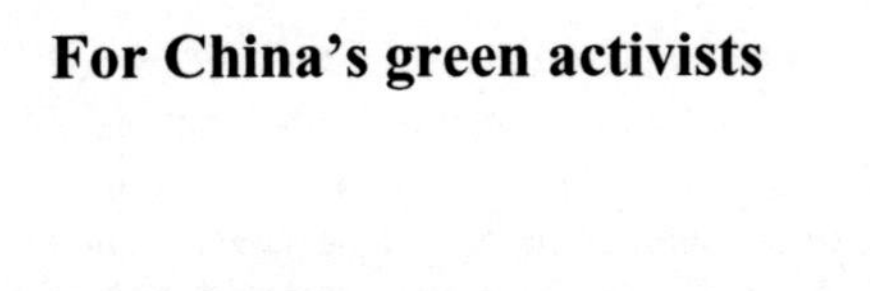

For China's green activists

There was once a man of Song who tilled his field. In the midst of his field stood the stem of a tree, and one day a hare in full course rushed against that stem, broke its neck and died. Thereupon the man left his plough and stood waiting at that tree in the hope that he would catch another hare. But he never caught another hare and was ridiculed by the people of Song. If, however, you wish to rule the people of today with the methods of government of the early kings, you do exactly the same thing as that man who waited by his tree.... Therefore affairs go according to their time, and preparations are made in accordance with affairs.

Han Feizi (233 BC)

Contents

Illustrations

Figures

Tables

Contributors

Anna Brettell has taught at Cornell University and the University of Vermont and is currently a Research Associate at the Harrison Program on the Future Global Agenda at the University of Maryland and a Program Officer at the National Endowment for Democracy. Her interests are Chinese and East Asian politics, international environmental politics and law, and comparative politics. She has published articles and chapters regarding the relationships among economic development, levels of pollution, and public participation in China; Chinese environmental groups; environmental justice and China's complaint and dispute resolution systems; and environmental cooperation in East Asia.

Craig Calhoun is President of the Social Science Research Council and University Professor of the Social Sciences at NYU. He is the author of *Neither Gods Nor Emperors: Students and the Struggle for Democracy in China* (University of California Press, 1994) and *Nations Matter* (Routledge, new edition 2007).

Richard Louis Edmonds is Visiting Professor in the Geographical Studies Program and Associate Member, Center for East Asian Studies, University of Chicago. His interests focus around historical geography and environmental studies. He was editor of *The China Quarterly* from 1996 to 2002. Dr Edmonds has written extensively on China, Japan, Taiwan, Macau, and Hong Kong.

Peter Ho is Professor of International Development Studies and Director of the Centre of Development Studies of the University of Groningen. He has published widely on issues of environmentalism and NGOs, risk and biotechnology, property rights and institutional change, and natural resource management in China and Asia. Over the period 1997 until 2002, he served as the personal Chinese interpreter for the Dutch Minister of Foreign Affairs. He is a frequent advisor for the Dutch and Chinese government, as well as the media, including the PBS and BBC.

Hong Jiang is an Assistant Professor of Geography at University of Hawaii – Manoa. Her research focuses China's human-environmental issues from cultural and historical perspectives. She has worked extensively on Inner Mongolia; her current project examines ecological and ethnic discourses of western China. Her work has appeared in a monograph (*The Ordos Plateau of China*) and numerous journal articles.

Katherine Morton is a Fellow in the Department of International Relations at the Australian National University. Her research interests include: China and the environment; the role and influence of NGOs; and human security. Her book on *International Aid and China's Environment* was published by Routledge in 2006.

Leonard Ortolano is the UPS Foundation Professor of Civil Engineering at Stanford University. Between 2003 and 2006, he was also director of the Haas Center for Public Service and before that he directed Stanford's Program on Urban Studies. His research and teaching focus on environmental planning and policy implementation.

Jiang Ru is an Environmental Specialist in the Rural Development, Natural Resources and Environmental Unit in the East Asia and Pacific Region at the World Bank. His interests focus on environmental management and public participation, environmental movement, implementation of international environmental treaties, hazardous waste management, and natural conservation. Dr Ru is currently working on a number of environmental investment projects in East Asia.

Yanfei Sun is a PhD student in Department of Sociology at University of Chicago. She has completed two articles on the Chinese environmental movement and is currently writing her dissertation on religions in contemporary China.

Maria Tysiachniouk is a chair of the Department of Environmental Sociology at the Center for Independent Social Research in St Petersburg. She is also teaching extensively in Europe and the United States, and has published more than 140 papers. Her interests are in environmental sociology and she is a member of the board of directors of the Research Committee of Environment and Society of the International Sociological Association. She is studying global environmental movements and transnational NGOs and in particular is interested in environmental movements in Russia and China

Guobin Yang is an Associate Professor in the Department of Asian and Middle Eastern Cultures at Barnard College and an affiliated faculty in the Department of Sociology at Columbia University. He has published on a broad range of issues in contemporary China, including the internet and civil society, environmental NGOs, the 1989 student movement, the Red Guard Movement, and collective memories of the Chinese Cultural Revolution.

Jiangang Zhu is an Associate Professor in the Department of Anthropology at Sun Yat-sen University and the director of the Institute for Civil Society. He has published the papers on China civil society, NGOs and urban communities.

Preface

This collection of papers is the final result of two capacity-building projects for environmental NGOs in China – the SENGO project (Strengthening Environmental NGOs in China), funded by Senter International of the Dutch Ministry of Economic Affairs, and the IMPACT project (Improved Participation of Civilians in Transitional China), financed by the Dutch Royal Embassy in Beijing. During the period from 2002 to 2004 when I ran these two projects, I have learnt most about Chinese environmentalism not so much by doing the research, but by living and working together with so many enthusiastic, and dedicated environmental activists. In retrospect, I realize that during these years I've actually been lucky enough to witness the emergence and growth of China's civil society.

The SENGO and IMPACT projects were no easy undertakings as they aspired to build a bridge between Chinese activists, officials and academics. I still remember clearly how difficult it was to start up a dialogue over environmental politics between the different project participants. Yet, the SENGO and IMPACT projects have been a highly rewarding achievement for doing what they intended: bridging people. During the Chinese delegation visit to the Netherlands in the fall of 2003, during the NGO training conducted by Both Ends in Amsterdam, and during the many, many discussions and project meetings, a sense of a common aim and endeavor gradually developed.

Many good and dynamic people have turned the SENGO and IMPACT projects, as well as this book, into a success. For this, I would like to express my heartfelt thanks to (in alphabetical order) Li Lailai of the Institute for Environment and Development, Liang Congjie of Friends of Nature, Huub Scheele and Wiert Wiertsema of Both Ends, Wang Canfa of the Centre for Legal Assistance to Pollution Victims, and Wang Ming of the NGO Research Centre of Tsinghua University. Special thanks go to Liu Guozheng, Deputy Director General of the Directorate General for Environmental Information of the State Environmental Protection Agency, Jiang Xiaoke, member of the Environmental and Natural Resource Committee of the National People's Congress, and Xing Jun, personal secretary of the Deputy Minister of the Ministry of Civil Affairs.

The papers that are brought together in this volume were presented at the final project conference entitled "Shifting Social Spaces in China's Transition? Environmentalism, Public Participation and Popular Protest" which was held at the NGO Research Center of Tsinghua University, Beijing, from 22 until 24 April 2004.[1] After the conference, my co-editor Richard Edmonds and I shaped the papers into their current form through several rounds of revisions with the authors. This book would not have been possible without his support and editorial hand, trained through many years when he served as editor of *The China Quarterly*. The book has also benefited greatly from the detailed and helpful written reviews by Elizabeth Economy, Takwing Ngo, Frank Pieke, Mario Rutten, Mark Selden, Julia Strauss, and two anonymous referees. Special thanks go to David Goodman, who not only commented on several chapters of the book but also supported its publication. Finally, I would like to express my gratitude to Christine Pirenne of the Ministry of Foreign Affairs and Arjen Schutten of Senter International of the Ministry of Economic Affairs for making this research possible.

Peter Ho
Groningen, 2008

Note

1 Some papers originally presented at the conference "Shifting Social Spaces in China's Transition? Environmentalism, Public Participation and Popular Protest" have not been included in this volume, but were – along with the introduction and conclusion – published in a special issue of *China Information*, Vol. XXI, No. 2 (2007), pp. 187–344.

Abbreviations

ASO	Association of Student Organizations
AST	Association of Science and Technology
BAST	Beijing Association of Science and Technology
BCAB	Beijing Civil Affairs Bureau
BROC	Bureau of Regional Public Campaigns (Russia)
CANGO	Chinese Association for NGO Cooperation
CAS	China Academy of Sciences
CAST	China Association of Science and Technology
CBIK	Centre for Biodiversity and Indigenous Knowledge
CCCCP	Central Committee of the Chinese Communist Party
CCP	Chinese Communist Party
CCTV	China Central Television
CITES	Convention on International Trade in Endangered Species of Wild Fauna and Flora
CPPCC	Chinese People's Political Consultative Conference
CYL	Communist Youth League
EPB	Environmental Protection Bureau
EPBHP	Environmental Protection Bureau of Heilongjiang Province
EVAs	environmental volunteer associations
EVAHP	Environmental Volunteers Association of Heilongjiang Province
FON	Friends of Nature
GGF	Global Greengrants Fund
GONGO	government-organized non-governmental organization
GSA	Green Students' Association
GVB	Global Village of Beijing
HIT	Harbin Institute of Technology
HUC	Harbin University of Commerce
ICF	International Crane Foundation
IFAW	International Fund for Animal Welfare
INGO	international non-governmental organization
IREX	International Research Exchange Board
ISAR	Initiative for Social Action and Renewal of the Far East
MOCA	Ministry of Civil Affairs

NFU	Northeast Forestry University
NGO	non-governmental organization
NPC	National People's Congress
NPO	not-for-profit organization network
PAN	Pesticide Action Network
PEAC	Pesticide Eco-Alternative Centre Yunnan Thoughtful Action
PERC	Pacific Environment
RFE	Russian Far East
ROLL	Reproduction of Lessons Learned
RRSSO	Regulation on Registration and Supervision of Social Organizations
SARS	Severe Acute Respiratory Syndrome
SEPA	State Environmental Protection Administration
SEU	Social-Ecological Union
SFA	State Forestry Administration
SNISD	South–North Institute for Sustainable Development
UYO	Upper Yangtze Conservation and Development Organization
USAID	United States Agency for International Development
VOOP	All-Russia Society for the Protection of Nature
WCS	Wildlife Conservation Society

1 Introduction

Embedded activism and political change in a semi-authoritarian context

Peter Ho

> For a Non-Governmental Organization (or NGO) to supervise the government it is necessary that it can see what the government is doing. At present, Chinese NGOs can only see the result of policies; they cannot see the process of policy-making. In that sense, they do not see much more than the common people. Moreover, does an NGO understand what it must do to influence policies? Even so, those who understand also have to consider the present situation. If you cannot survive because you get into conflict with the government, what use does that have? China's political system (zhengzhi tizhi) is different from that of the West. China has a one-party rule, and the development of NGOs and civil society is closely related to that.
>
> (Zhang Tianzhu, Co-drafter of the 10th National Environmental Five-Year Plan, 4 March 2004)

One of the great enigmas of China's reform experience is its economic metamorphosis without fundamental political change. The Catholic Church in Poland, the old Democratic Party in Hungary, and alliances of political and environmental activists in East Germany and Bulgaria, were instrumental in pushing for greater political freedom in East and Central Europe. Moreover, over the past years, broad movements of dissatisfied citizens toppled the governments of Ukraine, Kyrgyzstan and Georgia, popularly dubbed the "color revolutions." However, there is no successful Chinese equivalent for this. In fact, since the end of the Great Proletarian Cultural Revolution, social movements that tried to mobilize the Chinese masses have met a similar fate: suppression and delegitimization. The protesters of the Democracy Wall movement of 1979, the student movement at Tian'anmen Square in 1989,[1] the Chinese Democratic Party and Falun Gong Sect of the late 1990s, have without exception been repressed and/or forced into exile.[2] For many observers, this demonstrates the weakness of Chinese civil society. In 2005 a renowned columnist noted with disappointment that the lack of a nation-wide demand for political change dashes any hopes for democracy in China.[3]

However, the assumption that political change in China must result from a broadly supported social movement is actually a matter of "sexing up" the media debates. Indeed, a sole protester defying a Chinese tank and hordes of students surrounding a “Statue of Liberty” at Tian’anmen Square are forceful images that appeal to a widely imagined idea of how democratic change should come about. Yet, perhaps because of its recent turbulent past of mass campaigns and a longer period of political upheaval, China seems keen to avoid any new revolutions or commit itself to sudden political change. For one thing, it is obvious that since the start of the economic reforms in the late 1970s, profound changes in China’s polity and society have taken place. Many scholars have documented the increased importance of the People’s Congresses at various levels,[4] the development of village democracy,[5] the shifts in citizenship and its perceptions thereof,[6] and the burgeoning of civil society.[7] In this sense, the idea of democracy as multitudes of citizens taking to the streets may be attractive but simultaneously misleading, as it disregards the nature and course of political change that is taking place in China today: a gradual shift towards a polity adapted to an increasingly complex and pluralist society. Having said this, it is critical to ask ourselves the questions: what does the limited political space imply for the development and dynamics of a social movement in China? Is the possibility for, or the actual occurrence of, a social movement a precondition for political change and the development of civil society? Finally, what are the potential perspectives for the emergence of a social movement in China, and how would it relate to international environmental forces?

These are the questions that this edited volume will explore, and it will do so by focusing on one of the earliest and most active areas of civil society in modern China: the environmental realm. By narrowing down on the Chinese case, this volume takes an area studies approach in its theorizing efforts. However, we do not want to argue that the concepts developed here are restricted to China alone. Rather do we believe that China’s society – an apparently restrictive political environment in which rapid socio-economic and cultural changes are taking place – provides an interesting context for the study of social movements. However, further research on social movements in comparable settings such as the Chinese one will definitely be necessary. In this volume, we will argue that China’s semi-authoritarian constellation of formal state limitations on the freedom of association and speech, coupled with increased social spaces for civic and voluntary action has created a milieu in which social movement dynamics features in an embedded fashion.[8] Contrary to a “fully authoritarian” context in which the state wields virtually totalitarian control over society,[9] the semi-authoritarian environment is restrictive of, but paradoxically, also conducive to nation-wide, voluntary collective action with less risk of social instability and repression at the hand of the governing elite.

On the one hand, China’s restrictive political environment prompts social movements to be almost invisible, which they achieve through self-imposed censorship and a conscious de-politicization of environmental politics. Social movements[10] in China are forced to lead a low-profile or semi-clandestine existence.[11] For survival they rely heavily on the Party-state for legitimacy, which restrains them from developing all too intimate linkages with citizens and international donors. Activists need to adopt a non-confrontational strategy that cautiously evades even the slightest hint at organized opposition against the central Party-state. Note, however, that this is much less the case for the local state authorities. Regarding environmental protection, the central authorities and environmentalists often find themselves partners in the same struggle against local government.[12] On the other hand, however, the semi-authoritarian context has created an environment in which the divide between civic organizations, state, and Party is extremely blurred. Contradictorily, this context is conducive to green activism that can be seen from the rapid growth in environmental NGOs over the past years. By establishing informal organizations, façade institutions or "companies," environmental NGOs are capable of circumventing the stringent regulations for NGO registration. In addition, green activists make avid use of informal networking with Party and state officials. Through a web of informal ties, social structures can develop that are capable of effectively mobilizing resources, appealing to citizens' newly perceived or desired identities, and building up a modest level of counter-expertise against state-dominated information on social cleavages and problems – be it labor rights, gender issues, dam-building, or nuclear energy.

This contradictory duality – a semi-authoritarian setting that is restrictive and conducive at the same time – forms the essence of the embeddedness of Chinese social activism. Furthermore, this embeddedness might provide the basis for incremental political changes in China, rather than the overnight revolutions as have occurred elsewhere in the former socialist world. Apart from a discussion of the chapters in this volume, this contribution is divided into three main sections. In the first section below, we will review the literature on the relation between social movements and political change, with particular reference to green activism. The second part of this chapter discusses the historical development of social and green activism in China. This is followed by the final section, which explains the embeddedness of Chinese environmentalism by means of the main concepts on social and green movements.

Change through movement

The various contributions in this volume demonstrate that the dynamics of Chinese green activism and the means to take collective action diverge from what theory describes and predicts, and compel us to rethink certain concepts and ideas. This might be particularly true for the form in which social and green movements inject themselves – widely supported popular movements versus embedded, low-profile activism – as well as the type of political change these can effect – fundamental but abrupt and thus potentially destabilizing changes versus incremental but certain changes.[13] In the international literature on social movements and environmentalism, it is hypothesized – albeit at times not explicitly – that the force of broad social and green movements can impel politicians to adopt new governance styles.

Fundamentally different from a slow, bureaucratic politics of "muddling through," social movements are regarded as vested with the power to effect visible, fundamental change due to their non-institutionalized nature and a close proximity to their grassroots constituencies.[14] The ultimate consequence of social movements can, of course, be regime change, generally pictured as a change from an authoritarian setting to a pluralist, liberal-democratic polity.[15] After the fall of the Berlin Wall in 1989, a large body of literature appeared on the role of environmental movements in democratization processes in states-in-transition.[16] In this view, NGOs and the occurrence of popular movements are seen as a sort of index of the "healthy" development of civil society and an integral component for political change and, eventually, democratization. Writing about East and Central Europe, Cellarius and Staddon noted that "because environmental 'groups' and 'movements' – later environmental NGOs – played such a large role in bringing about the dramatic changes of 1989–1990, these organizations should then be considered the vanguard of democratic transition."[17]

As a variation on the theme of green social movements and NGOs as a lever for political change, we find the studies on environmental reforms at the local, national and transnational levels. In environmental studies of North and Western Europe it is said that the industrial pollution control measures of the 1960s and 1970s suffered from severe implementation defects because policies relied primarily on a "command-and-control" and "end-of-pipe" type of regulation. Instead of mobilizing and stimulating consumers and producers to deal with environmental problems from the grassroots, the state merely "ordered" sustainable development by imposing emission quotas, administrative regulations, and prohibitions. Moreover, sustainable development was regarded as simply a matter of installing environmental technologies at the end of the polluting pipe, rather than effecting environmental restructuring throughout the entire chain of consumption and production processes. Yet, as the environment deteriorated, the traditional top-down environmental policies, gradually yielded to a new and more effective kind of environmental governance.[18]

It is often claimed that because of its non-institutional nature, the environmental movement played a critical role in the shift from an incremental and polluting polity towards this novel governance structure that could better safeguard sustainable development. As Doyle and McEachern noted:

> Environmentalism in all its forms, was born in environmental movements.... They occupy a political terrain that is often quite separate from more established institutionalized political forms such as pressure groups, parties, and the administrative and parliamentary systems of the state. It was within these non-institutional, more informal realms of society and its politics that environmental movements emerged. It is safe to say that without the environmental movements there would be little or no "greening" of government and corporations.[19]

Also, the writings on a "global civil society" and "transnational environmentalism" that have forced large international corporations into positive action, such as happened over the controversy of the Brent Star oil platform owned by Royal Dutch Shell, are clear illustrations of a similar line of argumentation.[20]

However, in this volume we wish to steer away from the more normative assumptions on green social movements, NGOs and civil society. Instead we wish to go back to try and acquire an understanding of the process of how green social movements maneuver within differing political contexts. In a different wording, *whether* green social movements and NGOs succeed in bringing about political change or sustainable development is of less importance to us than *how* they attempt to do this. Before turning to an analysis of Chinese environmentalism along this line, it is necessary to provide some background information on its development in the near past.

Green activism in China

The collective period (1956–78) in Communist China was characterized by a virtual absence of civil society. The pre-revolutionary civic organizations – the guilds, the native-place associations, secret societies, clans, temple and peasant organizations in the countryside – were soon phased out after the Communist take-over in 1949. The "mass organizations" that were established in later years and appeared "non-governmental," such as the Women's Federation, the Communist Youth League, and trade unions, were in fact outposts of the Communist Party in society.[21] However, the economic reforms that started in the late 1970s unleashed great social changes initially unforeseen by the government.[22] Recently, a great variety of voluntary associations and non-governmental organizations (termed "social organizations" or *shehui tuanti* in Chinese) have sprung up in China. The number of officially registered social organizations rose from 100 national and 6,000 regional organizations in 1965 to over 1,800 national and 165,600 regional ones at the close of the past century.[23]

The explosive growth of social organizations is due to the cost-cutting retrenchment of the state from society. The rising complexity of the economy has prompted the government to cut back on its expenditures. Moreover, the rapid increase of the unemployed after a major restructuring of state industries and an axing of the state bureaucracy since the late 1990s have put an increasing strain on government resources. As a result, central and local state institutions have been privatized or disbanded altogether, opening up social spaces for voluntary civic action that were formerly in state hands, such as social welfare services, legal counseling, and cultural activities.[24] The central Party-state has also grown more sensitive of its own reach and capacity, and the need for civil society. High officials have called for "intermediary social organizations" to become active in the space between state and society.[25] However, at this point it is necessary to note that the Chinese state's notion of civil society and NGOs (or non-governmental organizations) is quite different from the Western concept of voluntary organizations that can protect citizens' rights or counterbalance state power. In fact, as Goldman noted, the Chinese government is willing "to grant political rights, not to recognize them. Moreover, political rights are to enable citizens to contribute to the state rather than to enable individuals to protect themselves against the state."[26] Notwithstanding this difference in the interpretation of civil society, many civic organizations with varying degrees of independence from the state have emerged in China today.

During the early 1990s it already had become painfully clear that without fundamental restructuring China would be heading towards an environmental debacle.[27] The high political priority given to environmental protection, lent environmental activism a certain legitimacy. Moreover, in contrast to civic associations organized along religious and ethnic – let alone political – lines, organizations that engaged in environmental protection were regarded as "politically less harmful." They could be employed to assist the state in achieving sustainable development goals. In some of the transition countries of East and Central Europe, environmentalism was allowed to flourish for similar reasons. Therefore, one of the earliest and also most dynamic sectors of civil society in China has been the environmental realm.[28] Against this backdrop, it is important to study Chinese civil society through the various forms in which environmentalism has manifested itself, including green NGOs, environmental protests, citizens' complaints, and residents' movements.

The critical question, naturally, is whether the hordes of protesters that demonstrated against environmental ills in East Germany, Hungary and Bulgaria can be likened to the semi-professional green organizations and loose, informal groups in China that engage in media-attractive but localized protests, voluntary tree-planting and waste collection campaigns. After all, the main feature of Chinese green activism that stands out to date is its absence of a broadly supported, popular *movement* against the established order. Even more so, organizations lack any desire to openly confront the central authorities, and instead remain relatively small, fragmentary and localized.

Environmental activism in China is not an activity with a fair degree of autonomy and self-regulation, but occupies a social space that is enmeshed in a web of interpersonal relations and informal/formal rules between political and social actors. Yet, different from a situation in which activism is merely repressed, the Chinese embedding conditions both limit formal environmental organization and make it possible. These embedding ties that can successfully cross the divides between the Party-state and society have enabled environmental activism to play an increasingly critical role in the greening of industries, the government and consumer life styles.[29] The various contributions in this volume are a clear testimony to this. In order to understand the embeddedness of green activism in the Chinese semi-authoritarian context, we need to analyze it in terms of its form of organization and the tactics it employs.

From grievances to mobilization

Many theorists have attempted to unravel the driving forces of social movements. The *primus inter pares* of social movement theorists is of course Karl Marx, concerned as he was with the mobilization of the proletariat against capitalist entrepreneurs. After him, many others have delved into the question of collective action in general, and social movements in particular.[30] In the proliferating literature about social movements, several recurring themes can be discerned: (i) the potential and current cleavages and conflicts in society that lead to popular dissatisfaction; (ii) the movement's overarching "ideology" and individuals' identification with that body of thought, beliefs and values; (iii) the organization of the movement and its capacity to muster financial, personnel and material resources; and (iv) the specific chances for collective or political action provided for the social movement – in other words, the "windows of opportunity." In the theories on social movements these issues are generally known under the notions of grievance, issue framing, resource mobilization, and political opportunities and constraints.[31]

Grievances

The first precondition for any activism is the presence of areas of social conflict, be it around economic, social, or environmental issues. One of the questions that has worried many foreign and domestic observers is the emergence of social cleavages in Chinese society. The economic reforms have given rise to a rapidly widening gap between poor and rich, a rising unemployment, and problems of governance as state capacity fails to keep pace with socio-economic developments. As a result, social conflict and popular protests have occurred more frequently over the past years.[32] A survey conducted by the China Statistical Bureau in 2002 found that the issues of greatest concern for Chinese leading officials included job lay-offs, corruption, education, income and housing, and ... the environment.[33]

China's economic boom has been accompanied by serious air, water, soil and noise pollution. Although the central government has been relatively quick in dealing with the environmental challenge, the initial positive effects of environmental policies have been completely offset against the sheer rise in the scale of production and consumption.[34] Major lakes, such as the Dianchi, Chaohu and Dongting lakes, have been heavily polluted by local industries and neighboring farms; in one-third of the main cities the air is seriously polluted; while the 24-hour economy and its construction activities frequently disrupt people's lives (particularly their night's rest). China's dismal environmental record is clearly reflected in the increase in popular discontent. For instance, Brettell shows in this volume that over the period 1990–2000, the number of people who complained about environmental problems rose 175 percent from 140,681 to 387,165. Over the same time span, the number of incidents which prompted complaints rose by almost 180 percent from 111,359 to 309,800.[35] For this reason, environmental activism finds fertile ground in China.

Issue framing

In the framing of environmental issues, there is one noteworthy feature of Chinese green activism that we might term the "de-politicization of environmental politics." This sounds like a contradiction in terms because if there is one thing that the East and Central European experiences have proven, it is that the environmental question is inseparably linked to politics. However, in China's semi-authoritarian context the overall majority of green activists stays clear from any suggestion that any political objectives are involved even though inevitably there is a political aspect within each movement. This non-confrontational strategy implies that: (i) green activists portray themselves as partners, rather than opponents of the central authorities; (ii) they skirt certain sensitive environmental questions, such as nuclear energy and agricultural biotechnology;[36] and (iii) they avoid any connotation with broad, popular movements, for instance, by reporting lower membership numbers to authorities than they actually have. Some observers dubbed this *self-imposed censorship* the "female mildness" of Chinese environmentalism. As one of China's foremost environmental activists, Liao Xiaoyi, the founder of the Beijing-based NGO Global Village, professed:

> We guide the public instead of blaming them and help the government instead of complaining about it. This, perhaps, is the "female mildness" referred to by the media. I don't appreciate extremist methods. I'm engaged in environmental protection and don't want to use it for political aims. This is my way, and my principle too.[37]

However, this self-imposed censorship does not really hamper green activism, but rather enables it. Put differently, as long as you don't openly oppose the central state, many things are possible in China. For one thing, environmental NGOs and activists are relatively successful in attracting the state's attention and getting certain issues on the political agenda. The vivid account of an urban residents' movement by Zhu and Ho in this volume demonstrate this political leverage. Of course, this is also due to the high political priority that the Chinese state attaches to solving the environmental question. In addition, Chinese citizens' environmental awareness and the perceived urgency for environmental protection is significant. In a survey conducted in 1999 it was found that environmental protection ranked fifth among issues of greatest concern by respondents.[38] Another survey, done in 2002, found that approximately 63 percent of Chinese respondents deemed environmental protection "extremely important."[39] The high environmental awareness among Chinese people can mostly be attributed to the government's vigorous campaigns. In addition, "the environment" has also become a high-profile issue in the media. In this volume, Calhoun and Yang note a proliferation of "greenspeak" in Chinese society, as can be witnessed in newspapers, television and radio, and particularly the internet. Green NGOs and voluntary groups make avid use of the popularity of the environment as a social issue, to push it further on the agenda of politicians and decision-makers.

In terms of framing, it is interesting to look at the potential links between environmentalism and traditional Chinese religion and culture. For instance, making reference to Chinese folk religion has greatly assisted environmental activists in furthering their political cause in Taiwan.[40] During the eventful year of 1987, environmental protests against the China Petrochemical Plant in Kaohsiung City clearly referred to Buddhist and Taoist symbols. Environmental activists claimed they had asked support from Guanyin, the Goddess of Mercy, and Shen Nong, the God of Agriculture, to bless their demonstrations. In addition, a traditional spirit altar had been erected at the gate of the factory. When the police arrived in an attempt to break the blockade, they were driven away by groups of traditional martial artists clad in traditional Chinese attire and armed with cudgels, broadswords, and spears. The alliance between environmentalists and religious institutions proved successful in many ways: the local temple associations helped to mobilize their people against the authorities, paid substantial sums of money to bail out those who had been arrested, and in fact greatly facilitated the protests in Taiwan, as "religious parades" usually receive formal approval in contrast to public demonstrations.[41] In the current context, it seems unlikely that activists in mainland China will make widespread use of religious symbols in the framing of environmental issues.[42] Particularly, the recent incidents around the Falun Gong sect and its subsequent repression have turned such actions into a sensitive issue. On the other hand, there are signs that certain green groups carefully organize themselves around traditional religious lines as well.[43]

Opportunities, constraints, and mobilization

The manner in which green NGOs can muster financial, material and personnel resources in the Chinese context is bound up with political opportunities and constraints. For this reason, we will deal with these parameters in a joint discussion. As is discussed in detail in the contribution by Ru and Ortolano in this volume, NGOs face a restrictive institutional and political environment in China. NGOs currently have to register with the Ministry of Civil Affairs or its subordinate institutions, as well as with a sponsoring institution (*zhuguan bumen*).[44] This two-tiered administrative system of control (in Chinese: *shuangchong guanli tizhi*) has often proven a bottleneck for NGOs and voluntary associations. Rather than going through these troublesome and protracted registration procedures, NGOs and voluntary associations opt to avoid registration or register as a legal entity, which they are not, such as a research institute or a company. Green NGOs that have failed to formally register are by law not regarded as a legal person, and thus not entitled to an independent financial account or to sign contracts on their own accord.

Furthermore, as shown elsewhere in this volume the recent "color revolutions" in the Ukraine, Georgia, and Kyrgyzstan, have led to an increase in state control over NGOs and voluntary organizations in China.[45] A recent illustration of this is the case of Green Watch, an environmental group founded by a certain Tan Kai in the municipality of Hangzhou, Zhejiang province. Green Watch was established as an informal organization after successful protests mounted by farmers against a polluting chemical plant in Huaxi, near Dongyang City. Tan Kai and five other activists were detained on 19 October after opening a bank account to collect funds for their cause. By doing so, Tan Kai had committed an illegal act because he is not allowed to fundraise unless the organization is officially registered, which requires a deposit of 100,000 yuan for a national and 30,000 yuan for a local organization – a substantial sum for most Chinese. According to Stacy Mosher, communications director of the New York branch of Human Rights in China, this poses a catch-22 situation as "unless you have very deep pockets to begin with, you have no way of reaching out to local or foreign organizations who might want to contribute."[46]

At the same time, however, "windows of opportunity" have clearly opened up for Chinese environmental groups. The China of the 1980s is definitely not the same China today, due, first, to what I have elsewhere termed the "greening of the Chinese state" over the past years – as is visible in the proclamation of an impressive body of environmental laws and regulations, and the strengthening of the environmental bureaucracy.[47] Green NGOs have strongly profited from this trend. Second, despite the fact that the Chinese Communist Party still rules supremely, many social areas that were closed off from political activities have gradually become accessible for citizens, including labor issues, poverty alleviation, and legal protection. Moreover, the internet revolution has heralded profound changes in the available channels for airing certain political views and popular discontent.[48] In this regard, the crucial question is how one can make use of these windows of opportunity, and here the issue of embeddedness surfaces once more.

The distinction between Party-state versus civil society in China is a blurred one, indeed. "True" environmental NGOs establish themselves as entities that they are not, such as companies, informal salons, or research institutes. In addition, the state also sets up its own environmental NGOs: the Chinese phenomenon of "government-organized NGOs" or GONGOs. Even more, as is demonstrated in the second chapter, some environmental groups are part of a nested structure of Party institutions that – if necessary – can exert strict control over their activities. Some note that

> the lack of separation between the government and society affects the development of grassroots organizations. To obtain the support or recognition from the government they readily accept administrative intervention. Many scholars have criticized that NGOs in China lack a non-governmental character and are not worthy of the name NGO.[49]

However, the blurred divide can also work to the advantage of NGOs if they know how to make use of it.

By developing informal ties with the central (and local) Party-state, environmental NGOs have managed to gain considerable political leverage and maneuvering space. At this point, we touch on the crucial Chinese notion of *guanxi* (literally meaning: "relation"), a complex Chinese notion that is inextricably connected to family ties, patron–client relations, and the art of gift-giving.[50] Through their *guanxi*, NGO leaders and politicians are tied together in a symbiotic situation in which they are mutually dependent on each other: the NGOs need the legitimacy and political influence granted by the central state, while the central state relies on the NGOs for their contacts with society,[51] and for exerting pressure on local authorities and polluting industries. Furthermore, in the case of the more prominent but also successful environmental NGOs, we often see that their leadership is part and parcel of a structure in which environmentalism is embedded in the Party-state. For example, Liang Congjie of Friends of Nature was until recently a member of the People's Political Consultative Conference; Wang Canfa of the Center for Legal Assistance to Pollution Victims is a frequent government advisor and co-drafter of national environmental legislation; and Jiang Xiaoke of the Beijing Environmental Protection Foundation was a member of the National People's Congress (NPC) and concurrent commissioner of the NPC Commission on Environmental and Natural Resources Protection. This commission is the highest state organ responsible for policies on environmental protection and natural resource management, supervises the State Environmental Protection Administration, and assists in the drafting of new environmental laws and regulations.

Rather than limiting their autonomy vis-à-vis the state, these embedding ties have enabled green activists to play an increasingly critical role in China's environmental governance. Against this backdrop, a scholar noted that even the "environmental Government-Organized NGOs have come to realize their own organizational missions, negotiate with the state for more self-governance, and facilitate trans-societal cooperation." In fact, they are "situated in between the state and society, and as a result they can influence the formation of new collective identities and political coalitions."[52]

The contributions

The collection starts with a contribution by Ho who examines the meaning of Chinese embedded activism by reviewing Party and state interventions, as well as the responses these incite by those attempting to explore the limits of the newly opened-up social spaces. The Party-state interventions have created a blurred and complex institutional environment in which authorities can by and large monitor and control civic activity if deemed necessary. A case in point are the green student associations at Chinese universities that are part and parcel of the Party system. Through an intricate system of Party and administrative cadres with double positions, green student associations are fully integrated in university Party structures. On the other hand, Ho demonstrates that environmental activists have gained firm ground through a combined tactics of self-imposed censorship, a conscious *de-politicization* of politics, and a reliance on informal strategies and relations. Ru and Ortolano examine the difference between the Chinese state's policy to control environmental NGOs, and the implementation of this policy. The authors study the state's intent to employ a rigorous corporatist structure based on granting monopoly representation to NGOs. Based on a study of a sample of eleven national and eleven Beijing municipal environmental NGOs, it is argued that state controls have been implemented as "agency control" or "no explicit control." In this context, agency control refers to control exerted by government agencies that sponsor and supervise NGOs. The difference between the state's intended and actual control over NGOs resulted because of a decrease in state administrative capacity, the self-interests of agencies implementing state controls, the existence of NGO leaders well connected to officials, and self-censorship employed by NGOs. This chapter is followed by a contribution by Calhoun and Yang who analyze one of China's foremost anti-dam campaigns. In early 2004, however, public controversies surrounding dam-building on the Nu River in southwest China prompted the government to halt a proposed hydropower project. The occurrence of such public debates indicates the rise of a critical green public for whom environmental non-governmental organizations are the main social actors in producing and disseminating their views. Calhoun and Yang describe how official mass media, the internet, and "alternative media" contribute to this green sphere in different ways.

In the fifth chapter we turn to the historical aspects of state–society relations with a contribution by Jiang, who moves ahead in time to Communist China's collective era (1958–late 1970s), where she takes a Mongolian community in Inner Mongolia as a case to explore local strategies employed by the Mongol herders in response to state grassland policies, arguing for the important role of local agency in such endeavors. Such agency was not expressed in the form of organized protests but through subtle resistance to and creative use of state grassland policies. Jiang affirms the use of local agency and a non-dichotomous approach to study state–society relationships. Her exploration not only places the post-Mao environmental actions in a historical context, but also helps inform our understanding of current grassroots environmental strategies in China.

Shifting back to modern times, Brettell tells us that Chinese citizen environmental complaints increased dramatically between 1990 and 2001. Her quantitative analysis of these complaints, aggregated at the provincial level, suggests that pollution, economic, and environmental awareness are correlated with variation across time and location, accounting for an average of 81 percent of the fluctuation in complaints. Brettell also includes a qualitative analysis which shows that official encouragement of citizen complaints and increasing official responsiveness have contributed to the rise in complaints. While encouraging citizens to complain, authorities have been able to channel dissent by standardizing and institutionalizing the complaint system, thereby bolstering legitimacy, reigning in errant local authorities, and heading off potential social instability. While grievances have increased, there is not a thriving anti-pollution movement – one of the most remarkable features of Chinese green activism. Instead, we see isolated, spontaneous, and highly localized protests that increasingly are turning violent but remain localized. Groups of aggrieved citizens file collective complaints, but they rarely work together or with NGOs, which, along with a government control of the media, stunts the growth of an anti-pollution movement.

Over recent years, Shanghai has grown into one of China's main economic centers. At the same time, the economic growth has also caused substantive changes in Shanghai's urban life and culture. As a result, there have been great shifts in citizenship and urban identities, which raises fundamental questions about the way green NGOs and activists can make use of social spaces. The chapter by Zhu and Ho provides a fascinating account of an environmental movement that grew out of tension between residents, state and real-estate developers. Through a detailed ethnography of activism in a Shanghai urban community, it is demonstrated that the privatization of housing, a growing ecological discourse, and the emergence of new civil organizations – notably the homeowners' committee – provided fertile grounds for a social movement. In their dispute with the local authorities, the residents soon found out that the de-politicized, environmental discourse, which had won them victory during previous confrontations, was no longer successful. For this reason, a new discourse had to be constructed which portrayed the residents as the central state's allies in controlling the violations of the law by the *local* government. This discourse was heavily contested and attacked by the local authorities. The activists' embedded strategy of "de-politicized politics" has thus itself become a stake in a state–society conflict.

The analytical scope of the book does not remain at the local level. In a rapidly globalizing world in which local politics and state autonomy is increasingly locked into transboundary relations, it is critical to assess Chinese environmentalism in a wider context as well. In recent years, there has been heated scholarly debate about the emergence of a "global civil society" and "global environmentalism" as certain NGOs have become powerful international players – such as Greenpeace and WWF (formerly known as the World Wide Fund for Nature). What impact do transnational NGOs have on the agenda-setting and activities of Chinese environmental activists? What cooperative links exist between global and Chinese NGOs, and how effective are they in taking up the domestic and international environmental challenge? These are some of the critical questions that are addressed in the final chapters of the book.

In Chapter 8, Sun and Tysiachniouk present an interesting case study from the Sino-Russian border area in former Manchuria. Despite grave cross-border environmental problems, Russian and Chinese NGOs have not formed effective networks. Sun and Tysiachniouk argue that there is an asymmetry in organizational capacity, structure, and the culture of Chinese NGOs and Russian NGOs in this border area, and that this asymmetry is one of the most important reasons for the lack of partnership. This asymmetry was molded by external forces such as Western charitable foundations, as well as different domestic opportunity structures in Russia and China.

Chapter 9 by Morton explores the emergence of a transnational sphere of environmental advocacy in China. By taking a bottom-up approach, Morton provides a preliminary analysis of the ways in which the expanding linkages between international and local NGOs are affecting democratic change at the grassroots level, especially in relation to public participation and policy reform. The observations offered by Morton suggest that international support can bring benefits to local NGOs by expanding the associational space within which they operate. However, the biggest risk is that strengthening civil society from the outside could undermine local political agency by stifling the potential of local NGOs to genuinely represent the interests of those whom they claim to represent. Tackling this dilemma presents a major challenge to the development of NGO cooperation across China's borders.

In the concluding chapter, Ho and Edmonds review the main theoretical implications of the various contributions in line with the overall research questions of this volume: what does the limited political space imply for the development and dynamics of a social movement in China? Is the occurrence of a green social movement a precondition for improved governance in terms of transparency and accountability? What are the perspectives for the emergence of a broadly supported environmental movement in China, and how would it relate to international environmental forces?

China's semi-authoritarian context is at the basis of the specific features of social activism today – a fragmentary, highly localized, and non-confrontational form of environmentalism. However, despite these features it would be a misconception to state that embedded environmentalism is a docile, and silenced movement. On the contrary, it is a continuously negotiated and therefore a highly effective adaptation to the current polity through which considerable political influence can be wielded. The main feature and success of China's reforms lies in a strategy of gradual change. In this respect, Ho and Edmonds argue, embedded environmentalism should be regarded as a transitory phase, a changing characteristic of an emerging civil society in a semi-authoritarian environment. Since environmental activism made its first appearance on the Chinese political stage, it has gained considerably in political leverage, and has developed increasing international linkages as a result of its embeddedness. The critical question that one should ask, therefore, is not whether the Chinese state should allow activists and NGOs to employ confrontational, radical, and mass mobilization tactics, but rather under what circumstances and at what time in the course of reforms.

Notes

1 See also Zhao Dingxin, *The Power of Tiananmen: State–Society Relations and the 1989 Beijing Student Movement*, Chicago: University of Chicago Press, 2001.

2 See for instance, Teresa Wright, "The China Democracy Party and the Politics of Protest in the 1980s–1990s," *The China Quarterly*, No. 172 (December 2002), pp. 906–26; Jan van der Putten, "*Chinese dissidenten opgepakt*" [Chinese dissidents arrested], *Volkskrant*, 1 December 1998, p. 4; Jan van der Putten, "*Bewind China*

vreest een naderende protestgolf" [China's regime fears an approaching gulf of protest], *Volkskrant*, 28 December 1998, p. 5; Jan van der Putten, "*De mens is nog nooit zo diep gezonken*" [Man has never fallen so low], *Volkskrant*, 27 April 1999, p. 5.

3 See Jonathan Power, "Will Democracy be Another 30 Years in Coming in China?" *Daily Times*, 3 June 2005, pp. 3–4; online, available at www.dailytimes.com.pk/default.asp?page=story_3-6-2005_pg3_4. This article also appeared in the *New York Times* of 4 June 2005.

4 For instance, Melanie Manion notes that although voters' choices for candidates for township People's Congresses are constrained by Party organizations, their role is not trivialized. She notes "the alignment of voter and Party committee preferences in these elections reflects a perspective quite different from an orthodox Leninist view of the appropriate relationship between Party and society, but also fundamentally different from pluralist visions of elections as instruments of democracy"; Melanie Manion, "Chinese Democratization in Perspective: Electorates and Selectorates at the Township Level," *The China Quarterly*, No. 163 (September 2000), p. 765. An excellent overview of the literature on People's Congresses is given in footnote 4.

5 See, for example, Wang Zhenyao, "Village Committees: the Basis for China's Democratization," in Eduard B. Vermeer, Frank N. Pieke and Woei Lien Chong (eds) *Cooperative and Development in China's Rural Development: Between State and Private Interests*, Armonk: M.E. Sharpe, 1998.

6 For information on citizenship and civil society, see also Merle Goldman and Elizabeth J. Perry (eds) *Changing Meanings of Citizenship in Modern China*, Harvard: Harvard University Press, 2002.

7 Robert P. Weller, *Alternate Civilities: Democracy and Culture in China and Taiwan*, Boulder: Westview Press, 2001; and Peter Ho, "Greening Without Conflict? Environmentalism, Green NGOs and Civil Society in China," *Development and Change*, Vol. 32 (2001), pp. 893–921.

8 Through which we trace back to the idea of embeddedness in the "Polanyian" interpretation of the contextualization of social action. See Karl Polanyi, *The Great Transformation*, New York: Holt, Rinehart, 1944. In the conclusion to this volume, a more detailed discussion of embeddedness in the social sciences literature is provided.

9 The fully authoritarian state is in fact an abstraction, as no state is capable of exerting complete totalitarian control. Also, in Maoist China this was not the case: see, for instance, David Crook and Isabel Crook, *Revolution in a Chinese Village: Ten Mile Inn*, London: Routledge and Kegan Paul, 1959; and Anita Chan, Richard Madsen and Jonathan Unger, *Chen Village under Mao and Deng*, revised edition, Berkeley, CA: University of California Press, 1992, pp. 32–3).

10 For a discussion of the concept of a social movement, see Doug McAdam, John D. McCarthy and Mayer N. Zald (eds) *Comparative Perspectives on Social Movements: Political Opportunities, Mobilizing Structures and Cultural Framings*, Cambridge: Cambridge University Press, 1996, pp. 13–17.

11 Some recent studies on social movements in China are Phillip Stalley and Dongning Yang, "An Emerging Environmental Movement in China?" *The China Quarterly*, No. 186 (June 2006), pp. 333–56; Feng Chen, "Privatization and its Discontents in Chinese Factories," *The China Quarterly*, No. 185 (March 2006), pp. 42–60. Both studies conclude that there is little likelihood of the emergence of a social movement in China, be it in the environmental field (Stalley and Yang) or as a labor movement (Chen). However, these studies fail to look at the embedded fashion of Chinese activism.

12 For instance, the local state generally tends to protect polluting industries because of hard-needed revenues and the employment of workers, whereas the central state and green activists are more likely to find themselves on the same side in environmental protection. As a result, it is often difficult to close down polluting factories. See also

Benjamin van Rooij, "Implementation of Chinese Environmental Law: Regular Enforcement and Political Campaigns," in Peter Ho and Eduard B. Vermeer (eds) "China's Limits to Growth: Prospects for Greening State and Society," special issue of *Development and Change*, Vol. 37, No. 1 (2006), pp. 1–271.

13 Broad social movements versus embedded movements, as well as sudden versus gradual changes, are of course ideal-types, as the forms and the types of political changes should be seen more as a continuum that can change over time and place.

14 See Charles Lindblom, "Still Muddling, Not Yet Through," *Public Administration Review*, No. 39 (November/December 1979), pp. 517–37; Charles Lindblom, "The Science of 'Muddling Through'" in Jay M. Shafritz and Albert C. Hyde (eds) *Classics of Public Administration*, third edition, Belmont: Wadsworth Publishing, 1992. About social movements as an instrument for new governance and politics, Jamison *et al.* wrote: "The social movements which emerged in all the countries of Western Europe and North America in the early 1970s had in common the intention of broadening the limits of existing national political cultures by creating a 'new politics'. By this was meant that these social movements were not primarily composed of new interest groups seeking incorporation into the established political discourse. Rather they were new social movements which sought to explode it: to expose its shortcomings and limitations, and to open up new spaces for political activity"; see Andrew Jamison, Ron Eyerman, Jacqueline Cramer with Jeppe Laessøe, *The Making of the New Environmental Consciousness: A Comparative Study of the Environmental Movements in Sweden, Denmark and The Netherlands*, Edinburgh: Edinburgh University Press, 1990, p. 192.

15 See, for instance, Chapters 4 and 5 in McAdam, McCarthy and Zald (1996) for sociopolitical change at the global level, see Margaret E. Keck and Kathryn Sikkink, *Activists beyond Borders*, Ithaca: Cornell University Press, 1998; John A. Guidry, Michael D. Kennedy and Mayer N. Zald (eds) *Globalizations and Social Movements: Culture, Power and the Transnational Public Sphere*, Ann Arbor: University of Michigan Press, 2000.

16 For instance, Susan Baker and Petr Jehlicka, "Dilemmas of Transition: The Environment, Democracy and Economic Reform in East Central Europe," *Environmental Politics*, Vol. 7, No. 1 (1998), pp. 1–26.

17 Barbara A. Cellarius, and Caedmon Staddon, "Environmental Nongovernmental Organizations, Civil Society and Democratization in Bulgaria," *East European Politics and Societies*, Vol. 16, No. 1 (2002), p. 183.

18 Albert Weale, *The New Politics of Pollution*, Manchester: Manchester University Press, 1992.

19 Timothy Doyle and Doug McEachern, *Environment and Politics*, London: Routledge, 1998, p. 55.

20 For a discussion of environmentalism at a global level, see Gareth Porter and Janet Welsh Brown, *Global Environmental Politics*, Boulder: Westview Press, 1996. However, whether such shifts should occur through radical "ecotage" and militant "ecological guerrilla warfare," or through consensual, tri-partite negotiations (state, businesses and environmentalists) remains a scholarly bone for contention. See also William M. Lafferty and James Meadowcroft (eds) *Democracy and the Environment: Problems and Prospects*, Cheltenham: Edward Elgar, 1996.

21 According to Whyte "by the mid 1950s an organizational system had begun to emerge that made it possible to control and mobilize the citizenry much more effectively than before. The building blocks of pre-1949 cities – guilds, native-place associations, clans, secret societies, neighborhood temple associations, and so forth – had either been eliminated or transformed. In their place a new urban infrastructure had been built that was firmly under Party control"; Martin King Whyte, "Urban Life in the People's Republic," in: Roderick MacFarquhar and John K. Fairbank (eds) *The Cambridge History of China: The People's Republic Part 2, Revolutions within the*

Chinese Revolution 1966–1982, Cambridge: Cambridge University Press, 1991, p. 697.

22 For a discussion of the early developments of civil society in China, see Gordon White, *Riding the Tiger: The Politics of Economic Reform in Post-Mao China*, London: Macmillan, 1993.

23 Statistics by Wu Zhongze – director-general of the Ministry of Civil Affairs – cited in Wang M., "Overview of China's NGOs," in Wang M. (ed.) *Zhongguo NGO Yanjiu* [Research on NGOs in China], no. 38, Beijing: UNCRD Research Report Series, 2000, pp. 15–16. According to the *1997 China Law Yearbook*, there were 1,845 national and 184,821 regional social organizations in 1996; see Zhongguo Falü Nianjian Bianji Weiyuanhui (1997: 1077).

24 According to a survey of 104 Beijing-based NGOs by Tsinghua University, 32 percent were active as trade and commercial associations, and study and students' societies; 7 percent were research and survey institutions; 6 percent provided social services; 6 percent were environmental NGOs; 5 percent focused on poverty alleviation; 4 percent engaged in international exchange; 27 percent belonged to other categories; and 13 percent did not respond. See Deng Guosheng, "Beijing NGO Wenjuan Diaocha Fenxi" [Analysis of the Questionnaires on NGOs in Beijing] in Wang M., *Zhongguo NGO Yanjiu* [Research on NGOs in China], No. 38, Beijing: UNCRD Research Report Series, p. 26.

25 At the Ninth People's Congress in March 1998, the secretary-general of the State Council, Luo Gan, declared that "government has taken up the management of many affairs which it should not have managed, is not in a position to manage, or actually cannot manage well" which has hindered the efficiency and effectiveness of the government. It was therefore necessary, said Luo Gan, to expand the activities of "social intermediary organizations"; see Anthony J. Saich, "Negotiating the State: The Development of Social Organizations in China," *The China Quarterly*, No. 161 (March 2000), p. 128.

26 Merle Goldman, "The Reassertion of Political Citizenship in the Post-Mao Era: The Democracy Wall Movement," in Merle Goldman and Elizabeth J. Perry (eds) *Changing Meanings of Citizenship in Modern China*, Harvard: Harvard University Press, 2002, p. 159.

27 See Richard L. Edmonds, *Patterns of Lost Harmony: A Survey of the Country's Environmental Degradation and Protection*, London: Routledge, 1994; and Vaclav Smil, *China's Environmental Crisis: An Inquiry into the Limits of National Development*, Armonk and London: M.E. Sharpe, 1993.

28 One of the earliest civil initiatives in environmental protection was undertaken by a dissatisfied official, who established the Association for the Research on Environmental Policy back in 1991. See Peter Ho, "Greening without Conflict? Environmentalism, Green NGOs and Civil Society in China," *Development and Change*, Vol. 32 (2001), p. 904.

29 For some interesting examples for Taiwan, where green NGOs have effected a clear "greening" of industries, see the articles on environmentalism by Tu Wenling and Lifang Yang in Peter Ho (ed.) "Greening Industries in Newly Industrializing Countries: Asian-style Leapfrogging?" *International Journal of Environment and Sustainable Development: A Special Issue*, Vol. 4, No. 3 (2005), pp. 209–26 (six articles in total).

30 See for instance, Theda Skocpol, *States and Social Revolutions: A Comparative Analysis of France, Russia and China*, New York: Cambridge University Press, 1979 (on the unintentionality of social movements); Neil J. Smelser, *Theory of Collective Behavior*, New York: Free Press, 1963 (Smelser distinguished six conditions underlying collective action: structural conduciveness; structural strain; generalized beliefs; precipitating factors; leadership and regular communication; and finally the operation of social control); Alain Touraine, *The Voice and the Eye: An Analysis of Social Movements*, New York: Cambridge University Press, 1981 (Touraine is known for his

concept of the "field of action" which refers to the connections between a social movement and the forces or influences against it. It is seen as a process of mutual negotiation among antagonists, that may lead to social changes).

31 See also the useful overview provided in Sidney Tarrow, *Power in Movement: Social Movements and Contentious Politics*, Cambridge: Cambridge University Press, 1998, pp. 10–25.

32 J. Elizabeth Perry and Mark Selden (eds) *Chinese Society: Change, Conflict and Resistance*, New York: Routledge, 2003, pp. 93–112.

33 China Statistical Bureau, "*Zhongguo Lingdao Ganbu zai 2002 Nian zui Guanxin de Shiqing*" [Issues of greatest concern for Chinese leading officials in 2002], *Zhong Guo Qing Nian Bao*, 18 December 2003, p. 14.

34 See Peter Ho and Eduard B. Vermeer (eds) *China's Limits to Growth: Greening State and Society*, Oxford: Blackwell, 2006; Richard Louis Edmonds, "China's Environmental Problems," in Robert E. Gamer (ed.) *Understanding Contemporary China*, Boulder: Lynne Rienner, 2003, p. 268.

35 The growth in complaints is not simply the result of population growth. In 1992, approximately 1.23 people out of 10,000 made a complaint. In 2001, 3.5 people out of 10,000 made a complaint. See the contribution by Anna Brettell in this volume.

36 For information on the political economy of biotechnology in China, see Jennifer H. Zhao and Peter Ho, "A Developmental Risk Society? Genetically Modified Organisms (GMOs) in China," *International Journal for Environment and Sustainable Development*, Vol. 4, No. 4 (2005), pp. 370–94; and Peter Ho, Eduard B. Vermeer and Jennifer H. Zhao, "Biotech and Food Safety in China: Consumers' Acceptance or Resistance?" in Peter Ho and Eduard B. Vermeer (eds) "China's Limits to Growth: Prospects for Greening State and Society," *Development and Change*, Vol. 37, No. 1 (2006), pp. 227–53.

37 Huang Wei, "Individuals Changing the World: An Interview with Sophie Prize 2000 Winner Liao Xiaoyi," *Beijing Review*, No. 33 (14 August 2000), p. 17.

38 Li Ningning, "*Huanbao Yishi yu Huanbao Xingwei*" [Environmental consciousness and environmental behavior], *Xue Hai*, No. 1 (2001), pp. 122–3. For a comparison with European attitudes about the environment, see European Opinion Research Group, *The Attitudes of Europeans Towards the Environment*, Eurobarometer 58.0, Brussels: Directorate-General Environment, European Union, 2002.

39 It is important to note that environmental awareness and understanding are two different matters. In the same survey, urban residents' average score on environmental knowledge issues was only 4.5 (full marks: 13 points), and rural residents scoring even lower with 2.4. This survey was carried out in 2002 by the Social Survey Institute of Systems Reform Commission; see Cui Shuyi, "*Gongzhong Huanjing Yishi Xianzhuang, Wenti yu Duice*" [Public Environmental Consciousness: Existing Situation, Problems, and Strategy], *Li Lun Xue Kan*, Vol. 4, No. 110 (July 2002), p. 87.

40 A description of environmentalism in Taiwan is provided in Li-Fang Yang, "Embedded autonomy, social movements and ecological modernization in Taiwan," in Peter Ho (ed.) "Greening Industries in Newly Industrializing Countries: Asian-style Leapfrogging?" *International Journal of Environment and Sustainable Development*, Vol. 4, No. 3 (2005), pp. 209–26.

41 See also Robert P. Weller, *Discovering Nature; Globalization and Environmental Culture in China and Taiwan*, Cambridge: Cambridge University Press, 2006, pp. 113–14; Robert P. Weller and Michael Hsiao, "Culture, Gender and Community in Taiwan's Environmental Movement," in Arne Kalland and Gerard Persoon (eds) *Environmental Movements in Asia*, Richmond: Curzon Press, 1998; and Michael Hsiao, On-Kwok Lai, Hwa-jen Liu, Francisco A. Magno, Laura Edles and Alvin Y. So, "Culture and Asian Styles of Environmental Movements" in Yok-shiu F. Lee and Alvin Y. So (eds) *Asia's Environmental Movements: Comparative Perspectives*, Armonk: M.E. Sharpe, 1999.

42 See also B. Pitman Potter, "Belief in Control: Regulation of Religion in China," *The China Quarterly*, No. 174 (June 2003), pp. 317–37.

43 For instance, Zhao has made note of a Daoist "Club for Green Civilization," which was set up in 1993 in Sichuan province. See Zhao Xiumei, *Fazhanzhong de Huanjing Baohu Shehui Tuanti* [Environmental Social Organizations in Development], Beijing: Tsinghua University (NGO Research Center Book Series), 1999. For more information on the relation between Daoism and nature, see N.J. Girardot, James Miller and Xiaogan Liu (eds) *Daoism and Ecology: Ways within a Cosmic Landscape*, Cambridge, MA: Harvard University Press, 2001.

44 There have been discussions to enable a more liberalized way of control, which in practice would imply that NGOs need not register under a sponsoring institution in the same field, nor in the same administrative area. In other words, it would then also become possible for NGOs, for instance, in Beijing to register under a local sponsoring institution in a different province (Wang M., oral communication, February 2004). However, the recent clampdowns on NGOs after the color revolutions have made it rather uncertain whether, and if so when and how, a reform in the registration procedures might be expected.

45 See the chapter by Ho.

46 David Eimer, "China Clamps Down on Environmental Monitoring Group," *Independent* (28 October 2005), pp. 1–2.

47 See the introduction in Peter Ho and Eduard B. Vermeer (eds) *China's Limits to Growth: Prospects for Greening State and Society*, London: Blackwell, 2006.

48 This, however, is not to say that the entire spectrum of environmental questions is open for civic action. For instance, biotechnology and nuclear energy still belong to social arenas that are largely restricted for green activism. See Peter Ho and Eduard B. Vermeer, "Biotech and Food Safety in China: Consumers' Acceptance or Resistance?" and Peter Ho, "Trajectories for Greening in China: Theory and Practice", both in Peter Ho and Eduard B. Vermeer (eds) *China's Limits to Growth: Prospects for Greening State and Society*, London: Blackwell, 2006; and Jennifer H. Zhao and Peter Ho, "A Developmental Risk Society? Genetically Modified Organisms (GMOs) in China," *International Journal for Environment and Sustainable Development*, Vol. 4, No. 4 (2005), pp. 370–94.

49 Wang M., "Overview of China's NGOs," in Wang M. (ed.) *Zhongguo NGO Yanjiu* (Research on NGOs in China), no. 38, Beijing: UNCRD Research Report Series, 2000, p. 20

50 Gary G. Hamilton (ed.) *Business Networks and Economic Development in East and Southeast Asia*. Hong Kong: Centre of Asian Studies, University of Hong Kong, 1991; Yan Yunxiang, *The Flow of Gifts: Reciprocity and Social Networks in a Chinese Village*, Stanford: Stanford University Press, 1996; Mayfair Yang, *Gifts, Favors and Banquets: The Art of Social Relationships in China*, Ithaca: Cornell University Press, 1994.

51 An article by Lo and Leung has shown that state environmental agencies face substantial problems in reaching out to the public, and actually need NGOs to do that for them. See Carlos Wing Hung Lo and Sai Wing Leung, "Environmental Agency and Public Opinion in Guangzhou: The Limits of a Popular Approach to Environmental Governance," *The China Quarterly*, No. 163 (September 2000), pp. 677–704.

52 Wu Fengshi, "New Partners or Old Brothers? GONGOs in Transnational Environmental Advocacy in China," *China Environment Series*, Issue 5 (2002), pp. 47 and 53.

2 Self-imposed censorship and de-politicized politics in China

Green activism or a color revolution?

Peter Ho

Introduction

The introduction to this volume has put forward the notion of embedded social activism to describe the specific setting and dynamics of social movements in China, with particular reference to environmentalism and green non-governmental organizations (NGOs). It was posited that embedded activism springs forth in a *semi*-authoritarian context. Different from an authoritarian context in which the state exerts virtually full control over society and has made large-scale voluntary action practically impossible,[1] the semi-authoritarian context features a formal structure of stringent state control that deviates from informal practices, which actually allow a fair degree of voluntary civic action. Against this backdrop, the current contribution addresses the meaning of embedded activism with particular reference to environmentalism: What is the institutional environment in which it takes place? How do green activists and NGOs respond to it? Lastly, can any shifts in the political opportunity structure for social activism be expected, and if so, in which direction? Apart from the introduction and conclusion, this contribution is divided into three separate sections.

The contribution will start by addressing the first question above: what is the environment of embedded activism? It will do so by reviewing the Party and state's interventions in civil society. To pinpoint the nature of Party power is an arduous undertaking as its structure is closely intertwined with administrative institutions, and what may appear as a civil organization might well be integrated with Party structures. To illustrate these blurred divides, the case of a green student association and its embeddedness in Party institutions will be discussed. In addition, at times of social crises, such as during the SARS outbreak or the demonstrations and subsequent repression of the Falun Gong sect, it is obvious that the Chinese Communist Party (CCP) forms the main source of political power in China today.[2] For NGOs and activists this became perfectly clear in the aftermath of the "color revolutions" in the Ukraine, Kyrgyzstan and Georgia, when the Chinese central authorities were quick to tighten control over civic organizations.

This, however, is not to say that the CCP exerts totalitarian control over society as the foreign media at times like to suggest. Whereas during the Maoist era the Party *was* the state and vice versa, there has been an increasing division between Party and state institutions since the start of the reforms in the late 1970s.[3] A majority of China-watchers have focused on the occurrence of free elections as a measure of democratization, and in their absence simply typified the People's Republic of China as an authoritarian state.[4] This neglects an underlying yet critical development: the separation of the three powers. Over the past decades, the judiciary and the legislative powers have increasingly claimed their own autonomous spheres from the executive branch. From being a mere "rubber stamp" in the 1980s, the National People's Congress, particularly its Standing Committee and special sub-commissions, have become actively involved in the initiation, drafting, and discussion of laws. Today it is no exception that bills are repeatedly sent back to the various ministries and experts for revisions,[5] while the legislature frequently reviews these very same laws after promulgation.[6] It is within this transitional, semi-authoritarian context that Chinese civic activism – be it for women's rights, the protection of factory workers or the environment – has found a certain yet embedded space.

The second section examines the response of green activism to the Party and state. It will be demonstrated that the relation between Party–state–society interaction is not a matter of "de-politicized and free" versus "politicized and repressed" public spheres.[7] Rather is the political context in which social activism finds itself a *paradoxical* one: restrictive but conducive. Indeed, the current Chinese political context limits the role and activity scope of environmental NGOs in society. It creates legitimacy problems and poses certain organizational constraints. Simultaneously, however, environmentalism has gained an increasing political leverage by avoiding any connotation with being a movement, by all means trying to appear small, low-key and localized, and acting as the state's partner rather than its adversary.[8] As much as it would be a fallacy to assume that China's on-going separation of the three powers can be reverted in an instant, it is a misconception to read in the authorities' cyclical policies of strict formal control a sure signal that Chinese civil society would be uprooted.

Against this backdrop, the final part delves into the remaining question: what future shifts in the political opportunity structure for social and green activism might be expected? This section underscores the idea that we need to view embeddedness as a transient phenomenon, a particular type of Party–state–society interaction shaped by the semi-authoritarian nature of Chinese society. A corroboration for such a hypothesis is found in the recent dam protests in southwest China. The anti-dam campaigns prove that the political opportunity structure for activists can change, and as it does, so do their strategies and interactions with the Party and state.

The nature of control: party and state

View and interventions: fearing color revolutions?

The position of green NGOs and activists in semi-authoritarian polities is to a great extent determined by the state's momentary political tendency. In this respect, the speech by President Hu Jintao when he presided over a collective study session of the Politburo on 27 December 2004 seemed significant. During the speech Hu stated that it was "important to stick to a people-oriented policy, strengthen the science and technology support to social and public welfare undertakings and technological planning in such fields as medicine and health, family planning, and environmental protection."[9] Observers were quick to read in this signs that the Chinese leadership would allow NGOs to flourish as the Party realized that it needed to rely on civil society groups. In fact, since the early 1990s the central government has repeatedly stressed the importance of "social intermediary organizations."[10] However, such government appreciation is confined to situations in which NGOs work as extensions of the government, or to specific niches of environmental management that the government cannot or wishes not to work in. In this respect, one of China's foremost environmental activists, the reporter Dai Qing, stated that "most officials are happy to support and encourage civilian organizations that carry out propaganda, education and other activities that do not offend the government."[11] On the other hand, the realm of policy-making is regarded as an exclusive state affair that should be left to environmental and political experts within state and Party circles.

In Western Europe, NGOs are part of a governance structure in which checks-and-balances are primarily imposed by a pluralist society.[12] Formal registration as a non-governmental entity is relatively easy, while the control of NGO activities is a measure of its legitimacy versus society that is generally built up in the years after their establishment. In China, we are dealing with a reversed situation where the state has professed to checking and supervising NGOs, instead of leaving it to self-regulatory mechanisms in society. Citizens face considerable problems when trying to obtain formal registration as a social organization. The current regulations limit horizontal linkages, favor groups with close government connections, and discourage grassroots' initiatives. A more detailed discussion of the Chinese state's regulations and control over social organization is provided in the following contribution by Ru and Ortolano.

Despite the restrictive nature of state regulations, it is strangely clear to both state and society that the former generally lacks the intention to effect totalitarian control. Instead, an ambiguous, symbiotic relation has developed that is continuously open for negotiation. On the one hand, the Chinese state controls civic activities by erecting barriers for civil organizations to obtain formal institutionalization. On the other hand, the state allows – or, better, tolerates – (as it could not do otherwise under the current socio-economic conditions) the emergence of social groups that increasingly escape their view and push at the state-imposed limits.[13] There are ample opportunities to circumvent the rules and become a legal body without direct registration under the Ministry of Civil Affairs, as the other authors in this volume demonstrate. The Chinese authorities are aware of the formation of such informal groups, but are not opposed to their establishment, provided that they do not challenge the political leadership. However, this seemingly tolerant government stance does not imply that the Chinese authorities are not concerned with the activities of NGOs, as a confidential government report illustrates:

> A critical link to secure the correct political direction of civil organizations and strengthen the management of civil organizations is the widespread establishment of Party organizations. Comrades stated that the weakening of Party construction [*dangjian gongzuo boruo*] is one of the crucial problems widely occurring among civil organizations.... This seriously influences the Party's implementation and enforcement of key policies for civil organizations.[14]

Moreover, new developments can also lead to the re-negotiation and redefinition of Party–state–society relations. The recent color revolutions in the Ukraine, Georgia and Kyrgyzstan have caused a rising weariness among government and Party officials. In order to strengthen control over civic groups, the central authorities ordered in March 2005 that all social organizations registered with industrial and commercial bureaux had to report promptly to the relevant civil affairs administration for a renewed review and approval of their registration status.[15] A month after, the government established the All-China Federation of Environmental Protection. Although officially announced as a non-governmental body to facilitate coordination among environmental NGOs, its establishment was viewed with suspicion by environmental activists. As a prominent green activist stated: "Environmental NGOs do not need such an organization. We already know who is who in the field, so why set up an organization for coordination? I think it is basically set up to increase government control over us."[16] The creation of the All-China Federation of Environmental Protection is in fact a clever move by the government to co-opt environmental civil groups. At its official launching on Earth's Day in April 2005, the inaugural ceremony was attended by Vice-Premier Zeng Peiyan, while all major figures within governmental and non-governmental circles had been invited.[17]

Furthermore, the Chinese government has formed a Leading Group to monitor social organizations, composed of officials from the State Environmental Protection Administration, the Ministry of Civil Affairs, and the International Department of the Central Committee of the Communist Party. The Leading Group has commissioned a nation-wide census of environmental organizations to be carried out by the All-China Federation of Environmental Protection. The website of the federation mentions that the census will collect information about NGOs engaging in environmental activities, unregistered groups and environmental groups at universities. By sending out over 10,000 questionnaires, and interviewing key-informants from NGOs, it is hoped more information can be gathered about the history, leadership, constraints and activities of civil groups. The census is expected to take four months and is done in cooperation with the Social Survey Center run by the *China Youth Daily* (in turn an extension of the Communist Youth League (CYL)).[18] According to a researcher at the Social Survey Center, the census aimed to find unregistered groups that were "operating under the water."[19]

Blurred divides: government-organized NGOs or embedded NGOs?

A social environment in which the state and Party pro-actively intervene in the actions and operations of civic and voluntary organizations has led to the somewhat ironic phenomenon of "government organized non-governmental organizations" (GONGOs). The term GONGO refers to the multitude of organizations that have in recent years been created at various administrative levels by and in support of the Party and state. There are different reasons for government institutions to establish "non-governmental organizations": first, the reallocation of government budgets and the related need for government departments to slim down;[20] second, the ability of GONGOs to operate more flexibly and adapt easier to market conditions than bureaucratic and unwieldy government institutions; third, the desire to attract (foreign) financial resources through GONGOs, as potential donors are generally more favorable to contribute to organizations independent from the state. The social organizations formally registered with the Ministry of Civil Affairs are generally established by those with stronger connections with the government. Many scholars have pointed to the drawbacks of GONGOs:

> The lack of separation between the government and society also affects the development of grassroots organizations. In order to obtain the support or recognition from the government they readily accept administrative intervention. Many scholars have criticized that NGOs in China lack a non-governmental character and are not worthy of the name NGO.[21]

It is important to note that the scholarly opinions on the negative effects by GONGOs on Chinese bottom-up activism are divided. Some observers have remarked that not only grassroots organizations but also GONGOs have a critical role to play in changing state–society relations and working towards environmental improvement.[22] According to Wu, some GONGOs are growing more autonomous from their "government-parent" and play increasingly independent roles in linking external and internal actors, state and non-state organizations in China. In this respect they also push for social and political change, but from a position closer to the political source of power. Lastly, similar to the "real" NGOs, the GONGOs frequently provide a solid training grounds for future NGO leaders.[23]

However, it is not the aim of this contribution to argue which of the two – NGOs or GONGOs – would be in a better position to strengthen civil society and environmental policies, as both probably do so in their own right and on their own terms. Rather, it is more important to note that the interventionist strategies of Party and state have created an environment of blurred divides, and embeddedness in Party-statist structures for NGOs and GONGOs. The precise nature of this embeddedness might differ. For NGOs embeddedness might be a matter of struggling through complicated registration procedures requiring personal relations in state and Party circles, while for GONGOs the proximity to Party-state institutions might more strongly affect the type of activities they can or decide to engage in. But the crux of embedded activism lies in the challenges it poses – to NGOs and GONGOs alike – in terms of legitimacy and the roles that social organizations can play in society.

The embeddedness for Chinese social organizations can go easily unnoticed, but can be far-reaching. For instance, although for outside observers GONGOs appear as voluntary groups, they are strongly embedded within Party institutions and state-supported mass organizations, such as the Communist Youth League, the Women's Federation, or the Workers' Union. A case in point are the Green Students' Associations that have sprang up at virtually every major academic institution or university in the country (see also the contribution by Sun and Tysiachniouk in this volume). Especially during the past few years since the late 1990s, Green Students' Associations have drastically increased, from a reported 22 in 1997 to over 150 in 1998. Most of the Green Students' Associations are located in major cities such as Beijing, Xi'an, Guangzhou and Tianjin.

Their establishment has been heralded with great enthusiasm about the voluntary force of China's educated youth. Yet, considering China's turbulent history of student movements as powerful forces pushing for political change,[24] it would be a sheer illusion to think that the CCP and the state would allow students to organize themselves in a fully uninhibited, and uncontrolled manner. This, of course, is not to say that Party control today is similar to that during the Mao era. Rather, it is a matter of having structures in place that could be mobilized to swing into action if deemed necessary. This becomes obvious by examining the case of the Green Students' Association (GSA or *Lüse Xuesheng Xiehui*) which was established at one of China's foremost universities in the early 1990s to "promote environmental protection at the campus and increase students' awareness for environmental problems."[25]

The GSA appears to be an association initiated "by students for students". In fact, it is embedded within the larger university Party structure (see Figure 2.1). All major decisions in Chinese universities regarding finances, positions and university policies have to be formally approved by the university Party Committee headed by the Party Secretary. Analogous to the Party-state structure, the Party-university structure has branches at every administrative level, down from the central university level, to the faculties and departments.[26] In order to understand how the political game is played, we have to turn our attention to the critical figure in this structure: the Party Secretary of the Communist Youth League (CYL). He is the person charged with the overall coordination of students' affairs within the university, and forms the node of communication and control

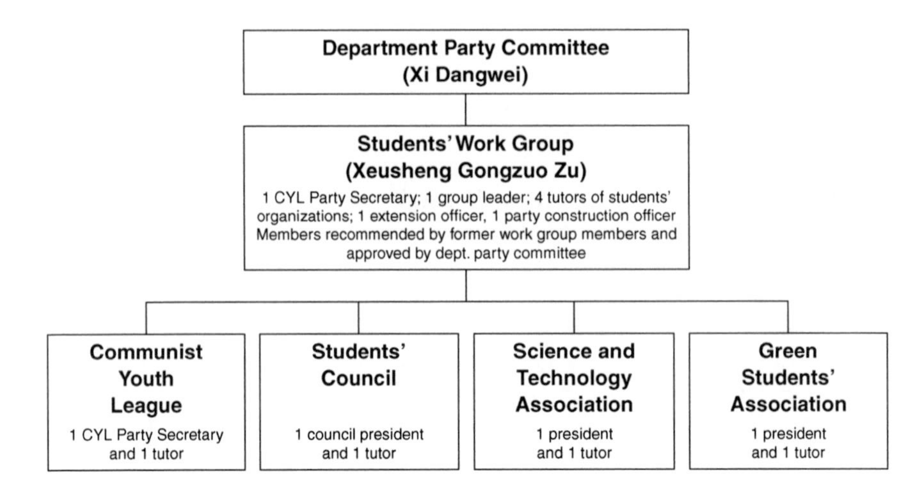

Figure 2.1 Embedded structure of the Students' Green Association (source: drawn by the author).

between students and the university's Party structure. On paper, the GSA is on a par with the CYL, the Students' Council (*Xuesheng Hui*), and the Science and Technology Association (*Keji Xiehui*). In this set-up, the presidents of the four student associations would be accountable to the group leader of the Students' Work Group (*Xuesheng Gongzuo Zu*).

In reality, however, the actual line of power runs through the Party Secretary, who heads the CYL and simultaneously sits in the Students' Work Group as the responsible person to report to the Department's Communist Party Committee (*Xi Dangwei*). This implies that not only the presidents of the students' associations are accountable to the CYL Party Secretary, but also the student tutors (*fudaoyuan*).[27] In the undergraduate phase, each of the four undergraduate years is headed by a tutor responsible for the management and supervision of the students in that year.[28] The tutor is concurrently a member of one of the students' organizations and of the Students' Work Team.[29] Through the weekly meetings of the Students' Work Group, the CYL Party Secretary is updated about the activities and events of the students and the four students' associations. The members of the Students' Work Group have the right to recommend new members, but their appointment has to be approved by the Department's Party Committee. Although the student associations have been officially registered, they do not have independent financial accounts and run their finances through the account of the Students' Work Group. It should be noted that this intricate structure described here counts only for a single university department, and is replicated for each department.

Embedded activism: impact and responses

Legitimacy and NGOs' role in society

Having reviewed the role of the Party and state in creating and upholding the embeddedness of social activism, it is now time to turn to the other side of the equation: the impact of and responses to embeddedness by NGOs and green activists. Formalized and institutionalized channels for public participation, providing NGOs with access to policy-makers, are virtually non-existent in China. However, diverting the focus from official procedures to the informal side of politics, it turns out that even without institutionalized channels for participation, China's environmental groups find ways, albeit incremental and *ad hoc*, to approach and influence relevant government departments with policy suggestions. Environmental NGOs in the sector of research and development occasionally submit policy and research papers to government departments or officials who work on specific environmental issues, sometimes on invitation, sometimes at their own initiative. In a survey of the NGO Research Center of Tsinghua University it was found that of approximately 1,500 polled NGOs, 58.7 percent had once or more provided policy advice to state institutions.[30] Another common channel created by NGOs to acquire recognition by and influence on policy-makers is the organization of conferences and workshops on specific environmental topics. On such occasions Chinese officials are invited, along with (inter)national experts who are sympathetic to the green case and who bestow a certain legitimacy to the event.[31]

An example of a more direct attempt to influence the authorities is the joint campaign of Beijing's three most prominent NGOs[32] to prevent the reconstruction of a 35-year-old channel. In order to prevent water losses, the channel that brings water from the Miyun reservoir into the capital city had to be straightened and lined with concrete. In March 2001 the campaign resulted in a meeting organized by the involved NGOs between the Beijing city government, including the vice-mayor, the greens and various experts and journalists. Although the campaign was not effective in altering the government's plans, the fact that high-level government officials agreed to meet NGOs in a forum of this kind made the meeting a remarkable event in itself. It needs to be mentioned, however, that the meeting was later described by officials as an attempt to "surround and attack" the vice-mayor and that local media were forbidden to make any mention of the event.[33] However, the issue of NGOs' political leverage remains a slippery one. A professor in environmental science at Tsinghua University and concurrent member of the Drafting Committee of the 10th National Environmental Five-Year Plan remarked:

> Talking about the political influence of NGOs implies they must be known by the people. But if you ask a passer-by on the street, almost nobody would know the "famous" Friends of Nature. This reflects NGOs' small influence. But this does not mean that NGOs don't have any influence.

> When president Clinton visited China, Global Village was invited by the US government to meet with him. The Chinese authorities did not object. In fact, they consider NGOs good for China's public image. China uses them to say to others: "Look, we have NGOs too!" Herein lies their potential power.[34]

This statement is illustrative for a potential problem in green NGOs' political leverage: the unfamiliarity of the larger public with their activities, which might limit their role as serious players in environmental politics. A nationwide survey carried out between July 1998 and July 1999 among 10,495 households provides some interesting insights in this respect. In response to the question which measures respondents considered most important in environmental protection, 23.7 percent mentioned strengthening environmental education and propaganda, 18.1 percent stated improving environmental laws and regulations, and 15.1 percent thought increased investments. The development of civil organizations only scored a mere 3.8 percent (see Table 2.1). In addition, most people display a strong reliance on the government for solving problems, rather than civil organizations. When asked to respectively rank the national government, local governments, companies, individuals and social organizations with respect to their responsibility in environmental protection, respondents gave the lowest ranking to social organizations.[35] From these data we can see that the low level of legitimacy of green NGOs and activists is a point of concern. The lack of grassroots support, materially and morally, is likely to hamper NGO development and weakens the role that NGOs can play in environmental politics.

To date the Chinese authorities have refrained from facilitating the creation of *formal* roles for NGOs, communities or other civil society groups in environmental policies.[36] This is also recognized by the public, as the results from the survey above demonstrate. Regarding the strengthening of public participation in environmental activities, almost 45 per cent of respondents thought the

Table 2.1 Which measures do you think are most important for environmental protection? ($n = 7,879$)

Measure	*Percentage*
Strengthen environmental education and propaganda	23.7
Improve environmental laws and regulations	18.1
Increase investments in environmental protection	15.1
Increase environmental law enforcement	13.6
Stimulate citizens' participation in environmental protection	10.6
Monitoring of pollution control by enterprises	9.7
Improve environmental technologies	5.4
Develop the role of civil environmental organizations	3.8
Total	100

Source: Translated and drawn by author on the basis of data from SEPA (ed.) *Quanguo Gongzhong Huanjing Yishi Diaocha Baogao* [Research Report on Public Environmental Awareness in China], Beijing: Zhongguo Huanjing Kexue Chubanshe, 1999, p. 22.

government's score was "bad" to "very bad," and in terms of stimulating the role of environmental NGOs approximately 40 percent found that the government had performed "badly" to "very badly."[37] However, the enigma of China's semi-authoritarian context is that its formal structures of stringent control still leave ample avenues open for the *informal* influencing of political decision-making. In this sense, civil organizations find themselves embedded in a Party-statist structure of control, as well as in an intricate web of personalized relations and informal politics. Those attempting to find the key to successful environmentalism can only do so by adhering to the principles of any other successful social activist: "de-politicize politics" and self-impose censorship. That is the reason Chinese green activism is a "movement without a movement," lacks any urgency to openly confront central authorities, and remains localized, fragmentary, and small-scale. This becomes obvious by looking at its specific activities.

Green activism: forced into de-politicized spaces?

This sub-section sketches some of the main activities that domestic green NGOs – and GONGOs, for that matter – are engaged in today.[38] It is by no means meant to be comprehensive or complete, as other contributions in this volume describe NGO activities in greater detail. The overview here is meant to facilitate the understanding of the roles that green groups currently play in environmental politics. Chinese environmentalism features those activities, which it learned or copied from its green pendants in Western Europe and North America: awareness-raising, environmental education, research, and advocacy. Yet typical of Chinese green activism might be what is absent: a broadly, supported popular movement. Moreover, the Chinese embedded context forces green organizations and activists in the role of a partner of the state, rather than a representative and defender of certain interests. As a result, environmental activists take all necessary means to avoid being regarded as, or even having the potential for, growing into a full-fledged, green social movement. To stay out of political harm's way, they generally engage in specialized and media-attractive yet politically innocent activities. It forms the essence of a conscious strategy to "de-politicize environmental politics."

Environmental education and awareness-raising

Environmental education and awareness-raising constitute the main focus for the majority of China's green NGOs. The range of activities lumped together under this rubric is relatively varied. It includes teaching environmental classes in schools and universities, instructing journalists on environmental news coverage, and organizing bird-watching trips for NGO members. Over time, some environmental NGOs have become quite professional in this field. For example, Friends of Nature has composed a comprehensive textbook for environmental education in primary schools, now widely in use throughout China. The Institute for Environment and Development (an NGO in the disguise of a research institute) implements the Chinese component of the international LEAD program (Leadership for Environment and Development Program), which is sponsored by the Rockefeller Foundation. The program offers training to small groups of environmental professionals.[39] Many more examples of awareness-raising activities could be given here. At times these activities seem to serve no other purpose than taking city-dwellers outside to bring them into contact with "nature" (or rather what activists perceive as such). Activities by "Green Earth Volunteers" – a loose network of nature-conservationists and bird-watchers in Beijing – is a typical example of this. At times, environmental education is catering for specific target groups and for a specific purpose. An interesting example of this are the "Mother's Environmental Protection Volunteers" in Shaanxi province. This environmental group is linked to the Provincial Women's Federation, and provides training to female farmers to promote organic agriculture and improve women's participation in agricultural operation.

Although this partly overlaps with education, *participatory approaches and community building* also make up for a substantive share of NGO activities in China. Common examples are the organization of the collection of waste batteries in residential areas, tree planting, and campaigns against the sending of greeting cards during the Chinese New Year. A more institutionalized public activity is the "Green Community" project initiated and carried out by Global Village of Beijing.[40] This NGO co-operates with residents of selected urban communities and their neighborhood committees. Various measures are promoted to achieve a green lifestyle, such as separate waste collection, recycling, water- and energy-saving, and green consumption patterns. The project attempts to secure the participation of citizens as well as local authorities (such as the neighborhood committees and district councils). In this sense, the project moves from mere awareness-raising to trying to change lifestyles by addressing issues of environmental technology, mobilization of communities, and the necessary logistics. Although results varied, the Green Community project was one of the first of its kind at the time of its inception. The Green Community project has been "adopted" by the Beijing Environmental Protection Bureau that has made the project part of its campaign for a "Green Olympics, Green Beijing."

Research for development and advocacy

Whereas the NGO activities above are directed at the general public, there is also a range of environmental activities that take place outside the public's view. A number of NGOs work on specialized issues demanding more technical expertise, such as the Pesticide Eco-Alternative Center in Yunnan province (see the contribution by Morton elsewhere in this volume), or the Hebei Research and Service Center for Environment and Resource Law in Shijiazhuang city, consisting of a group of environmental lawyers.[41] An interesting case is the South–North Institute for Sustainable Development (SNISD) in Beijing. It is one of the few domestic NGOs that has taken up the advocacy and development of renewable energy in China as its main focus. The mission of SNISD is to conduct research on energy laws, regulations and policies and to promote the use of renewable energy sources.[42] In rural areas SNISD has cooperated with local governments and farmers to stimulate sustainable energy, such as through integrated biogas systems. In urban areas the main project activities of SNISD include feasibility studies on fuel cells for vehicle use in Shanghai, and a project in Beijing focused on wind energy use by Beijing-based companies. After a survey of companies' willingness to purchase wind energy at slightly higher costs, SNISD is currently lobbying with the government to establish a market-mechanism in co-operation with all major stakeholders.[43]

Constraints: resources, functioning and leadership

How does a semi-authoritarian context of formal control and informal opportunities for social activism affect the mobilization of resources, organizational functioning and leadership? In a survey of 1,540 NGOs, the respondents ranked the lack of resources in terms of finance, material, and personnel respectively first, second and third as problems that they generally encounter (see Table 2.2). These problems are similar to those experienced by emerging civil groups in other parts of the world.[44] However, to a great extent they also result from a restrictive environment that frustrates formal registration. In absence of formal registration, green groups are not a legal person, and thus are not entitled to have a separate financial account, which hampers financial transparency and accountability.[45] In addition, solving the *hukou* or household registration in hiring personnel might also constitute a problem for unregistered NGOs. In the cities the *hukou* entitles citizens entry to the urban job market, public health facilities, the social welfare system, and education. Those with a rural *hukou* are excluded from these services, as a result of which the *hukou* system can strictly control

Table 2.2 Main problems encountered by social organizations (note: more than one answer possible)

Problem type	*Percentage*	*Ranking*
Lack of funding	41.4	1
Lack of office space and supplies	11.7	2
Lack of qualified personnel	9.9	3
Other	9.3	4
Insufficient government support	8.5	5
Internal management problems	7.5	6
Lack of information exchange and training opportunities	5.2	7
Deployed activities lack social response	3.6	8
Relevant laws and regulations undeveloped	3.4	9
Lack of projects	3.0	10
No problems	1.8	11
Too much government intervention	1.1	12

Source: Translated and drawn by author on the basis of data from NGO Research Center (G.S. Deng, "A Preliminary Analysis of the NGO Survey" in M. Wang (ed.) *Zhongguo NGO Yanjiu* [Research on NGOs in China], no. 43, Beijing: UNCRD Research Report Series, 2001).

rural–urban migration. Over the years, however, government control on migration has relaxed.[46]

Most environmental groups are permanently short of financial resources, struggling to make ends meet. Domestic fund-raising is problematic for several reasons: the absence of government funding for citizen initiatives; unfavorable tax regulations on donations for charity; government regulations on limits to membership and membership fees for social organizations; and several cultural obstacles to fund-raising among the general public. As a result, Chinese grassroots NGOs – or at least those with no or weak government connections – are almost completely dependent on foreign donor money, obtained either through donations or through grants for specific projects or activities. Of all funding sources, international funds are least available to NGOs in China. As shown in Table 2.3, the greater part of funds for social organizations – which includes GONGOs as well as NGOs – comes from government subsidies, with a mere 1.6 percent coming from international donors such as the Ford Foundation, WWF and the Asian Development Bank. In particular the grassroots organizations rely on foreign funding, and have little to expect from the government. For instance, of the total budget over 1996–9 of Global Village of Beijing 85 percent came from international sources, while this amounted to 53 percent in the case of Friends of Nature.[47]

Related to the dire financial situation is the lack of human resources. Many NGOs are shorthanded and lack staff with environmental expertise, management experience, and the English-language proficiency needed to communicate with potential donors. In the aforementioned survey it was found that among NGO staff, only 28 percent had an undergraduate degree, while 43 percent had finished high school or an equivalent education, and 27 percent higher vocational

Table 2.3 Funding structure of social organizations in China

Type of funding	*Percentage*	*Ranking*
Government subsidies and grants	50.0	1
Membership fees	21.2	2
Income through commercial activities	6.0	3
Project funding or grants by companies	5.6	4
Other	4.0	5
Project funding by the government	3.6	6
Collection funding	2.2	7
Donations by individual members (excluding membership fee)	2.0	8
Surplus funding from previous year	1.8	9
Project funding or grants by international organizations and governments	1.6	10
Capital turnover	1.2	11
Project funding or grants by domestic foundations	0.5	12
Loans (commercial and non-commercial)	0.3	13

Source: translated and drawn by author on the basis of data from NGO Research Center (G.S. Deng, 'A Preliminary Analysis of the NGO Survey' in Wang M. (ed.), *Zhongguo NGO Yanjiu* (Research on NGOs in China), no. 43, Beijing: UNCRD Research Report Series, 2001).

education.[48] Environmental groups seem trapped in a vicious circle: they are unable to secure funding because they lack qualified staff, which they cannot hire in any case, since they cannot afford their salaries. At the same time, however, green NGOs do everything they can to appear small in the eyes of the state, specifically in their numbers of associated volunteers or members. This self-imposed censorship is most obvious in what is perhaps China's largest grassroots NGO, Friends of Nature. When Liang Congjie, the director of Friends of Nature, was once asked how many members his organization counted, he mentioned "a couple of hundred."[49] Yet, even before the height of the organization's popularity in the beginning of the twenty-first century, a conservative calculation estimated that the organization counted over 500 individual members and 1,300 collective members through institutional membership.[50]

The majority of NGOs also considered poor internal management to be among the main constraints in their operations. The lack of a clear and democratic structure for decision-making is perceived as a vital problem in this respect. It is not uncommon for all major and minor decisions to be taken by the NGO leader, with or without consultation of subordinate staff. Moreover, should the leader for some reason become incapacitated, the organization is generally hamstrung until leadership is resumed. This top-down and commandist structure of Chinese NGOs is intertwined with the phenomenon of *guanxi* or interpersonal relations (see the introduction of this volume). Contemporary Chinese society is still strongly influenced by the social obligations originating from the Confucian value system.[51] Although Guthrie predicted that the importance of *guanxi* might decline as socio-economic development progresses,[52] it is overly obvious that linkages of NGOs with the state still exhibit a strongly personalized rather than an institutionalized character. Political influence is mostly gained through individual connections, personal prestige and the networks of one particular person within an NGO (usually the leader), as a result of which there is a certain degree of detachment from the organization itself. A drafter of environmental regulations stated: "When people from NGOs are asked to participate in political discussions, it is not by asking their organization to participate, but a particular individual from that NGO."[53] A short look at a few environmental NGOs makes this perfectly clear.

For instance, the political influence of Friends of Nature is a direct function of its director, Liang Congjie, member of the Chinese People's Political Consultative Conference (CPPCC) and grandson of the famous early twentieth-century reformer Liang Qichao. Similarly, the founder of the South–North Institute of Sustainable Development, Yang Jike, was the former governor of Anhui province, a current member of the Standing Committee (the daily committee) of the Chinese People's Political Consultative Conference and Deputy Director of the Population, Resources and Environment Committee of the CPPCC. The female founder of the Beijing Environmental Protection Foundation, Jiang Xiaoke, is a member of the National People's Congress (NPC) and concurrent commissioner of the NPC Commission on Environmental and Natural Resources Protection. This commission is the highest state organ responsible for policies on environmental protection and natural resource management in China. Moreover, it supervises the State Environmental Protection Administration and its subordinate institutions, and assists in the drafting of new environmental laws and regulations.[54] A similar situation counts for smaller, local NGOs outside the Chinese capital.[55] It should not come as a surprise that many green NGOs have been established by strong personalities who are extremely well connected to Party and state institutions. Particularly during the early days of Chinese environmentalism, only those with a large "backstage" (*houtai*) were willing and able to go through the complicated and protracted procedures to establish a civil organization. One might therefore conclude that although the Chinese NGOs are "non-governmental," they are – particularly the well-established ones – also part of an elite hierarchy.

Shifting opportunities: disputed dams and organized movements?

One might wonder whether in the long run the Communist Party will remain distrustful about any larger organized movements. Allowing organized opposition is also a matter of confidence in society, trusting that its institutions and populace are sufficiently developed and adapted to deal with a pluralist polity. After more than a quarter-century of successful economic reforms, it is stating the obvious that the modern People's Republic of China is unlikely to revert to the past Maoist, totalitarian state. In this sense, one might need to view embeddedness as a transitory stage in China's developmental experiment. An illustration of this is the central authorities' concern with building an environment that will not yield merely obedient supporters of the state but, ultimately, well-functioning, non-governmental entities. At present, the Ministry of Civil Affairs is revising the regulations for social organizations. Some of the expected amendments will prove favorable for NGOs, as the Ministry plans to improve the working conditions for NGO staff,[56] and to clarify the issues of fund-raising and tax payment. In September 1999, the central government adopted the Law for Donations for Public Welfare to alleviate financial pressures on NGOs. The law stipulates that donors can enjoy tax breaks for money donated for the public good. However, the law merely defines a basic framework for charity donations, while it is still uncertain how it needs to be implemented. For this reason, new regulations on how tax breaks can be enjoyed and safeguarded, and how donations for the public good are defined, need to be addressed.[57]

More importantly, the question of *guanxi* and intertwined Party, state and civil organizations, as discussed in the previous section, have attracted the attention of the CCP Politburo. In a confidential notice, it was written that

> in Party and government organizations there is still a relatively large number of county and higher leading Party and government cadres who hold concurrent leading posts in social organizations. To adjust to the needs of reforming our nation's economic system and polity, as well as [to the

> needs] for organizational restructuring; to speed up the change of government functions; to develop the role of social intermediary organization [shehui zhongjie xuzhi] that social organizations should have; and after the approval of leading comrades of the Politburo and State Council [it has been decided that] leading cadres at county level and above in the Party organizations, People's Congresses, government organizations, bodies of the Chinese People's Political Consultative Conference, judiciary bodies, procuratorial departments, as well as their subordinate departments, should not hold concurrent posts in social organizations.[58]

At this point, it is of importance to note the recent changes in the tactics of Chinese environmental activism. Probably the most eye-catching development is the increased contention over dam-building. In the environmental arena, dams are one of the likely contenders to stir up strong popular and political controversy around the globe. We see this in the vehement protests ranging from movements against the Narmada Dam in India, the Danube Dam in Hungary, and the Glen Canyon Dam in the USA.[59] The construction of the world's largest dam – the Three Gorges (Sanxia) Dam in Hubei – incited social conflict from the very beginning. Although the Three Gorges Dam contained all the necessary ingredients to evolve into China's first national figurehead for environmental protest, it actually did not.[60] However, since construction on this mega-dam has started, the conditions for environmental activism have changed. Not only did the Chinese state undergo an overt "greening" as a result of which "the environment" landed high on the political agenda,[61] but also the specific politics around large dams had shifted towards the end of the 1990s. Notably the replacement of Premier Li Peng, the greatest proponent of the Three Gorges Dam, with the more progressive Premier Zhu Rongji proved critical in this regard.[62]

The shifting political opportunity structure for environmental activists is apparent in the eruption of anti-dam movements in various localities. Amongst these, one of the most powerful movements occurred around the Nu River (Nujiang) Dam construction in Yunnan province (a detailed study of this campaign is provided by Calhoun and Yang in this volume).[63] Initially, the campaign sprang forth from an informal "journalists' salon" (*jizhe shalong*) which was hosted by two committed activists, Wang Yongchen of Green Volunteers, a Beijing-based NGO, and Yu Xiaogang of Green Watershed in Kunming. The proclamation of the Environmental Impact Assessment Law on 1 September 2003 meant an expansion of the political space for green activism, and cleared the way for a broad coalition between scientists, journalists, officials,[64] and NGOs (including the influential Friends of Nature and Global Village of Beijing). The ensuing public debate and the participation by a wide variety of different stakeholders implies a marked change in decision-making processes in China. A few months later, the coalition had already led to numerous conferences and workshops, petition letters, news reports, on-line publications, and exhibitions.[65] In April 2004, a first unexpected success was won, when Premier Wen Jiabao suspended the dam construction pending further investigation:

> We should carefully consider and make a scientific decision about major hydro-electric projects like this that have aroused a high level of concern in society, and with which the environmental protection side disagrees.[66]

During a gathering of the journalists' salon in the summer of 2004, it was decided to set up a formal structure to coordinate the anti-dam campaign: the China River Network (CRN). At present, CRN is well established with its own office space in the office of Friends of Nature, one full-time staff member and a number of volunteers, a website (www.chinarivers.ngo.cn), and financially supported by eight institutional members: Green Watershed, Green Volunteers, Friends of Nature, Global Village of Beijing, the Institute for Environment and Development, Tianxia Xi Education Institute, Green Han River, and Green Island.[67] Although at the time of writing, the struggle over the Nu River Dam continues in full force,[68] the campaign has proven two important matters: (i) the Chinese political context changes, and (ii) as it does, so do the social spaces and strategies that are available to activism. In that sense, today's embedded activism might well hide various grassroots' movements that might one day venture into the open.

Concluding observations: embeddedness as a negotiated symbiosis

Various scholars and observers have attempted to typify the newly emerging social spaces in China, as well as the dynamics of civil society that tries to fill these. For instance, Howell argued that

> there is a tension in contemporary China between the emergence of a depoliticized public sphere concerned with philanthropy, service and public goods, and a politicized public sphere that resonates more closely with liberal-democratic versions of civil society as the site for promoting individual rights and checking state power, and critical traditions of protest in China.[69]

This volume, as the other contributions will also demonstrate, argues that Party–state–society interaction is characterized not so much by a tension between depoliticized versus politicized public spheres, but rather by a paradox: a restrictive yet simultaneously conducive environment. *That* is the essence of what we here defined as "embedded social activism": the resultant of a semi-authoritarian political constellation in which the Party-state imposes formal measures of strict control, while still leaving open ample informal avenues for political action by non-governmental groups and activists. A case in point is the development of Chinese environmentalism.

This contribution sought to examine the meaning of embedded environmentalism in China's semi-authoritarian context. As embeddedness is a "two-way affair," both sides of the Party-state and society divide have been reviewed here.

In an embedded, semi-authoritarian context the rules of engagement are negotiable, but it is the state – and particularly the Chinese Communist Party – that ultimately sets the limits of the game. It does so by consciously co-opting critical players in the non-governmental field, and by creating an institutional environment of blurred structures ("government-organized NGOs") between Party-state and society. The case of the Green Students' Association at one of China's leading universities, which is embedded within the Communist Youth League, is a clear illustration of these mechanisms. Moreover, the Chinese authorities do not hesitate to occasionally flash the threat of a clampdown on civil groups. The State Council recently mentioned that "the democratic management and financial administration of some social organizations is abnormal ... which influences China's political and social stability."[70] It is in this light that we must also read the renewed tightening of control over Chinese NGOs in the aftermath of the "color revolutions" in the Ukraine, Georgia, and Kyrgyzstan.

On the other hand, however, as long as environmental NGOs and green activists at least profess their allegiance to the Party and state, and act as a partner instead of an adversary, a substantive deal of maneuvering space and leverage will be their share. Through tactics of "de-politicized politics," self-imposed censorship, and avoiding suggestions of being able to mobilize, a broad-based grassroots movement, green NGOs and activists have managed to secure a well-established societal position. Embedded activism in the Chinese semi-authoritarian context is a negotiated symbiosis between Party, state, and society, whereby the Party-state draws certain boundaries, while NGOs and activists continuously attempt to test the political waters. In this process, the political opportunity structure continuously shifts over time. Whereas anti-dam protests leading to a suspension of construction and renewed environmental impact assessments were a sheer impossibility in the 1990s, this has already happened in China today. In this respect, Chinese embedded activism is not merely a strategy for survival, but more a model of development. Activists' dual strategy of relying on the state for legitimacy, while wooing and linking with citizens for resources, has caused the development of social structures that are successful in mobilizing resources, in framing environmental issues that fit in with citizens' newly perceived identities, and in accumulating environmental expertise versus the state and Party – be it on biotechnology, nuclear energy, or dam-building.

Notes

1 However, it is important to note that a total state control over society has never been the case in China, neither during the early period of reforms (e.g. E. Croll, *From Heaven to Earth: Images and Experiences of Development in China*, London: Routledge, 1994, pp. 112–13; and P.P.S. Ho, *Institutions in Transition: Ownership, Property Rights and Social Conflict in China*: Oxford: Oxford University Press, 2005, Chapter 5) nor during the collectivist period from 1956 to 1976 (e.g. D. Crook and I. Crook, *Revolution in a Chinese Village: Ten Mile Inn*, London: Routledge and Kegan Paul, 1959; A. Chan, R. Madsen and J. Unger, *Chen Village under Mao and Deng*, revised edition, Berkeley: CA: University of California Press, 1992, pp. 32–3).

2 In contrast to the idea of the declining authority of Party power over the period of reforms, other scholars maintain that instead of a decline it has rather changed in nature. See for instance, J. Tong, "An Organizational Analysis of the Falun Gong: Structure, Communications, Financing," *The China Quarterly*, No. 171 (September 2002), pp. 636–60. John Burns demonstrated that throughout the 1990s the CCP has maintained tight control over the institutions and processes for creating and deleting official posts, specifically in order to preserve political patronage and social stability: J.P. Burns, "'Downsizing' the Chinese State: Government Retrenchment in the 1990s," *The China Quarterly*, No. 175 (September 2003), pp. 775–802.

3 See also S. Zheng, "The New Era in Chinese Elite Politics," *Issues and Studies*, Vol. 41, No. 1 (March 2005), pp. 190–203; K. Lieberthal, *Governing China: From Revolution through Reform*, New York: W.W. Norton, 1995, pp. 208–14. In addition, various scholars have also recorded the decline of Party authority over the past years: K. Lieberthal and M. Oksenberg, *Policy Making in China: Leaders, Structures and Processes*, Princeton: Princeton University Press, 1988; and K. Lieberthal and D.M. Lampton (eds) *Bureaucracy, Politics and Decision Making in Post-Mao China*, Berkeley and Los Angeles: University of California Press, 1992.

4 See J. Power, "Will Democracy be Another 30 Years in Coming in China?" *Daily Times*, 3 June 2005, pp. 3–4; online, available at www.dailytimes.com.pk/default.asp?page=story_3-6-2005_pg3_4. This article later appeared in the *New York Times* of 4 June 2005.

5 See K.J. O'Brien, "Agents and Remonstrators: Role Accumulation by Chinese People's Congress Delegates," *The China Quarterly*, No. 138 (1994), pp. 359–80; M. Manion, "Chinese Democratization in Perspective: Electorates and Selectorates at the Township Level," *The China Quarterly*, No. 163 (2000), pp. 764–82.

6 An interesting case of the review of the Land Administration Law and related national regulations is described in Chapter 2 of P.P.S. Ho, *Institutions in Transition*. See also S.B. Lubman (ed.) *China's Legal Reforms*, Oxford: Oxford University Press, 1996.

7 As Judd Howell maintains; see J. Howell (ed.) *Governance in China*. Boulder, CO: Rowman and Littlefield, 2004, p. 163.

8 This is different from some of the ex-socialist states in East and Central Europe, such as Hungary, East Germany and Bulgaria, where the environmental movement emerged as a broad popular movement that pushed for democratization. See, for instance, B. Jancar-Webster, "Environmental Movement and Social Change in the Transition Countries," *Environmental Politics*, Vol. 7, No. 1 (1998), pp. 69–90.

9 Qiu Xin, "China Curbs Civil Society Groups," *Asia Times Online*, 19 April 2005, pp. 1–3.

10 J. Duo (ed.) *Shehui Tuanti Guanli Gongzuo* [Work on the Administration of Social Organizations], Beijing: Zhongguo Shehui Chubanshe, 1996, p. 9.

11 D. Qing and E.B. Vermeer, "Do Good Work, but Do Not Offend the 'Old Communists'" in W. Draguhn and R. Ash (eds) *China's Economic Security*, Curzon Press: Richmond, 1999, p. 144.

12 Complicating this governance model is the fact that the regulatory framework *after* registration is unclear and incomplete. This relates to a wide range of issues varying from regulations on fund-raising to requirements for the auditing of income and expenditures of social organizations.

13 See also P.P.S. Ho, 'Greening without Conflict? Environmentalism, NGOs and Civil Society', *Development and Change*, Vol. 32 (2001), pp. 893–921.

14 Ministry of Civil Affairs (ed.) *Minjian Zuzhi Guanli Zuixin Fagui Zhengce Huibian* [Compilation of the Newest Regulations and Policies for the Management of Civil Organizations], Confidential material [*neibu ziliao*]: Beijing, speech by Deputy Minister Xu Ruixin (8 December 1999), 2000, pp. 141–2.

15 Qiu Xin, "China Curbs Civil Society Groups," pp. 1–3.

16 Anonymous oral communication, April 2004. The respondent is the director of one of China's foremost environmental NGOs.

17 In addition, the federation was also supplied with a substantive "dowry" of 1.29 million RMB allocated by the Ministry of Finance for its daily operations and activities. See Qin Chuan, "Government Turns up NGO Volume," *China Daily*, 26 April 2005, p. 5; online, available at www.chinadaily.com.cn.

18 The *China Youth Daily* was set up in 1951 by the Communist Youth League. Ultimately, the *China Youth Daily* falls under the authority of the Propaganda Department of the CCP.

19 J. Ma, "Green Groups Fall under Microscope in Beijing," *South China Morning Post*, 18 August 2005, p. 12. According to Lin, the Chinese government has recently also stepped up control over international NGOs, fearing that they might destabilize Chinese society; see Y. Lin, "China Worries that Foreign NGOs are Importing a 'Color Revolution'," Sun/Central News Agency, Hong Kong, 6 August 2005, available at www.chinalaborwatch.org.

20 Many GONGOs serve as reservoirs for officials laid off during the reforms, enabling them to continue their work – in their new capacity as 'NGO-staff' – more or less autonomously from their previous government-employer.

21 M. Wang, "Overview of China's NGOs" in M. Wang (ed.) *Zhongguo NGO Yanjiu* [Research on NGOs in China], no. 38, Beijing: UNCRD Research Report Series, 2000, p. 20. In the same vein, Bryant claims that: "NGO-mimics, can be an unwelcome and potentially corrosive force eating away the good reputation of real grassroots NGOs.... The establishment of NGOs-mimics is a tactic by which individuals or groups seek to acquire some of the moral respectability that NGOs possess in order to advance their own self-serving agendas"; R. Bryant, "False Prophets? Mutant NGOs and Philippine Environmentalism," *Society and Natural Resources*, No. 15 (2002), pp. 629–39.

22 See F. Wu, "New Partners or Old Brothers? GONGOs in Transnational Environmental Advocacy in China," *China Environment Series*, No. 5 (2002), pp. 45–58. An example of a GONGO that fulfills a critical role in Chinese civil society, the Beijing Environmental Protection Foundation established by Jiang Xiaoke, is given in P.P.S. Ho, 'Greening without Conflict?', pp. 893–921.

23 In the case of the government-organized student associations, Lu writes, "After years of working to strengthen their groups and develop creative activities, student environmental associations leaders have acquired not only strong organizational skills but also environmental literacy and passion that they will integrate into their work as China's new generation of officials, teachers, reporters and NGO leaders"; H. Lu, "Bamboo Sprouts after the Rain: The History of University Student Environmental Associations in China," *China Environment Series*, No. 6 (2002), pp. 55 and 57.

24 For instance, the May 4th movement in 1919 which started as over 5,000 students of Peking University demonstrated against the ceding of Shandong province to Japan, or the Tian'anmen uprising in 1989 when students from a great variety of universities protested against the Communist government.

25 For reasons of privacy, the respondent from which this account originates has asked the author not to reveal the name of the university. The respondent was a doctoral student and Party Secretary of the CYL at this university (oral communication, April 2004).

26 The Communist Party controls and directs the government apparatus through an interlocking system of Party Committees parallel to that of the People's governments at all levels. It falls outside the scope of this contribution to describe the general workings of the Party and its interaction with the government. For more details see J.C.F. Wang, *Contemporary Chinese Politics: An Introduction*, 3rd edition, New Jersey: Prentice-Hall Editions, 1989, pp. 130–54; K. Lieberthal, *Governing China*, pp. 208–14.

27 There is also switching of positions. For instance, it is noteworthy that the former president of the Green Students' Association is currently one of the tutors in the Students' Work Team.
28 Moreover, each year is divided into three classes (*ban*), with each class managed by a student representative (*banzhang*) and one of the teachers as class head (*banzhuren*).
29 Members of the Students' Work Team are paid a monthly allowance of several hundred yuan per month by the Department's Party Committee depending on whether they are PhD or graduate students.
30 Of these 5.6 percent indicated that their advice had not been taken into account, while 8.6 percent stated that they were uncertain whether their policy advice had been considered. Some successful policy advice in the environmental field are the Green Community Project of Global Village of Beijing and the Green Energy Program of the South–North Institute for Development that both received positive government attention; G.S. Deng, "A Preliminary Analysis of the NGO Survey" in M. Wang (ed.) *Zhongguo NGO Yanjiu* [Research on NGOs in China], no. 43, Beijing: UNCRD Research Report Series, 2001, p. 23.
31 A recent example is a training course in environmental law for lawyers and judges organized by the non-governmental Center for Legal Assistance to Pollution Victims, for which they invited lawyers and legal experts from around the nation.
32 Friends of Nature, Global Village of Beijing and Green Earth Volunteers.
33 N. Young, "Green Groups Explore the Boundaries of Advocacy," *China Development Brief*, Vol. IV, No. 1 (2001), pp. 3–4.
34 Zhang Tianzhu, oral communication, 4 March 2002.
35 Organization with highest responsibility: national government 29.3 percent; local governments 22.3 percent; companies 19.3 percent; individuals 16.5 percent; and social organizations 11.2 percent; SEPA (ed.) *Quanguo Gongzhong Huanjing Yishi Diaocha Baogao* [Research Report on Public Environmental Awareness in China], Beijing: Zhongguo Huanjing Kexue Chubanshe, 1999, pp. 28–9.
36 M.T. Rock, "Integrating Environmental and Economic Policymaking in China and Taiwan," *American Behavioral Scientist*, Vol. 45, No. 9 (2002), p. 1449.
37 Organization with highest responsibility: national government 29.3 percent; local governments 22.3 percent; companies 19.3 percent; individuals 16.5 percent; and social organizations 11.2 percent; SEPA (ed.) *Quanguo Gongzhong Huanjing Yishi Diaocha Baogao* [Research Report on Public Environmental Awareness in China], Beijing: Zhongguo Huanjing Kexue Chubanshe, 1999, pp. 22 and 28–9.
38 Activities by transnational NGOs are dealt with in the contributions by Morton, and Sun and Tysiachniouk.
39 During a two-year period participants receive a total of 16 weeks' training. Under the auspices of the overarching institution, LEAD International, the Institute for Environment and Development has trained over 100 young professionals from public and private sectors. See also www.enviroinfo.org.cn.
40 The Global Village of Beijing Environmental Culture Centre was established in 1996 by the charismatic Liao Xiaoyi, a famous TV personality. The fame of its leader has helped Global Village of Beijing to become relatively well known among the Chinese public.
41 The Hebei Law Support Office was set up by Cao Huanzhong, the director of Hebei Century United Law Firm in the fall of 2003 with a similar mission and range of activities as the famous Centre for Legal Assistance to Pollution Victims (CLAPV) in Beijing. All seven experts connected to the center work on a voluntary basis. If necessary, pollution victims will be introduced to cooperating attorneys in Hebei Century United Law Firm, who can provide legal services on a no-cure no-pay basis. Obtaining independent registration, as an NGO is extremely hard within China's current legal and political context. Therefore the environmental law support center has been set up under the Hebei Research Center of Environment and Resources Law, which is

registered as an NGO or social organization under the Hebei province Social Science Academy.

42 SNISD was founded in 1998 by Chen Qing and Yang Jike. SNISD's activities include research, training and consultancies, vis-à-vis concrete environmental protection and sustainable development activities at the regional, national and international level.

43 C. Yin, *Women and Energy: A Story in Baima Snow Mountain Reserve*, South–North Institute for Sustainable Development: Beijing, 2002, pp. 2–3. Currently green energy is not yet available in Beijing. The only state-owned energy company does not supply wind energy, although it is produced in sufficient quantities in Inner Mongolia; see www.snisd.org.cn.

44 R. Dwivedi, "Environmental Movements in the Global South. Issues of Livelihood and Beyond," *International Sociology*, Vol. 16, No. 1 (2001), pp. 11–31.

45 Among 1,564 NGOs it was found that only a small percentage (14.7 percent) made financial reports that had been checked by registered accountants; see G.S. Deng, "A Preliminary Analysis of the NGO Survey" in M. Wang (ed.) *Zhongguo NGO Yanjiu* [Research on NGOs in China], no. 43, Beijing: UNCRD Research Report Series, 2001, p. 21.

46 See also T. Cheng and M. Selden, "The Origins and Social Consequences of China's Hukou System," *The China Quarterly*, No. 139 (September 1994), pp. 644–69; and M. Selden, "Household, Cooperative, and State in the Remaking of China's Countryside" in E.B. Vermeer, F.N. Pieke and W.L. Chong (eds) *Cooperative and Collective in China's Rural Development: Between State and Private Interests*, New York: M.E. Sharpe, 1998.

47 For Global Village of Beijing the budget over 1996–9 amounted to US$390,474; for Friends of Nature this was 252,600 RMB; see L. Feng, "Cultural Center of the Global Village of Beijing" in M. Wang (ed.) *Zhongguo NGO Yanjiu* [Research on NGOs in China], no. 38, Beijing: UNCRD Research Report Series, 2000. However, the time frame of this budget has not been specified; see W.A. Hu, "Friends of Nature," in M. Wang, "Overview of China's NGOs" in M. Wang (ed.) *Zhongguo NGO Yanjiu* [Research on NGOs in China], no. 38, Beijing: UNCRD Research Report Series, 2000, p. 175. For more details on the sources of income of various NGOs, see also N. Young, "250 Chinese NGOs; Civil Society in the Making," *China Development Brief*, Beijing, special edition, 2001.

48 In addition, 2 percent had a master's degree; G.S. Deng, "A Preliminary Analysis of the NGO Survey" in M. Wang (ed.) *Zhongguo NGO Yanjiu* [Research on NGOs in China], no. 43, Beijing: UNCRD Research Report Series, 2001, p. 18.

49 Personal communication, April 2002.

50 In 1999, the largest institutional members included the Green Students' Association of Peking University (approximately 150 members); the Association for Environment and Development of Peking University (460 members); the Mountain's Promise Society of the Beijing Forestry University (180); and the Environment and Development Society of the China People's University (60). At present, Friends of Nature also counts many environmental NGOs among its members, such as Global Village, the Institute for Environment and Development, and the South–North Institute for Sustainable Development; X. Zhao, *Fazhanzhong de Huanjing Baohu Shehui Tuanti* [Environmental Social Organizations in Development], Tsinghua University, Beijing: NGO Research Center Book Series, 1999, p. 46.

51 The importance of *guanxi* in Chinese society can hardly be overstated, and Yang even goes as far as describing the coalescence of *guanxi* networks in China as a kind of nascent civil society. Yang states in this respect that *guanxi*, as the realm of people-to-people relationships, exists separately from formal bureaucratic channels, is long realized in social practice and now beginning to emerge in institutionalized form; see M. Yang, *Gifts, Favors and Banquets: The Art of Social Relationships in China*, Ithaca: Cornell University Press, 1994; and C. Hann and E. Dunn, *Civil Society, Challenging Western Models*, London: Routledge, 1996, p. 201.

52 D. Guthrie, "The Declining Significance of *Guanxi* in China's Economic Transition," *The China Quarterly*, No. 154 (June 1998), pp. 254–82; and T. Gold, D. Guthrie and D. Wank (eds) *Social Connections in China: Institutions, Culture and the Changing Nature of Guanxi*, Cambridge: Cambridge University Press, 2002.
53 Zhang Tianzhu, oral communication, 4 March 2002.
54 For a description of this body, see M. Palmer, "Environmental Regulation in the People's Republic of China: The Face of Domestic Law," *The China Quarterly*, No. 156 (December 1998), p. 793.
55 For instance, Green Han River, an environmental NGO established in 2002 in Xiangfan city in Hubei province, is headed by a female president, Ms Yun, who has served on the municipal government for over 20 years. In 1993, she was selected as a candidate to be mayor of Xiangfan. The executive director of this NGO, Mr Li, is a permanent member of the Xiangfan City Legislative Committee. Oral communication, December 2003.
56 Employment in China is supposed to carry with it welfare benefits, such as health care, pensions and the household administration, or *hukou.* Presently it is impossible both legally and financially for NGOs to provide such benefits to their staff, which makes working in an NGO comparably unattractive.
57 M. Wang, "Overview of China's NGOs," in M. Wang (ed.) *Zhongguo NGO Yanjiu* [Research on NGOs in China], no. 38, Beijing: UNCRD Research Report Series, 2000, p. 18.
58 Chinese Communist Party Politburo, "*Zhonggong Zhongyang Bangongting Guowuyuan Bangongting guanyu Dangzheng Jiguan Lingdao Ganbu bu Jianren Shehui Tuanti Lingdao Zhiwu de Tongzhi*" [Notice by the CCP Politburo and the State Council regarding the fact that leading cadres in Party and government organizations should not hold concurrent leading posts in social organizations], Beijing, No. 1998/17, 2 July 1998, pp. 1–2.
59 For controversies around dams, see B.R. Taylor (ed.) *Ecological Resistance Movements: The Global Emergence of Radical and Popular Environmentalism*, Albany: State University of New York Press, 1995, pp. 90–4; W.M. Adams, *Green Development: Environment and Sustainability in the Third World*, 2nd edition, London and New York: Routledge, 2001, p. 376.
60 J. Jing, "Rural Resettlement: Past Lessons for the Three Gorges Project," *The China Journal*, No. 38 (July 1997), pp. 65–94; J. Jing, "Environmental Protest in Rural China" in E.J. Perry and M. Selden (eds) *Chinese Society; Change, Conflict and Resistance*, London: Routledge, 2000.
61 P.P.S. Ho and E.B. Vermeer (eds) *China's Limits to Growth: Greening State and Society*, Oxford: Blackwell, 2006.
62 The change in politics is apparent in the greater sensitivity to environmental impacts of large dams. See, for instance, G. Heggelund, "Resettlement Programs and Environmental Capacity in the Three Gorges Dam Project" in P.P.S. Ho and E.B. Vermeer (eds) *China's Limits to Growth.*
63 Similar protests have occurred against the dam that threatens to inundate the famous Dujiangyan project – a 2,200-year-old engineering marvel in Sichuan province.
64 Two officials of the State Environmental Protection Agency were close supporters of the campaign against the Nujiang River Dam: Mu Guangfeng, the former director of the Environmental Impact Assessment Department of SEPA (and a close friend of environmental activist Wang Yongchen), and Pan Yue, the Vice Minister of SEPA; see Y. Lu, *Environmental Civil Society and Governance in China*, London: Chatham House, 2005, p. 2.
65 R. Litzinger, "Damming the Angry River," *China Review Magazine*, Issue 30 (November 2004).
66 Wen Jiabao quoted in J. Yardley, "Beijing Suspends Plan for Large Dam," *International Herald Tribune*, 8 April 2004, p. 1, available at www.irn.org/programs/nujiang/.

67 See also M. Buesgen, *The Campaign Against the Nujiang Dam*, The Hague: Institute of Social Studies, Research Paper, 2005, pp. 26–36.

68 J.T. Shi, "Watchdog Clears the Way for Power Plant Construction to Restart," *South China Morning Post*, 17 February 2004, available at http://www.scmp.com/; J. Yardley, "Beijing Suspends Plan for Large Dam," *International Herald Tribune*, 8 April 2004, available at www.irn.org/programs/nujiang/index.asp?id=041304_ihtnyt.html; J. Yardley, "Dam Building Threatens China's 'Grand Canyon'," *The New York Times*, 10 March 2004; and International Rivers Network (ed.) *Human Rights Dammed Off at Three Gorges: An Investigation of Resettlement and Human Rights Problems in the Three Gorges Dam Project*, Berkeley, CA: IRN, 2003.

69 J. Howell (ed.) *Governance in China*, p. 163. Saich has mentioned that state–society interaction is more a matter of negotiation, and in that sense comes closer to the concept of "embeddedness"; T. Saich, "Negotiating the State: The Development of Social Organizations in China," *The China Quarterly*, No. 161 (March 2000), pp. 125–141.

70 State Council Legal Affairs Office, *Shehui Tuanti Dengji Guanli Tiaoli Shiyi* [Interpretation of the Administrative Regulations on the Registration of Social Organizations], Beijing: Zhongguo Shehui Chubanshe, 1999, p. 11.

3 Corporatist control of environmental non-governmental organizations

A state perspective

Jiang Ru and Leonard Ortolano

Since the 1990s, scholars have examined the relationship between the state and society in China by examining the development of associations. To make a contribution to this ongoing discussion of state–society relations in contemporary China, we focus on environmental non-governmental organizations (NGOs), one of the most developed forms of social associations in China.[1] The concept of "embedded activism" as set out in the introduction of this volume describes state–society relations in China as a paradox: restrictive and conducive at the same time. Embeddedness refers to the nature of state control, as well as the specific reaction it triggers among civil society actors. In this contribution, we will explore in more detail the first aspect of embeddedness: the nature of state control. For this purpose, we first highlight measures employed by the state to control NGOs, and then investigate how these control measures have been implemented based on experiences from 11 national and 11 Beijing municipal environmental NGOs. Finally, we explore factors that have shaped patterns of state control experienced by the studied NGOs. In this paper, the term "state" refers specifically the Central Committee of the Chinese Communist Party and the State Council.

Based on evidence collected from 22 domestic environmental NGOs in Beijing,[2] we argue that the concept of corporatism (using the liberal democratic definition) is useful to understand state–society relationships in China. Even though the state has developed a complicated and rigorous set of measures to promote corporatism, state controls have been implemented in practice as *agency control*; i.e. control by government agencies that sponsor and supervise NGOs. Apart from its influence in writing regulations and implementing the NGO registration process, the control exerted by the civil affairs *xitong* (literally: "system") – the agent representing the state in matters related to NGOs – has been minimal. Furthermore, we find that the extent to which state control measures have been implemented is significantly affected by the administrative capacity of the state, the interests of agencies involved in implementing the state's control measures, the influence of NGO leaders who were personally or politically well connected to government officials, and NGOs' self-censorship.[3] Based on our finding that state-authorized supervisory agencies have dominated the NGO control process and that Chinese citizens have organized and operated some environmental NGOs with minimal state interference because of self-censorship, we propose to use the concept of *agency corporatism* to describe the relationship between environmental NGOs and the state at the beginning of the twenty-first century.[4]

State control measures: a corporatist structure

Since the late 1980s, the Chinese government has developed a complex set of NGO regulations to control the development and activities of social organizations in China. These regulations are implemented in the context of a multi-level registration system, which requires NGOs to register at the Ministry of Civil Affairs (MOCA) or local civil affairs offices based on the geographic scope of the NGOs' activities. In addition, a dual surveillance system is employed: an NGO must accept supervision from a civil affairs office and the NGO's supervisory organization.[5] A supervisory organization, referred to as a "mother-in-law organization" or "sponsoring organization" by some scholars, is a state-authorized organization that sponsors an NGO's registration application prior to and during the registration process, and then supervises the NGO's activities after the NGO registers with the appropriate civil affairs office.[6]

A core feature of the state's multi-level registration system and the dual surveillance system is a state-granted monopoly of representation: only one NGO may work on a particular set of issues in one administrative area. This is established in Article 13.2 of the 1998 Regulation on Registration and Supervision of Social Organizations (hereafter the 1998 RRSSO). The state has adopted three distinct categories of control measures – registration, routine supervision, and coercive measures – to ensure that all NGOs in China are following the state's NGO regulations and policies. These control measures can be interpreted as a *corporatist structure* that the state uses to control NGOs.[7]

Registration control measures

To be legally recognized as an NGO, an organization must go through a rigorously controlled process that ends with the organization being registered with a civil affairs office at the appropriate level of government.[8] This process includes six control measures: monopoly of representation, restricted geographic scope of activities, approval from a supervisory organization, compliance with state policy, ban on regional chapters, and minimum funding requirements. According to the monopoly of representation, a prospective NGO cannot be registered if another NGO has already registered to conduct a similar set of activities in the same geographic area.[9] The control measure on restricted geographic scope of activities reinforces this monopoly by requiring the prospective NGO to act within a restricted geographic boundary. These two measures are consistent with the corporatism concept, which emphasizes an association's representational monopoly over a particular type of mission in a given area.

To become registered, a prospective NGO must find a supervisory organization to sponsor its application for registration at a civil affairs office. Without such support, the prospective NGO is not permitted to initiate activities to prepare for registration. The prospective NGO and its proposed activities must comply with state policy and pose no political threat to the state. In addition, the prospective NGO is not allowed to establish any regional chapters. Thus, a national NGO can conduct activities anywhere in China, but it is not allowed to create a national network of local chapters. Moreover, an NGO registered with a particular municipal civil affairs office is not allowed to conduct activities outside of that municipality. The last registration control measure concerns funding. Except for foundations, the funding requirement is a minimum of 100,000 RMB in operating funds for national NGOs, and a minimum of 30,000 RMB in operating funds for local NGOs. All foundations, whether they are national or local, must have at least a two million RMB endowment in addition to a minimum of 100,000 RMB in operating funds.[10]

Routine supervision measures

After a prospective NGO is registered, the state employs the following routine supervision measures: membership dues control, three Chinese Communist Party (CCP) members rule, routine activity and financial supervision, and annual check. Under the dual surveillance system, supervisory organizations work with civil affairs offices to oversee activities of NGOs registered with civil affairs offices. Each NGO (except foundations which, by definition, are not membership-based NGOs) must comply with the membership dues limits set up by MOCA in 1992.[11] As of 2002, an NGO could only set its annual membership dues for individual members at a maximum of 10 RMB (or about US$1.2) and for institutional members at levels calculated based on the nature of the institution. For an NGO that has more than three CCP members in its secretariat, the NGO must create a CCP group within its secretariat.[12]

All other routine supervision measures involve day-to-day activities. In particular, an NGO must have its activities and financial affairs overseen by its supervisory organization, which is authorized to *guide* the major activities of the NGO.[13] In addition, at the end of every year, each NGO must go through an annual check procedure performed by a civil affairs office; this provides still another opportunity for the government to scrutinize an NGO's activities and financial performance.[14] If an NGO fails the annual check, it is given a grace period to make required changes.[15] If an NGO commits wrongdoings, it may face coercive control measures.

Coercive measures

If an NGO registered with a civil affairs office fails to comply with the registration and routine supervision measures, the relevant civil affairs office can force the NGO to correct its wrongdoings by issuing a warning, suspending the NGO's activities, asking the NGO to replace culpable leaders, confiscating its illegal income, collecting fines, and, finally, revoking the NGO's registration.[16]

These measures target NGO misdeeds as defined in the 1996 Ad Hoc Measures for NGO Annual Check and the 1998 RRSSO. In addition, civil affairs offices can use the 2000 Measures for Illegal NGOs to declare any NGO that is either unregistered or improperly registered to be illegal, and then confiscate its assets and ban it.

Implementation of state control measures

Between the late 1980s and 2002, agencies that implemented the state's NGO control measures carried out their work in three ways: regular NGO surveillance, reactive repression, and cleanup-and-consolidation campaigns. Regular surveillance refers to the control measures implemented routinely through processes linked to NGO registration, activity and financial supervision, and annual checks. An example of this is the increased control over NGOs in China shortly after the color revolutions occurred in the Ukraine, Kyrgyzstan and Georgia (see the contribution by Ho). "Reactive repression" refers to activities taken against NGOs that civil affairs offices or other state agencies view as troublesome. There have been sporadic cases of repression in response to NGO activities: for instance, the state's actions against the Falun Dafa Research Society (i.e. Falun Gong) in 1999.[17]

The state uses cleanup-and-consolidation campaigns to overhaul its NGO policies and regulations in response to unanticipated consequences that arise as NGOs develop and as new regulations are issued. Between 1978 and 2002, the state conducted two cleanup-and-consolidation campaigns: one from 1988 to 1991, and another from 1997 to 2002. During such a campaign, civil affairs offices scrutinize activities of NGOs, correct perceived wrongdoings, and "clean up" NGOs that have caused trouble for the state. On the other hand, NGOs with similar missions and activities, yet which are found in line with state policies, may find themselves being consolidated. In the first campaign, the state made civil affairs offices the sole agency in charge of NGO supervision.[18] In addition, the state issued three new NGO regulations to strengthen its control over NGOs.[19]

Take the 1997–2002 cleanup-and-consolidation campaign as an example. The state initiated this campaign after it observed difficulties linked to its control over NGOs.[20] In response to these problems, the state clarified that only MOCA and local civil affairs offices were authorized to register NGOs.[21] The state also delineated carefully the responsibilities of supervisory organizations and civil affairs offices involved in the dual surveillance system.[22] In addition, the State Council promulgated the 1998 RRSSO and MOCA promulgated the 2000 Measures for Illegal NGOs. MOCA and local civil affairs offices also issued more restrictive national and local lists of organizations that could serve as NGO supervisory organizations. In 1999, an incident involving the previously mentioned Falun Gong motivated the state to tighten its control over NGOs involved in *qigong*-practices (a traditional kind of Chinese yoga), and this effort prolonged the entire cleanup-and-consolidation process.[23] As a result of the

1997–2002 campaign, the number of national NGOs dropped from 1,845 in 1996 to 1,687 in 2001, and the number of local NGOs fell from over 184,000 to about 127,000 during the same period.[24]

Experience of environmental NGOs in Beijing

The state's registration control measures (especially the monopoly of representation) and the drastic cuts in the number of registered social organizations as a result of two cleanup-and-consolidation campaigns can be understood as the state's effort to promote corporatism by imposing rules to ensure that relations with NGOs are non-threatening and carefully controlled. But has the state achieved this goal? To examine this question, we investigated the impact of state control measures on environmental NGOs.

Sample environmental NGOs

In designing our study, we intended to examine state control over environmental NGOs at both national and local levels. Because of the need to rely heavily on interviews with staff members of NGOs and their supervisory agencies and because of time and budget constraints on our research project, we conducted our data-gathering in one place – Beijing. This allowed us to examine NGOs that operated nationally as well as NGOs with activities confined to being within the boundaries of a single municipality. At the same time, however, we are aware of the fact that Beijing – as the nation's political centre – undoubtedly has its specific influence on the functioning and activity scope of social organizations. It might be expected that state regulations will be more rigorously enforced in Beijing than in other localities. For a discussion of the activities of green NGOs in other locations, we refer to the contributions by Sun and Tysiachniouk (Heilongjiang), and Morton (Qinghai and Yunnan). For NGOs operating nationally, the NGO registration process is handled by MOCA, whereas for those operating only within Beijing, registration is through the Beijing Civil Affairs Bureau (BCAB).

As of 2002, 63 environmental NGOs operated nationally, and 37 environmental NGOs had operations restricted to being within Beijing municipality.[25] Among these 100 domestic environmental NGOs, 58 out of the 63 national NGOs were registered with MOCA, and 28 out of the 37 Beijing NGOs were registered with BCAB. Fourteen of the 100 national or Beijing municipal environmental organizations were not registered with civil affairs offices, and all 14 had been *voluntarily* organized by citizens. Three of these 14 NGOs elected to register as for-profit companies with bureaus of industrial and commercial administration. Four were registered with existing organizations. The remaining seven environmental NGOs were unregistered, even though being unregistered violates regulations.

We label all environmental NGOs that did not register with civil affairs offices as *de facto* environmental NGOs. The NGOs registered with civil affairs offices were officially classified as follows: industrial associations, academic societies, professional associations, foundations and federations (*lianhe xing shetuan*).[26] To further narrow the scope of our study, we selected the subset of the 100 environmental NGOs that were concerned with wildlife conservation. We made this choice for three reasons: wildlife conservation is a major area of activity for many Chinese environmental NGOs, NGOs have been working in this field since the early 1980s, and the activities of many NGOs that work on wildlife conservation have been well documented.

Once we chose to focus on wildlife conservation, the set of NGOs we needed to consider was reduced from 100 to 22. This reduced set of NGOs included three of the five types of NGOs that register with civil affairs offices – academic societies, professional associations, and foundations. The set also included de facto NGOs (i.e. unregistered NGOs, and NGOs registered with either existing organizations or with industrial and commercial administration bureaus).[27]

Table 3.1 summarizes the 22 environmental NGOs working on wildlife conservation. Because the discussion would be awkward and unnecessarily complex if we used the full names of the NGOs, we use code names for the NGOs. The code names in Table 3.1 are based on the NGO's geographic scope (the first letter of the names) and type (letter or letters after the first letter of the names). Thus, all NGOs with "N" as the first letter are national, and all with "B" as the first letter are Beijing municipal NGOs. All academic societies have "AS" as the second and third letters of their code names, all professional association have "PA," and all de facto NGOs have "DF." All foundations have an "F" as the second letter of their names. For example, NAS-1 in Table 3.1 is one of the national academic societies included in our study. Among the 22 environmental NGOs we studied, the following nine were voluntarily organized by citizens: BAS-4, NPA-3, BPA-3, BPA-4, and all five de facto NGOs.

Organizations supervising the studied NGOs

According to the law, two types of organizations – civil affairs offices and supervisory organizations – are responsible for overseeing the activities of the

Table 3.1 Case study environmental NGOs

Types	*Level*	*Code name*
Academic society	National	NAS-1, NAS-2, NAS-3
	Beijing	BAS-1, BAS-2, BAS-3, BAS-4
Professional association	National	NPA-1, NPA-2, NPA-3
	Beijing	BPA-1, BPA-2, BPA-3, BPA-4
Foundation	National	NF-1, NF-2
	Beijing	BF-1
De facto NGOs	National	NDF-1, NDF-2, NDF-3
	Beijing	BDF-1, BDF-2

studied environmental NGOs. A supervisory organization may request another type of organization, an "affiliated organization," to assist by hosting the NGO that is to be supervised. In this context, "hosting" consists of providing offices and other facilities for the NGO. By definition, de facto NGOs are not registered with civil affairs offices, and thus they have no supervisory or affiliated organizations defined by the 1998 RRSSO. How ever, de facto NGOs registered with industrial and commercial administration offices or existing organizations are subject to supervision by either the industrial and commercial administration offices or the existing organizations with which they are affiliated.

At the national level, the NGO Supervision Bureau within MOCA is responsible for registering and supervising national NGOs (including environmental NGOs). The Bureau is also responsible for proposing and implementing guidelines, policies and regulations for the registration and supervision of social organizations. As is the case for other central government agencies, MOCA (and its NGO Supervision Bureau) maintains a professional relationship (*yewu guanxi*) with local civil affairs offices (and their NGO supervision divisions). MOCA guides and supervises the practice of local civil affairs offices, but local governments control the personnel and budgets of local civil affairs offices. Eight of the 11 national NGOs in our study are registered with MOCA.

The 11 Beijing municipal NGOs in our study are subject to control by BCAB's NGO Supervision Office, but only nine of them are registered with BCAB. The Office has duties similar to those of the NGO Supervision Bureau of MOCA, but those duties are carried out within the administrative boundaries of the Beijing Municipal People's Government. Tables 3.2 and 3.3 list the supervisory and affiliated organizations for the 17 environmental NGOs registered with civil affairs offices at the national and Beijing municipal levels, respectively. As indicated, several of these organizations are supervised by associations of sciences and technology (AST). The China Association of Science and Technology (CAST) and its local ASTs are quasi-government agencies led and fully funded by the CCP.[28] CAST and local ASTs are the designated supervisory organization for NGOs engaged in physical sciences and technology.[29] CAST requires NGOs supervised by either CAST itself or local ASTs to have affiliated organizations that support their operations, and this is not the case for NGOs linked to the other supervisory organizations listed in Tables 3.2 and 3.3.[30] Two types of organizations serve as affiliated organizations of NGOs supervised by CAST and the Beijing Association of Science and Technology (BAST): government agencies and research institutes. Affiliated organizations host the secretariats of the NGOs affiliated with them, but they have no legal responsibility to control NGOs. Normally, affiliated organizations provide office space, staff members, and even some operating funds for the secretariats of the NGOs.

Experience of environmental NGOs with state controls

Not all the state's control measures influence NGOs' operations in the same way. Interviewees from the 22 environmental NGOs examined in our study

Table 3.2 Supervisory and affiliated organizations of national environmental NGOs in Beijing[a]

NGO type	*NGO*	*Supervisory organization*	*Affiliated organization*
Academic societies	NAS-1	China Association for Science and Technology (CAST)	State Environmental Protection Administration (SEPA)
	NAS-2	CAST	Institute of Zoology, China Academy of Sciences (CAS)
	NAS-3	CAST	Research Center for Ecological Environment, CAS
Professional associations	NPA-1	CAST	State Forestry Administration (SFA)
	NPA-2	Ministry of Construction	–
	NPA-3	Ministry of Agriculture	–
Foundations	NF-1	SEPA	–
	NF-2	CAST	Beijing Science and Technology Commission

Note
a This table concerns only NGOs registered with civil affairs offices that are examined in this chapter.

Table 3.3 Supervisory and affiliated organizations of Beijing municipal environmental NGOs[a]

NGO type	*NGO*	*Supervisory organization*	*Affiliated organization*
Academic societies	BAS-1	Beijing Association for Science and Technology (BAST)	Beijing Environmental Protection Bureau (EPB)
	BAS-2	BAST	Beijing Nature Museum
	BAS-3	BAST	Beijing Institute of Botany, CAS
	BAS-4	Beijing EPB	–
Professional associations	BPA-1	Beijing Forestry Bureau	–
	BPA-2	Beijing Forestry Bureau	–
	BPA-3	BCAB	–
	BPA-4	BCAB	–
Foundations	BF-1	Beijing EPB	–

Note

a This table concerns only NGOs registered with civil affairs offices that are examined in this chapter.

reported a wide variety of experiences with state controls. Based on information provided by interviewees from these NGOs, state control measures have either minimal impacts or effects that can be grouped into three categories of constraints: financial, managerial, and legitimacy. In this context, "minimal impact" does not mean that a control measure has no influence; instead, it implies that the NGO can comply with the control measure easily. Financial constraints limit the funds available to an NGO to conduct its activities, managerial constraints restrict the autonomy of an NGO, and legitimacy constraints keep an NGO from being legally recognized. In addition to these impacts, state controls have indirect, implicit effects via NGO self-censorship, a subject we treat in a later section.

Table 3.4, which is based on our interview data, summarizes control measures that influence the operations of NGOs. Because de facto NGOs are not registered with civil affairs offices, they are not subject to routine supervision measures. In addition, because foundations are not membership-based NGOs, they are not subject to membership dues control and the ban on regional chapters. As shown in Table 3.4, only five of the many control measures actually constrained the operation of the NGOs included in our study. Moreover, the patterns of influence of various control measures were generally similar for both the national and Beijing municipal NGOs.

Of the five control measures that had a constraining influence, two involved registration measures (i.e. approval of supervisory organizations and monopoly of representation) that imposed legitimacy constraints. None of the academic societies and foundations experienced difficulties in gaining approval of supervisory organizations, and none of them had any difficulties with legitimacy. However, three out of seven professional associations and all five de facto NGOs had trouble obtaining the approval of supervisory organizations and thus they had problems in gaining legitimacy.[31] The supervisory organizations that refused to sponsor the three professional associations all gave the NGOs the same reason: they had already sponsored similar NGOs, and thus according to the control measure of monopoly of representation they could not offer their support.

The remaining three control measures that imposed managerial and financial constraints concerned routine matters (i.e. routine financial and activity supervision, a ban on regional chapters, and membership dues control). Seven NGOs were managerially constrained by routine supervision of finances and activities.[32] A common mechanism for imposing managerial constraints involved the selection of NGO leaders. Indeed, leaders of each of the seven managerially constrained NGOs were nominated (and their staff members were overseen) by their supervisory organizations (or affiliated organizations in the case of academic societies). The five academic societies and two professional associations that were financially constrained by membership dues control felt their members would have been willing to pay higher dues, and they could have put the additional revenue to productive use.[33] Finally, two national academic societies were financially constrained by the ban on regional chapters because they could not recruit local members directly.[34]

Table 3.4 Impacts of state controls on NGO operations

Control measures	*Impact on NGO operations*	*Academic societies*		*Professional associations*		*Foundations*		*De facto NGOs*	
		National	*Beijing*	*National*	*Beijing*	*National*	*Beijing*	*National*	*Beijing*
Approval of supervisory organizations	Legitimacy constraints			NPA-3	BPA-3, BPA-4			NDF-1, NDF-2 NDF-3	BDF-1, BDF-2
Monopoly of representation	Legitimacy constraints			NPA-3	BPA-3, BPA-4				
Routine financial and activity supervision	Managerial constraints	NAS-1	BAS-1	NPA-1, NPA-2	BPA-1, BPA-2	NF-1		Not applicable	
Ban on regional chapters	Financial constraints	NAS-2, NAS-3				Not applicable		Not applicable	
Membership dues control	Financial constraints	NAS-2, NAS-3	BAS-1, BAS-2, BAS-3		BPA-1, BPA-2	Not applicable		Not applicable	

Nine NGOs managed to evade or violate one or more state controls without penalty. Examples of evasion are as follows: NAS-3 evaded the ban on regional chapters by recruiting local members directly; and BPA-3 and BPA-4 evaded the control measures of monopoly of representation by successfully registering with BCAB (which has nothing to do with their scope of activities) even though both BPA-3 and BPA-4 engaged in similar activities (i.e. doing animal protection work in Beijing). Examples of NGOs that violated state controls without penalty are as follows: NPA-3 violated the ban on regional chapters by setting up a regional chapter in Tianjin in 2001; NAS-3, NPA-3, BPA-3 and BPA-4 violated membership dues control by setting their individual membership dues at 20, 60, 100, 200 RMB/year, respectively; and BPA-3 conducted a conservation activity outside Beijing (in Xinjiang Uighur Autonomous Region), and thereby violated the control measure that restricts the geographic scope of its activities. In addition, de facto NGOs violated NGO registration regulations. According to regulations, they should not have been able to act as NGOs without registering with civil affairs offices.[35]

Civil affairs offices play a minimal role in an NGO's affairs once the NGO has become registered. Neither MOCA nor BCAB exerted any control over the NGOs' routine operations. Instead, they depended on supervisory organizations to exert post-registration controls. As reported by interviewees from the 22 environmental NGOs, these two agencies had little impact on the operations of the NGOs, regardless of the NGOs' registration status. Indeed, interviewees from the 17 registered NGOs reported that MOCA and BCAB, as agencies in charge of NGO affairs, exerted controls only during the NGO registration process. As of 2002, none of the 22 case study NGOs (including the de facto NGOs) was subjected to coercive control measures exerted by MOCA or BCAB. Moreover, neither MOCA nor BCAB took actions to correct the regulatory violations committed by NAS-3, NPA-3, BPA-3 and BPA-4, and the two agencies did nothing to repress any of the five de facto NGOs included in our study. Even during the 1997–2002 cleanup-and-consolidation process, MOCA and BCAB did not bother any of the 22 studied environmental NGOs. The agencies did take action against two national and two Beijing municipal environmental NGOs (not examined herein) by canceling their registration. However, the reasons for doing so were unrelated to the imposition of punishment for violating regulations.[36]

Some of our findings are particular to ASTs. When they acted as supervisory organizations, neither CAST nor BAST exerted any significant controls. In contrast, however, the government agencies that served as affiliated organizations to the NGOs supervised by CAST or BAST (see Tables 3.2 and 3.3) exerted controls over the NGOs with which they were affiliated, even though affiliated organizations are not legally responsible for NGO supervision. This was observed for NAS-1 (with the State Environmental Protection Administration (SEPA) as the affiliated organization), NPA-1 (with the State Forestry Administration (SFA) as the affiliated organization), and BAS-1 (with the Beijing Environmental Protection Bureau (EPB) as the affiliated organization). In these cases, SEPA, SFA, and the Beijing EPB constrained the operations of the NGOs with which they were affiliated by nominating NGO leaders and by supervising staff members. The only exception to this pattern for ASTs is NF-2, a national foundation supervised by CAST and affiliated with the Beijing Science and Technology Commission. In this case, the affiliated organization did not exert managerial control over the NGO. This lack of control may have resulted because several influential, high-ranking governmental officials were involved directly in managing NF-2. In contrast to government agencies that are affiliated organizations, the research institutes serving as affiliated organizations for NAS-2, NAS-3, BAS-2 and BAS-3 had minimal influence on the operations of those NGOs.

A final set of findings concerns correlations between organizational characteristics of the 22 case study NGOs and the effects of state control measures on NGO operations.[37] We observed that only three out of the 11 organizational characteristics in our correlation study – extent of government intervention, degree of autonomy, and existence of influential leaders – were correlated significantly with the three types of constraints: legitimacy, managerial and financial. Here, extent of government intervention is defined as the percentage of full-time staff in an NGO that are government officials (including retired officials); degree of autonomy is defined by the extent to which an NGO is free to select its leaders and conduct its activities; and existence of influential leaders is defined as the presence of NGO leaders who are personally or politically well-connected to government officials.

Our correlation analysis indicated that NGOs with a significant presence of government staff (from either supervisory or affiliated organizations) had relatively little autonomy, but no problem with legitimacy. Seven of the 13 NGOs with government officials in their secretariats were tightly controlled by their supervisory (or affiliated) government agencies.[38] None of these 13 NGOs had difficulty in getting support from their supervisory (or affiliated) agencies. In contrast, the nine NGOs with no government officials as full-time staff members (i.e. BAS-4, NPA-3, BPA-3, BPA-4 and all five de facto NGOs) were highly autonomous in terms of making management decisions. However, eight of them (BAS-4 being the exception) experienced legitimacy constraints because they had problems gaining the sponsorship of a supervisory organization. NF-2, BF-1 and BAS-4 presented exceptions to the above-noted pattern related to autonomy and legitimacy, but these exceptions can be explained by the presence of influential leaders. NF-2 and BF-1 were run by former leaders of the affiliated organization of NF-2 and of the supervisory organization of BF-1, respectively. In addition, BAS-4 was established and managed by an eminent environmental economist who had sustained a strong working and personal relationship with leaders of the supervisory organization of BAS-4, the Beijing EPB. Thus the backgrounds of the leaders of NF-2, BF-1 and BAS-4 were exceptional. Because of their influential connections and particular backgrounds, these individuals managed their NGOs without interference from their supervisory organizations (or from an affiliated organization in the case of NF-2).

Observed patterns of control: agency control or no explicit control

The above findings demonstrate that while the state has developed a corporatist structure with a set of measures to control tightly the operations of NGOs, the influence of these measures depends on the characteristics of both the NGOs and the organizations involved in implementing the state's NGO regulation. Our explanation of how the state exerts control over NGOs makes a distinction between the agencies responsible for designing and enforcing state control policies – MOCA and local civil affairs offices – and the organizations involved in implementing those controls; i.e. NGOs' supervisory and affiliated organizations. Table 3.5 summarizes the previously discussed findings using a format that highlights the influence of civil affairs offices, and supervisory and affiliated organizations on NGO operations.

Our analysis employs the term "agency control" to refer to the control exerted over NGOs by their supervisory or affiliated organizations. We use the term "state control" to refer exclusively to the state controls implemented by MOCA and local civil affairs offices. Some citizen-organized de facto NGOs are subject to neither state nor agency control. We observed that NGOs can evade interference by civil affairs offices or other agencies as long as they are not identified as threats to either the state (as represented by civil affairs offices) or other agencies.

MOCA and BCAB, as the designated agencies in charge of NGO registration and supervision, have had a limited impact on the 22 NGOs except in the context of registration. Indeed, MOCA and BCAB have not taken action against either the six unregistered NGOs or the seven NGOs registered with existing organizations or as for-profit companies with bureaus of industrial and commercial administration. These NGOs were illegal according to the 1998 RRSSO and the 2000 Measures for Illegal NGOs, and thus they could have been disbanded, but they were not. In addition, MOCA and BCAB did not take action against the nine of 22 environmental NGOs whose activities violated one or more state control measures.

As in the case of MOCA and BCAB, neither CAST nor BAST exerted significant controls on NGO operations. Serving as supervisory organizations, these two quasi-governmental agencies barely intervened in the operations of NGOs, and thus they did not fulfill their official duties. In addition, research institutes serving as affiliated organizations of the studied NGOs imposed no formal constraints on the operations of NGOs with which they were affiliated. As summarized in Table 3.5, government agencies acting as NGO supervisory or affiliated organizations were the only organizations to exert strict control on the studied environmental NGOs. Even though these agencies did not correct the wrongdoings of the NGOs they supervised (or with which they were affiliated), the agencies controlled the personnel arrangements of the secretariats of NGOs related to them. An exception occurred in the cases where the NGOs had influential leaders in charge of their operations. Furthermore, agencies serving as

Table 3.5 Impact on NGO operations of state controls

Organization exerting state controls		*Type of NGO*			
		Academic societies	*Professional associations*	*Foundations*	*De facto NGOs*
MOCA and BCAB		Minimal impacts, except strict control over registration			Minimal impact
Supervisory organizations	CAST and BAST	Minimal impacts on the operations of NGOs			Legitimacy constraints due to inability to get support from a supervisory organization
	Government agencies	Managerial constraints, unless the NGO has influential leaders in charge of its operations			Legitimacy constraints due to inability to get support from a supervisory organization
Affiliated organizations	Government agencies	Managerial constraints, unless the NGO has influential leaders in charge of its operations			Not applicable
	Research institutes	Minimal impacts		Not applicable	Not applicable

supervisory organizations have the unique capability to determine whether a prospective NGO will be able to even begin the registration process with a civil affairs office.

So why have civil affairs offices and supervisory (and, in some cases, affiliated) organizations implemented state controls over NGOs in the manner identified above? And why haven't civil affairs offices and supervisory (and affiliated) organizations pursued NGOs that violated state controls? To answer these questions, we examine the capacity constraints and incentives faced by civil affairs offices and supervisory (and affiliated) organizations and the strategic behavior of the 22 NGOs.

Why MOCA and BCAB depend heavily on other agencies

All interviewees from the 22 studied NGOs claimed that MOCA and BCAB rarely interfered with their NGOs. They felt the requirements of MOCA and BCAB were merely procedures they had to follow. Why have MOCA and BCAB been so assiduous about issuing regulations and policies and so apparently disinterested in disciplining or banning environmental NGOs engaged in problematic activities or with improper registration status? According to officials from civil affairs offices, three factors help to answer this question: the shortage of resources at civil affairs offices, the interests of other government agencies, and the increasing demands of citizens for rights to join associations.

Because of *limited resources at civil affairs offices*, those offices are unable to keep track of every NGO, and then they cannot identify and penalize all violations by NGOs. Several interviewees from MOCA and BCAB stated that civil affairs offices lacked the budgets and staff members to implement NGO control measures.[39] The NGO Supervision Bureau of MOCA, which had only 25 staff members, had difficulties with its NGO control tasks, given that there were over 1,600 national NGOs and the Bureau had other duties, such as registering and supervising private non-enterprise entities.[40] BCAB also lacked the resources to implement NGO control measures.[41] According to one MOCA official we interviewed, the 1997–2002 cleanup-and-consolidation campaign aggravated the constraints on staff by forcing civil affairs offices to put more of their resources into NGO registration; that left fewer resources for NGO surveillance.[42] He felt this resource shortage problem was faced by civil affairs offices throughout China.

According to interviewees from MOCA and BCAB, the control exerted by civil affairs offices was also affected by *interests of other government agencies*.[43] Since 1978, agencies have created, staffed, and financed many NGOs. As detailed in the next section, sponsoring agencies have used NGOs for many purposes, such as raising funds and offering jobs for retiring agency staffs. Before the 1989 RRSSO, government agencies had complete responsibility for controlling the NGOs they created. Civil affairs offices once considered eliminating the requirement that every NGO should have a supervisory organization.[44] However, out of respect for the prior relationships between government agencies and the

NGOs they had created, and considering the limited resources to supervise NGOs, civil affairs offices decided to maintain the role of supervisory (or, in some cases, affiliated) organizations in implementing post-registration NGO control measures.[45]

The absence of adequate resources for NGO supervision presents civil affairs offices with another challenge: how to address the *increasing demands of individuals to form associations*. The number of registered NGOs in China increased by about 30,000 between 1992 and 1996.[46] According to an interviewee from BCAB, it would be a practical impossibility for civil affairs offices to register all prospective NGOs; nor was it necessary, given that many such organizations would disappear soon after they had been established as a result of limited funds or diminished interest among members.[47] Interviewees from MOCA and BCAB argued that to encourage voluntary activities as a way to advance the well-being of society, and recognizing the limited resources of civil affairs offices, their agencies would not go after an NGO (whether it was registered or not) as long as the NGO had not committed any financial misdeeds or posed any political threats.[48] This focus on misdeeds and threats together with resource constraints explains why civil affairs offices ignored many de facto NGOs.

Concerns and interests of supervisory and affiliated organizations

Supervisory organizations (and, in some cases, affiliated organizations) have played a significant role in implementing state control measures. At the end of 2002, a supervisory organization could either refuse to sponsor a potential NGO or drop the sponsorship of an existing NGO without giving any reason to the organization. An NGO cannot appeal a decision made by a supervisory organization.[49] (It could, however, look for a different supervisory organization.) On numerous occasions, supervisory and affiliated organizations far exceeded their legally defined roles of *guidance* and *supervision* as articulated in the 1998 RRSSO. For example, some of these organizations have nominated or appointed NGO leaders.

In the context of supervisory organizations, two key questions arise: *why* does a potential supervisory organization choose to sponsor an NGO? And *what* determines how strictly a supervisory organization controls an NGO? We argue that a supervisory organization's decision to sponsor an NGO is based on an implicit (and sometimes explicit) weighing of the associated benefits, costs and risks to the supervisory organization. And a similar calculation of benefits, costs and risks determines the strictness of control exerted by a supervisory (or affiliated) organization. To develop this argument, we first analyze the concerns of supervisory and affiliated organizations, and then detail the incentives that organizations have to supervise NGOs.

Concerns of supervisory and affiliated organizations

NGOs included in this study have two types of supervisory organizations (ASTs and government agencies) and two types of affiliated organizations (government agencies and research institutes). CAST and local ASTs have official responsibilities to develop NGOs linked to science and technology. However, because of their limited resources, CAST and local ASTs ask affiliated organizations to provide resources to these NGOs and to supervise their activities.[50] Apart from CAST and BAST, the agencies and research institutes that serve as supervisory (or affiliated) organizations of NGOs investigated herein have no official obligations to sponsor NGOs. Supervisory and affiliated organizations *voluntarily* incur the costs and risks of their involvement with NGOs. Assuming they behave as rational actors, members of these supervisory and affiliated agencies would obviously not risk their careers for an NGO with the potential to commit financial or political wrongdoings.[51] Moreover, they would not invest resources in an NGO unless they expected to receive commensurate benefits. Thus, an NGO must motivate a potential supervisory organization to support its application for registration by ensuring the potential sponsor that there is a potential net gain and no real risk.

Let us, for example, consider SEPA. Even though SEPA has been urging NGOs to participate in environmental protection for many years, it serves as the supervisory organization of only four national NGOs, and it is the affiliated organization of one other.[52] Each of these five NGOs was either founded or sponsored by SEPA, and each has personnel and financial ties with SEPA. Moreover, SEPA treats each of the NGOs it supervises as a subsidiary. Compared with other central government agencies examined herein, SEPA sponsors few NGOs.[53]

Interests of supervisory and affiliated organizations

Experience of the 22 environmental NGOs in this study shows that supervisory (and affiliated) organizations have vested interests in NGOs supervised by (or affiliated with) them. CAST and BAST are officially responsible for supporting the development of NGOs in science and technology. In a sense, the very existence of CAST and BAST is based on sponsorship of these NGOs.[54] For research institutes that are the affiliated organizations of NAS-2, NAS-3, BAS-2 and BAS-3, the link with these academic societies gives the institutes enhanced stature and prestige. Indeed, the affiliated research institutes often find it advantageous to arrange their activities in the names of academic societies.[55] By doing so, relatively large numbers of participants from different sectors and places are likely to attend activities organized by affiliated organizations of the academic societies, and consequently the affiliated research institutes can gain greater influence within the academic community.[56] In addition, affiliated research institutes can cooperate with academic societies on projects, and thereby gain enhanced publicity.[57] However, research institutes typically do not to obtain net financial gains from their work with academic societies.[58]

Enhanced prestige and publicity are not the only incentives for supervisory (or affiliated) organizations. NGOs with government agencies as their supervisory or affiliated organizations have been used by agencies for the following purposes:

1 to arrange jobs for officials retired or removed from the agencies;[59]
2 to engage in cooperative programs with international NGOs;[60]
3 to raise funds;[61] and
4 to provide services to their supervisory or affiliated organizations.[62]

Risk control strategies

How can an organization supervising an NGO determine that the NGO will not pose risks by committing financial or political misdeeds? Of the 22 NGOs examined herein, the four NGOs that were affiliated with research institutes – NAS-2, NAS-3, BAS-2, and BAS-3 – were viewed as low risk because they were working on natural sciences and were managed by eminent scientists. These academic societies have reputations for being objective and independent and affiliated organizations had no legal authority to intervene in the operations of NGOs associated with them. Therefore, the affiliated research institutes tended not to interfere with their activities.

The supervisory (and affiliated) organizations of seven of the 13 NGOs supervised by (or affiliated with) government agencies (i.e. NAS-1, NPA-1, NPA-2, NF-1, BAS-1, BPA-1 and BPA-2) minimized their risks by controlling the personnel arrangements within the NGOs' secretariats. For example, leaders of NAS-1, NPA-2, BAS-1, BPA-1 and BPA-2 were nominated by the personnel departments of the agencies that served as their supervisory (or affiliated) organizations, and approval of these nominations by NGO congresses has typically been a formality. For the remaining six of the 13 environmental NGOs supervised or affiliated with government agencies (i.e. NPA-3, NF-2, BAS-4, BPA-3, BPA-4, and BF-1), their supervisory (or affiliated) organizations agreed to support them because leaders of these six NGOs used their personal relationships to surmount the difficulties posed by lack of incentives for supervisory (or affiliated) organizations and their concerns with risks. For example, founders of both NF-2 and BF-1 were retired high-ranking government officials.[63] Because they had been highly respected leaders in their government agencies, each of these NGO founders still had influence within their former agencies. They used their personal connections to persuade their former agencies to support their NGOs. In a sense, the supervisory (or affiliated) organizations of these six NGOs gave up their concerns about risks and supported these NGOs based on their trust in NGO leaders of these six NGOs. The existence of this trust also explains why these supervisory (or affiliated) organizations exert only weak control over the NGOs.

NGO strategies for dealing with state controls

The absence of repressive action against the subset of the 22 NGOs that violated registration requirements is explained partly by the conditions faced by civil affairs offices and NGOs' supervisory and affiliated organizations, but it also results, in part, because of actions taken by the NGOs. Indeed, some NGOs have developed countermeasures to circumvent control measures, and in some cases NGOs have deliberately violated control measures to create a better context for conducting their operations. A key strategy we identified from interviews with the 22 NGOs is *self-censorship* exerted by NGO leaders, which according to the previous contribution is one of the many strategies employed by activists to adapt to an embedded context.

Although some of the 22 NGOs violated one or more formal control measures, leaders of these NGOs were aware of the limits on how far they could go in violating controls without attracting negative attention from the state. This awareness was particularly important to citizen-organized de facto NGOs and NGOs that are not strictly controlled by their supervisory (or affiliated) organizations. As one interviewee put it, an NGO that is not involved in financial or political wrongdoings can operate without being regulated by a civil affairs office, but an NGO that has made financial or political mistakes will have its every move scrutinized.[64] Leaders of the 22 NGOs censored themselves by avoiding actions that would bring their organizations to the attention of civil affairs offices. Concerns about avoiding any possibility of being accused of financial misconduct were mentioned consistently by our interviewees.

Conclusions

Before summarizing the principal findings of our work, we provide a basis for putting them in a broader, national context. While our conclusions are restricted to the 22 environmental NGOs included in our study, they may have broader implications. The pattern of results on effects of state controls shown in Table 3.5 indicates that, with few exceptions, the types of effects observed for the 11 national NGOs are the same as those for the 11 Beijing municipal NGOs. Throughout our study, we found no substantive basis for disaggregating the findings into separate units – one for the national NGOs and the other for the municipal NGOs.[65]

The state's NGO regulations give the impression that the state, acting through the its civil affairs *xitong*, is determined to control strictly the development and activities of NGOs by employing the control measures discussed herein. However, based on the examination of the way the state's policies have been implemented in the context of 22 NGOs, we find that the state has limited ability to constrain NGOs. The state has not prevented citizens from forming NGOs and it has shown little interest in acting against the de facto NGOs and registered NGOs (considered herein) that have violated state control measures. These results suggest that state corporatism, which is characterized by the state granting a representational monopoly to a group in exchange for the group's acquiescence to state controls, may not be appropriate as a model describing the realities of contemporary China with respect to NGOs.

Our analysis highlights the importance of distinguishing between the civil affairs *xitong* and state agencies involved in state control measures. Agents of the state have acted differently based on their own calculation of benefits, costs and risks associated with supporting NGOs. Indeed, these interests have sometimes been competitive among different agents. For example, BAST was in dispute with the Beijing EPB over the right to supervise BAS-1 for about ten years.[66] The resource constraints faced by civil affairs offices together with vested interests of NGO supervisory (and, in some cases, affiliated) organizations have played key roles in determining how state control measures are implemented. In their roles as NGO supervisory (and affiliated) organizations, state agencies have granted representational monopolies to NGOs that satisfy their interests, which include raising funds, arranging jobs for retired or downsized government officials, and even nurturing personal relationships with influential individuals. These interests determine how supervisory and affiliated organizations control NGOs. Given that supervisory (and affiliated) organizations have the power to control NGOs associated with them, these agencies (regardless of whether they are central or Beijing municipal government agencies) have taken advantage of the corporatist structure created by the State Council and the Central Committee of the CCP to reduce their risks and extract net benefits from NGOs when possible. To get official recognition, a prospective NGO must adapt to this corporatist structure. We label the control practiced by supervisory (and affiliated) agencies as "*agency corporatism*."

An important conclusion of our research is that a majority of those examined NGOs that were organized voluntarily by citizens actually operated with a high degree of autonomy. This finding supports the overall argument of this volume that an embedded context is not merely a restrictive context, but also leaves substantial maneuvering space for voluntary action. Such social groups can be seen as critical elements of an embedded, but increasingly liberal, civil society in China. However, these citizen-organized NGOs (as well as other NGOs in our study) were cautious about avoiding confrontations with the state. Indeed, based on our interviews, each of the 22 NGOs demonstrated a keen appreciation of the capability of the state to repress any NGO that poses threats. Each of these NGOs was also aware of the limits on what it could do without attracting negative attention and repressive actions from the state. Our findings led us to characterize environmental NGOs in the late twentieth and early twenty-first centuries in China as containing elements of an embedded activism of which self-censorship is an integral component. Clearly, self-censored does not necessarily mean "cooperative."

Our results also highlight the importance of NGO leaders who have personal and political influence. These leaders were able to obtain sponsorship from supervisory agencies, and because of their personal ties they were able to operate their NGOs with a high degree of autonomy. As also noted in the previous contribution:

> It should not come as a surprise that many green NGOs have been established by strong personalities who are extremely well-connected to Party and state institutions.... One might therefore conclude that although the Chinese NGOs are "non-governmental," they are – particularly the well-established ones – also part of an elite hierarchy.[67]

Based on evidence collected from the 22 NGOs, we conclude that even though regulations were designed to have the civil affairs offices control NGOs strictly, the state controls have been implemented as "agency control" over the operations of these NGOs, and the influence of these controls on NGOs' routine operations are described as being minimal by the NGOs. The declining capacity of the civil affairs *xitong* to implement the state's control measures, variability in the interests of state agencies, the existence of NGO leaders who were well connected to government officials, and the self-censorship strategies of the NGOs have contributed to this political outcome. We foresee that it is probable that the civil affairs *xitong* will continue losing its control over NGOs because its capacity will continue to decline as the economic and political reforms progress. However, this does not mean that NGOs will be less controlled. As long as the ongoing reforms do not prevent agencies that serve as supervisory or affiliated organizations from extracting benefits from the NGOs with which they are associated, the practice of agency corporatism over these NGOs will be sustained. In the short term, at least, environmental NGOs will continue using self-censorship as it allows NGOs to meet their environmental protection goals without being subjected to repressive actions by the state.

Notes

1 For criticisms and suggestions on earlier drafts of this paper, we thank Ruth Greenspan Bell, Baruch Boxer, Elizabeth Economy, Richard Louis Edmonds, Xin Gu, Lei Guang, Peter Ho, Jean Oi, Mark Selden, Yang Su, Jennifer Turner and Fengshi Wu. We also wish to acknowledge the sources of financial support for the study: the UPS Foundation and the Morrison Institute for Population and Resources Studies at Stanford University. Clearly, we take sole responsibility for the views expressed in this paper. For a detailed discussion of the concept of NGOs and its application in China, see J. Ru, "Environmental NGOs in China: The Interplay of State Controls, Agency Interests and NGO Strategies," PhD diss., Stanford University, 2004, pp. 16–41.

2 We conducted 33 interviews in 2001 and 19 interviews in August 2002 in China. Interviewees included 45 NGO staff members, two NGO scholars, and five government officials working with NGOs. For reasons of space, methodological information on the sampling of the NGOs, the sampling criteria, sampling biases, interview methods, and the interviewees who provided data for their analysis are not included here. This information is available from the authors.

3 See also J. Unger, "'Bridge': Private Business, the Chinese Government and the Rise of New Associations," *The China Quarterly*, No. 47 (September 1996), pp. 795–819.

4 Earlier studies described agencies mainly as agents of the state without identifying the interests of these agencies in sponsoring and supervising NGOs affiliated with them. See A. Chan, "Revolution or Corporatism? Workers and Trade Unions in Post-Mao China," *Australia Journal of Chinese Affairs*, No. 29 (January 1993), pp. 31–61.

5 These surveillance mechanisms were introduced in the 1989 Regulation on Registration and Supervision of Social Organizations (RRSSO).

6 MOCA promulgated the Notice on Re-confirming Supervisory Organizations of Social Organizations [*Guanyu chongxin queren shehui tuanti yewu zhuguan danwei de tongzhi*] (MOCA, File No. 2000–41), which outlined a list of organizations that could function as supervisory organizations of NGOs.

7 Schmitter defines corporatism as "a system in which the state grants social groups deliberate representational monopolies in particular sectors in exchange for the groups' positive response to various state controls." See P.C. Schmitter, "Still the Century of Corporatism?" in F.B. Pike and T. Stritch (eds) *The New Corporatism: Social-Political Structures in the Iberian World*, Notre Dame: University of Notre Dame Press, 1974, pp. 93–4.

8 All NGOs in China are required to register with civil affairs offices. Registration with a civil affairs office allows an NGO to obtain an officially acknowledged stamp, open a bank account, and purchase official financial receipts.

9 See Article 13 of the 1998 RRSSO.

10 Funding limits are detailed in the 1988 Management Measures for Foundations, and the Notice on Further Strengthening the Management of Foundations [*Guanyu jingyibu jiangqiang jijinhui guanli de tongzhi*] (Office of State Council and the People's Bank of China, 1995).

11 On 1 August 2003, MOCA and the Ministry of Finance (MOF) modified the 1992 rule (on a trial basis) to allow NGOs to decide independently on the levels of their membership dues (MOCA and MOF, Notice on Issues Related to Adjusting Policies related to Membership Dues of Social Organizations [*Mingzhengbu caizhengbu guanyu tiaozheng shehui tuanti huifei zhengce deng youguan wenti de tongzhi*], 2003).

12 The Organization Department of the Central Committee of the CCP introduced this rule in 1998.

13 According to the 1998 RRSSO, the NGO must obtain approval from its supervisory organization before making major changes (such as making a change of leaders) in its organization and before seeking approval for such changes from its civil affairs office.

14 Under the Ad Hoc Measures for NGO Annual Check (MOCA, 1996), each year an NGO must submit required information (e.g. audited financial report) to the civil affairs office at which it is registered.

15 The 1996 MOCA document stipulates conditions under which an NGO may fail the annual check.

16 We only focus on wrongdoings that violate NGO regulations and state control measures. An NGO can also violate other laws or regulations, such as a fire protection code, and be subject to administrative or judicial punishments.

17 See J. Pomfret and M. Laris, "China Confronts a Silent Threat," *Washington Post*, 30 October 1999; the Decision to Clamp Down on the Falun Dafa Research Society [*Minzhenbu guanyu qudi falundafa yanjiuhui de jueding*] (MOCA, 22 July 1999); and the Announcement of the Ministry of Public Security [*Zhonghua renmin gongheguo gonganbu tonggao*] (Ministry of Public Security, 22 July 1999).

18 Before 1989, NGOs were regulated by state agencies, and sometimes by non-enterprise entities and other NGOs.

19 These regulations are the 1988 Management Measures for Foundations (September 1988), the 1989 Regulation on Registration and Supervision of Social Organizations (October 1989, hereafter referred to as the 1989 RRSSO, and later revised as the 1998 RRSSO), and the 1989 Ad Hoc Regulation on Management of Foreign Trade Unions (April 1989).

20 For examples of difficulties, see the Notice on Strengthening the Supervision of Social Organizations and Private Non-enterprise Entities [*Guanyu jianqiang shehui

tuanti minban fei qiye danwei guanli gongzuo de tongzhi] (Office of CCCCP, File No. 1996–22).

21 Ibid.

22 Ibid. The 1996 Notice on Strengthening the Supervision of Social Organizations and Non-Enterprise Entities asked the supervisory organizations to be responsible for NGOs' registration applications, ideology work, CCP development, financial activities, personnel management, workshops, international cooperation, donations and other important activities.

23 X. Zhong, "Qigong Social Organizations Will Be Cleaned Up to Prevent the Public from Feudalistic Superstition and Pseudoscience" [*Zhizhi xuanchuan fengjian mixin he wei kexue, qigong shetuan Jiang bei qingli*], *Fuzhou Evening News* [*Fuzhou wanbao*], 21 January 2000.

24 MOCA, *Civil Affairs Statistics Yearbook of China: 1996* [*1996 nian mingzheng tongji nianjian*], Beijing: Ministry of Civil Affairs, 1997; and MOCA, *Civil Affairs Statistics Yearbook of China: 2001* [*2001 nian mingzheng tongji nianjian*], Beijing: China Statistics Press, 2002.

25 For a complete list of these NGOs, see J. Ru, "Environmental NGOs in China," pp. 40–1.

26 The conceptualization of "federation" reflects official language in the Notice on Issues Related to the Regulation on Registration and Supervision of Social Organizations (MOCA, 30 December 1989).

27 The other two types of NGOs that register with civil affairs offices – industrial associations and federations – have no activities in wildlife conservation at the national or Beijing municipal levels. The term "existing organizations" refers to governmental agencies (other than civil affairs offices and industrial and commercial administration bureaus), enterprises, non-enterprise entities, or registered NGOs

28 C. Gao, G. Ying, and G. Tian, *Study of Chinese Association of Science and Technology* [*Zhongguo kexie xue*], Beijing: China Science and Technology Press, 1992, p. 86.

29 In addition to academic societies, CAST also supervises professional associations and foundations. As of 2002, CAST supervised 184 national NGOs, about 11 percent of all national NGOs registered with MOCA in 2002; and BAST supervised 142 NGOs, about 15.1 percent of all NGOs registered with BCAB.

30 S. Wang, Q. Shen and Z. Gao, *China Association of Science and Technology* [*Zhongguo kexue jishu xiehui*], Beijing: Contemporary China Press, 1994, pp. 512–13; and C. Gao and Y. Zhang, *Concise Dictionary for Associations of Science and Technology* [*Kexie gongzuo jianming cidian*], Beijing: China Science and Technology Press, 1997, pp. 110–11.

31 These NGOs were NPA-3, BPA-3 and BPA-4.

32 These NGOs were NAS-1, BAS-1, NPA-1, NPA-2, BPA-1, BPA-2 and NF-1.

33 These NGOs were NAS-2, NAS-3, BAS-1, BAS-2, BAS-3, BPA-1 and BPA-2.

34 These two national academic societies were NAS-2 and NAS-3.

35 T. Saich suggested that the 1998 RRSSO has, in principle, banned NGOs registering as business operations. However, the three environmental NGOs registered in Beijing as for-profit companies experienced no interference from any agencies as of the end of 2002. See T. Saich, "Negotiating the State: The Development of Social Organizations in China," *The China Quarterly*, No. 161 (March 2000), pp. 124–41.

36 One national NGO and one Beijing municipal NGO had their registration cancelled because they had been inactive for many years. A second national NGO was eliminated because the NGO was asked to incorporate into a national NGO by its supervisory organization. In addition, a second Beijing municipal NGO was disbanded because its leaders were unable to sustain the NGO's operations (Interview Q2).

37 For details of the correlation study, see J. Ru, "Environmental NGOs in China," pp. 120–5.

38 These seven NGOs were NAS-1, NPA-1, NPA-2, BAS-1, BPA-1, BPA-2 and NF-1.

39 Interviews P7, Q0, Q1, and Q3.
40 The number of officials was attained from Interviews Q0 and Q1. For the number of NGOs, see MOCA, *2001 Civil Affairs Statistics Yearbook*, pp. 102–5.
41 Interview C8.
42 Interview P7.
43 Interviews P7, Q0, Q1 and Q3.
44 T. Saich, "Negotiating the State," pp. 130–1. MOCA proposed elimination of the requirement for supervisory organizations in the amended Supervision Measures for Foundations (Interviews P7 and Q1).
45 Interviews P7, Q0, Q1 and Q3.
46 MOCA, *2001 Civil Affairs Statistics Yearbook*, pp. 102–5.
47 Interview Q3.
48 Interviews Q0 and Q3.
49 In contrast, an NGO does have the right to make an administrative appeal of a negative decisions of the civil affairs office at which it attempted to register by following appeal procedures in the 1990 Administrative Procedure Law of the People's Republic of China.
50 C. Gao, G. Ying, and G. Tian, *Study of Chinese Association of Science and Technology*, p. 110.
51 T. Saich, "Negotiating the State," pp. 129–30.
52 According to SEPA's administrator, Xie Zhenghua, the agency's sponsorship tends to compromise the independence of NGOs, and he would prefer to have NGOs in a position where they can freely criticize SEPA (Speech at the Woodrow Wilson International Center for Scholars, Washington, DC, 9 December 2003). An alternative explanation offered by Shin in this collection is that China, "as an environmental latecomer," has created an environmental institutional framework that discourages civil networks and environmental NGOs.
53 For example, the Ministry of Construction, the Ministry of Agriculture and the Ministry of Culture sponsor 31, 56 and 85 NGOs, respectively.
54 Interviews C8 and F2.
55 Interview P2.
56 Interview C8.
57 Interviews F1 and F2.
58 Interview P2.
59 These NGOs were NPA-1, NPA-2, BAS-1, BPA-1 and BPA-2.
60 These NGOs were NAS-1, NPA-1, NPA-2, and NF-1.
61 These NGOs were NPA-1, NPA-2 and NF-1.
62 All registered NGOs in our study provided services to their supervisory or affiliated organizations.
63 Interviews D4, D5, F3 and F4.
64 Interview F3.
65 Although we do not report results herein, as part of a broader study we also interviewed staff members from eight additional environmental NGOs, and they were located in three different provinces: Qinghai, Sichuan and Yunnan. The general patterns of effects of state controls on environmental NGOs reported herein apply also in the case of these eight NGOs. An aspect of the larger investigation, which is only partially reflected in this paper, also suggests that our findings may apply beyond the 22 environmental NGOs. As another part of the larger investigation, we conducted three case studies of particular wildlife conservation efforts. These concern NGO involvement in efforts to preserve giant pandas in Sichuan, Yunnan golden monkeys in Yunnan, and Tibetan antelopes in Qinghai. Our interviews at the eight NGOs and the findings in our three case studies were generally consistent with the findings reported herein and summarized below.
66 Interview C8.
67 See the contribution by Ho elsewhere in this volume.

4 Media, civil society, and the rise of a green public sphere in China

Guobin Yang and Craig Calhoun

Introduction

According to media reports, China's State Council halted the hydropower project being planned on the Nu River in Yunnan province in April 2004. The decision came after months of intense public debates. China's Premier Wen Jiabao reportedly cited "a high level of social concern" as an important reason for suspending the dam-building project (*Ming Pao Daily*, 2 April 2004). Such a reversal after public criticism has hardly been typical of the Chinese government – nor was the nature of the public criticism typical. In the first place, the public debate addressed policy. By contrast to the more common pattern, it was not simply the exposure of corruption or the suggestion that local officials deviated from the goals of the central Party and government. Second, a broad range of participants was involved in public discourse. This differentiated it from the "reportage" literature through which criticism flourished in the 1980s, for example, which typically required a strong individual personality, such as Liu Binyan, willing to focus on broader concerns in his or her writing.

How did the public debates about the Nu River happen? Who was involved? What media were used? Whereas the previous contributions have focused more on the Party–state–society interactions and the way these are shaped by an embedded context, the current contribution explores civil activism against the background of a newly emerging public sphere, as reflected in the media: television and radio programs, newspapers, magazines, and the internet. We argue that the articulation of "a high level of social concern" depended on a public sphere of environmental discourse in China – a green public sphere. Communication and debates in the public sphere channeled citizen opinions to influence government policies.

A green public sphere fosters political debates and pluralistic views about environmental issues, and for this reason is intrinsically valuable.[1] The rise of a Chinese green public sphere commands special attention, however. First, it is exemplary of a variety of new forms of public engagement in contemporary China. These include, for example, feminist activism, cyber-activism, HIV/AIDS activism, and rights activism more broadly. Second, with environmental issues as its central concern, the green public sphere represents the emergence of an issue-specific public. The differentiation of issue-specific publics is a relatively new development in China. Third, the transnational dimension of the Chinese green sphere indicates still another new trend, namely its transnationalization.[2] In effect, then, a Chinese green public sphere is significant in terms of both content and form. With respect to its formal attributes, the green sphere is distinctive because it engages politics and public policy without being primarily political.[3] Carving out a space for "nonpartisan" advocacy is a new development in China. It is also distinctive because of its reliance on a range of media and organizational forms, including traditional press, the internet, "alternative media," as well as environmental NGOs.

This article delineates the main features of the green public sphere, analyzes the main factors that have contributed to its emergence, and explores its functions. We argue that the emerging green sphere consists of three basic elements: (1) an environmental discourse or greenspeak; (2) publics that produce or consume greenspeak; and (3) media used for producing and circulating greenspeak. First, we show that one major indicator of the rise of a Chinese green public sphere is the proliferation of environmental discourse – a greenspeak. Contrary to an earlier Maoist and Marxist view of the human conquest of nature,[4] this new discourse warns about the dangers of irresponsible human behavior toward nature and calls for public action to protect the environment. Appearing in television programs, radio programs, newspapers, magazines, leaflets, flyers, posters, and on the internet, this blossoming discourse represents a participatory conversational situation, one "of seven mouths and eight tongues" (*qizui bashe*), as a Chinese folk saying would have it.

Second, we argue that environmental NGOs provide the pivotal organizational basis for the production and circulation of this greenspeak. A greenspeak that promotes a new environmental consciousness does not fall from heaven, but has its advocates and disseminators. We focus on environmental NGOs both because they are relatively new and because they play a central role in producing greenspeak. Third, we analyze the media of the green sphere. Distinguishing among mass media, the internet, and "alternative media," we argue that because these different types of media differ in social organization, access, and technological features, they influence the green sphere differently. Finally, we return to the case of the Nu River to illustrate the dynamics and functions of the emerging green sphere. In the conclusion, we discuss the implications of our analysis for understanding the sources of political change in China.

Green public sphere and greenspeak in China

"Public sphere" is a controversial concept. Jürgen Habermas initially defined it as "A domain of our social life in which such a thing as public opinion can be formed." Access to this domain is "open in principle to all citizens," who may "assemble and unite freely, and express and publicize their opinions freely."[5] Critics were quick to point out that Habermas's version of the public sphere was a bourgeois sphere that in reality excluded certain categories of people (such as women) and was fraught with problems of social, economic, cultural, even linguistic inequality.[6] In response, Habermas later recognized the internal dynamics of the public sphere, the possibility of multiple public spheres, as well as the conflicts and interactions among them.[7]

In China studies, the concept of the public sphere has similarly been controversial. It was initially used to explain the rise of the student movement in 1989.[8] Then a symposium on "public sphere"/"civil society" in 1993 introduced an influential debate. Some scholars in the debate find civil society and public sphere in late imperial China;[9] others argue that these concepts are too value-laden and historically specific for understanding Chinese realities.[10] More recently, there has been a revival of interest among China scholars in civil society and public sphere. For example, it has been argued that these categories are pertinent to China, because they emerged out of experiences of modernity, which transformed China no less than the West.[11] Others use a relaxed notion of the public sphere, adopting more neutral terms such as "public space" or "social space" or focusing on publics rather than the public sphere.[12] One reason why we continue to use the concept of the public sphere is that Chinese intellectuals themselves have come to embrace it. A recent Chinese book on green media, for example, focuses on the building of a "green space for public opinion" (*lüse gonggong yulun kongjian*), alluding directly to the Habermasian concept.[13]

Recognizing the historical baggage of the Habermasian concept, however, we maintain a broad conception of the public sphere as spaces for public discourse and communication. It consists of discourse, publics engaged in communication, and the media of communication. The emerging green sphere in China thus has the following basic elements: (1) an environmental discourse or greenspeak; (2) publics that produce or consume greenspeak; and (3) media used for producing and circulating greenspeak. "Public" is a broad and loose concept. By the publics of China's green public sphere, we refer specifically to individual citizens and environmental NGOs directly engaged in the production and consumption of greenspeak.

A main indicator of an emerging green public sphere in China is the proliferation of a greenspeak. "Greenspeak" refers to the whole gamut of linguistic and other symbolic means used for raising awareness of environmental issues.[14] The Chinese greenspeak includes recent neologisms in the Chinese language such as sustainable consumption, white pollution, eco-centricism, endangered species, animal rights, global warming, desertification, deforestation, biodiversity, bird-watchers, and more. An entire dictionary of greenspeak can now be compiled.

Different social actors use greenspeak for different purposes. Business corporations, for example, may use a green language as a way of "greenwashing" its interests.[15] We focus on greenspeak produced by civil society actors. This civic greenspeak has several features. First, it is tacked onto the mainstream global discourse of sustainable development. The popularity of such terms as "one world," "common earth," "holistic approach," "global village," "Earth Day," and of course "sustainable development" attests to the global dimension of this discourse. Following the 1992 Rio Earth Summit, moreover, the Chinese government published its strategies for sustainable development in a "China Agenda 21" white paper issued in March 1994, thus legitimating an official discourse of sustainable development in China.

Second, greenspeak expresses the tension between environmentalists and economic actors. Similar to new social movements elsewhere in the world, the environmental movement in China attracts some sections of the population and not others.[16] Its main constituency consists of students, intellectuals, journalists, professionals, and other types of urbanites. For example, university student environmental associations are a main part of the movement. A non-student member-based organization like Friends of Nature also draws its membership mostly from these urban groups. Greenspeak gives these people, who tend to have more cultural capital than economic capital, a rhetoric for identifying themselves and their concerns by contrast to the dominance of a more crass economic rhetoric in society at large.

Third, in response to the ascendance of consumerism and materialism, greenspeak promotes new moral visions and practices. A central moral message is that environmentalism must be practiced as a new way of life. It promotes a new understanding of the relationship between humans and their natural environment, one that stresses human–nature harmony. A practical corollary of this view is that humans must treat nature and its flora and fauna with respect and kindness.[17] It also promotes the vision of a new personhood. Practicing a green consciousness must start with oneself. If each and every person lives an environmentally friendly life, then the earth might be saved. A common personal practice is to reject the use of disposable consumer products (such as disposable chopsticks and shopping bags). These practices convey some sense of religiosity. They embody a search for more spiritual "meanings" in life – again, something that tends to be opposed to sheer economism. In this sense, the green discourse continues an opening up to "expressive individualism" in China over the last 20 years.

Yet the relationship between economism and environmental protection is also a source of dilemmas and personal perplexity. Reflecting the growing richness and diversity of current environmental discourse, these dilemmas are often openly discussed and shared. A good example comes from the publications of Green Camp, an unregistered NGO in Beijing. Each year since 2000, Green Camp has produced an informal publication featuring personal essays by participants in that year's green camp activities. These personal stories demonstrate how China's young environmentalists develop a deeper understanding of environmental issues by experiencing them firsthand and talking about their experience. One essay in the 2000 volume describes a small incident that happened to its author when he and a few other green campers were studying the feasibility of eco-tourism in Changbai Mountain. He heard the following conversation between a fellow camper and an employee of a local nature reserve's protection station:

CAMPER: What do you think is most needed for developing eco-tourism?
LOCAL EMPLOYEE: Money.
CAMPER: If you see people doing damage here, would you intervene?
LOCAL EMPLOYEE: Yes!
CAMPER: Why?
LOCAL EMPLOYEE: Because this is my home.
CAMPER: If you lose your job and have no income, would you go into the mountains to stealthily gather mountain stuff?[18]
LOCAL EMPLOYEE: Yes!

After hearing this conversation, everyone became silent. This was a transformative moment for this individual. He realized that between reality and young environmentalists' idealism, there was a vast gap. He continues:

> After the field trip to Changbai Mountain, I came to a deeper understanding of the difficulties of environmental protection in China. To transform economy and environment into a virtuous relationship of mutual benefit and embark on a road of sustainable development – this is still a dream. It will take arduous efforts to turn the dream into a reality.[19]

Part of the arduous efforts is to promote environmental consciousness and citizen participation. These efforts betray a fourth feature of the greenspeak – its political thrust. The Chinese greenspeak emphasizes participation and volunteerism. While recognizing that environmental problem-solving depends on the joint efforts of government, citizens, and NGOs, the greenspeak emphasizes the role of citizens and the importance of developing an NGO culture. In addition, greenspeak is a veiled way of talking about many other things, including making criticisms of government policies. One good example is a speech delivered by the representative of an environmental NGO from Qinghai province at an NGO workshop in Beijing in October 2002. Referring to the central government's ambitious plan to develop the western regions, the speaker argued that in western minority regions such as Qinghai province, the protection of biodiversity of the natural environment should be integrated with the protection of cultural diversity and that local communities should be involved in the decision-making process.[20] Greenspeak thus can be political, though as Peter Ho suggests in his contribution to this volume, this may be a de-politicized politics.[21]

Environmental NGOs: the discourse-producing publics of the green sphere

Michael Warner describes a public in the following terms:

> A public is a space of discourse organized by nothing other than discourse itself. It is autotelic; it exists only as the end for which books are published, shows broadcast, Web sites posted, speeches delivered, opinions produced. It exists *by virtue of being addressed* [original emphasis].[22]

Warner's description captures only half of what "a public" means. A public is not just an addressee, but also an addressor. It not only reads books, watches shows, reads internet posts, listens to speeches, and receives opinions, but also publishes books, produces shows, writes or responds to internet posts, delivers speeches (or conversations for that matter), and expresses opinions.

As mentioned above, the publics of China's green sphere consist of all citizens and civil society organizations involved in the production and consumption of greenspeak. Here we concentrate on the discourse-producing publics. These are again diverse and may include scientific communities, educational institutions, and a broad array of old and new social organizations. We focus on environmental NGOs because they are the most distinctive and novel organizational base for the green public sphere.

In one of the first systematic analyses of this topic, Peter Ho shows that environmental NGOs in China cover a wide spectrum, from more or less independent NGOs to government-organized NGOs (GONGOs), student environmental associations, unregistered voluntary organizations, and NGOs that are set up "in disguise" in order to bypass the registration requirements and hide their true nature from the government's view.[23] A survey of university student environmental associations shows that as of April 2001, there were 184 student environmental associations.[24] Non-student grassroots environmental NGOs numbered at about 100 as of 2003, not including the numerous GONGOs.[25] According to a recent Chinese news release, this number reached about 200 toward the end of 2006.[26]

Environmental NGOs are engaged in a broad range of activities, from public education and community building to research and advocacy. In these activities, they resort to all forms of media and public forums, including television, radio, newspapers, magazines, web sites, exhibits, workshops, and salons. As a result, these organizations become an important institutional base for bringing green issues into the public sphere. Many organizations publish newsletters and special reports in print or electronic form. Some produce television programs and publish books. For example, Global Village of Beijing (GVB) has an ongoing project to produce environmental television programs. Between 22 April 1996 and March 2001, GVB produced 300 shows for its weekly television program *Time for the Environment* on CCTV-7. It also publishes books on environmental issues which may be ordered via its web site. Current titles include *Citizen's Environmental Guide*, *Children's Environmental Guide*, *Green Community Guide*, and *Environmental Song Book*.

Two environmental organizations in Beijing have been organizing public forums on a regular basis. Tian Xia Xi Education Institute, an educational and environmental NGO founded in 2003, organizes forums on topics ranging from dam-building to citizenship education and public health. Green Earth Volunteers, an unregistered NGO based in Beijing, has been organizing monthly environmental salons for journalists since 1997. Featuring guest presentations on various environmental issues, these salons aim to provoke broader discussions among Beijing's unofficial environmental circles and help journalists to write more and better environmental stories. They cover a wide range of topics. At a salon event in June 2002, a retired worker who introduced himself as an environmental volunteer made a slide presentation about the desertification of the grassland in China's western regions. Then the founder of Save the South China Tiger, a non-profit organization registered in Great Britain, spoke about the protection of the tiger. Another event, held on 19 March 2003, featured three guest speakers who spoke respectively on issues of animal protection and reproduction in nature reserves, the new challenges facing wildlife protection in the development of China's western regions, and the import and export of medicinal ingredients made from wild animals. Table 4.1 lists the topics of the journalists' salons from January 2003 through July 2004.

One effective tactic used by NGOs to publicize environmental issues is campaigns. These campaigns help to concentrate public attention on specific issues by creating media visibility and public discussions. Some have directly influenced policies. One of the first public campaigns was organized in 1995 to stop the felling of an old forest in Yunnan in order to protect the indigenous golden monkeys. Since then, not a year has passed without some kind of such environmental campaign. In 1997, a group of college students organized a campaign to promote recycling on university campuses in Beijing, while another group ran a campaign to boycott disposable chopsticks. The most important recent campaign was launched in 2003 to

Table 4.1 Topics of journalists' salons, Beijing, January 2003–July 2004

Dates	*Topics*
22 January 2003	Genetic modifications and ecological safety
19 February 2003	Sand storms and their management
19 March 2003	Animal protection and reproduction; wildlife protection and western development; import and export of medicinal ingredients made from wild animals
3 July 2003	Water system design
22 July 2003	The roles of government, NGOs, business, and media in environmental policy-making in the USA
13 August 2003	Water system design
17 September 2003	The drying up of Chagannuo'er Lake and salt storms
29 October 2003	Urban transportation
11 November 2003	The public supervision of life sciences
17 December 2003	The ecological functions of rivers and streams
15 January 2004	World Dams Conference in Thailand and the impact of dams on economy, society, and culture
11 February 2004	Half a century of changing urban space in Beijing
24 March 2004	Global old-growth forests crisis and China's ocean ecology
18 May 2004	Preliminary studies about guiding ocean water to Beijing; old-growth forests and wetlands
16 June 2004	Energy for vehicles; the development of geothermal energy resources
21 July 2004	Present conditions of World Heritage Sites in China

Source: Green Earth Volunteers web site at www.chinagev.org/.

stop dam-building on the Nu River and is still under way as of this writing. We will come back to this case later in this article. As Table 4.2 shows, the scope of these campaigns is wide-ranging. They include both moderate educational campaigns such as promoting "Earth Day" activities and more confrontational campaigns to boycott commercial products and challenge industrial projects.

To understand why environmental NGOs can organize such a broad range of activities in China's constraining political context, it is essential to analyze why they have developed in the first place. We underscore three conditions. First, the growth of environmental NGOs is part of a larger "associational revolution" in China.[27] This has to do with many factors, including state decentralization and the government's recognition of a third sector. Second, the Chinese state is taking on shades of green, as reflected by the large body of environmental laws and regulations it has promulgated.[28] This "greening of the state" provides a favorable condition for environmental NGOs. Third, the growth of environmental NGOs takes place in a dynamic context of multiple social actors and complex social relationships. As the concept of embedded environmentalism proposed by the editors of this volume indicates, environmental NGOs are embedded in social relations that may enable as well as constrain their growth. One enabling type of social relations, for example, is the growing ties between domestic NGOs and international organizations. Ranging from financial support to professional exchanges, these ties have promoted the visibility of Chinese NGOs both in China and in the international arena.[29]

The media of the green sphere: official, "alternative," and the internet

We distinguish three types of the media of China's green public sphere, namely, mass media, alternative media, and the internet. Because they differ in their

Table 4.2 Selected environmental campaigns in China, 1995–2003

Year	*Campaign themes*
1995–6	Campaign to protect the golden monkey
1996	Campaign to boycott disposable chopsticks
1997	Campaign to promote campus recycling
1998	Campaign to guard the wild geese in Purple Bamboo Park, Beijing
1998–9	Campaign to protect the Tibetan antelope
2000	"Earth Day" publicity campaign
2000	"Save the Tibetan Antelope Website Union" campaign
2001	Campaign to boycott "wild tortoise" medicinal products
2001–3	Campaign to protect the Jiangwan wetlands in Shanghai
2002	Internet petition to stop the building of an entertainment complex near the wetlands in suburban Beijing
2003	Public campaign to fight SARS
2003	Campaign to protect the Dujiangyan Dam
2003	Campaign to stop proposed dam-building on the Nu River

relationship to the state and in technological features, they are not equally accessible to Chinese environmentalists and they influence China's green sphere differently.

Mass media

Mass media – newspapers, television, and radio – have enormous influence on the green public sphere. Since the 1990s, mass media coverage of environmental issues in China has greatly increased. Surveys conducted by Friends of Nature find that the average number of articles on environmental issues published in national and regional newspapers was 125 in 1994. This number rose to 136 in 1995 and 630 in 1999.[30]

The public campaigns listed in Table 4.2 were all covered by the mass media. The bigger campaigns such as those to protect the golden monkey and the Tibetan antelope and the current campaign against dam-building on the Nu River generated intense media publicity. Even a small campaign to guard two hatching wild geese in a public park in Beijing caught media attention.[31]

In principle, Chinese mass media are the organs of the state. How to explain the growing media coverage of sometimes very contentious environmental issues? First, mass media have undergone de-ideologization, differentiation, and commercialization in the reform era.[32] Party organs such as *People's Daily* and commercial papers, for example, are not subject to the same degree of state control. The increasing dependence on commercial revenues gives the mass media more latitude in covering issues of broad social interest. Growing environmental problems such as pollution are issues of great concern.

Second, the Chinese government has supported media coverage of environmental issues by launching its own environmental media campaigns. The most ambitious project is the "China Environment Centennial Journey" (*zhonghua huanbao shiji xing*). Funded by the government and led by a commission composed of high-level officials from various ministries, the campaign was inaugurated in 1993. Each year since then, the commission has sponsored mass media institutions to send journalists out into the field to do investigative reporting on a selected environmental theme. The theme for 1993, for example, was "Fighting Environmental Pollution," and for 2005, "Clean Drinking Water." From 1993 to 2005, 50,000 journalists from across the country participated in the project and produced 150,000 reports on environmental issues.[33]

Third, Chinese environmentalists attach great importance to mobilizing the mass media and have been remarkably successful in this respect. Besides the structural changes in the media and a favorable political context, a major reason for their success is that many environmentalists and even leaders and founders of environmental NGOs are themselves media professionals. Green Camp, Green Earth Volunteers, Green Plateau, Tianjin Friends of Green, and Panjin Black-Beaked Gull Protection Association are all led by journalists or former journalists. Friends of Nature has some influential journalists in its membership.

These environmentalist media professionals serve as direct linkages between the mass media and the environmentalists.

Alternative media

"Alternative media" refers to the informal and "unofficial" publicity material produced and disseminated by NGOs, such as newsletters, special reports, brochures, flyers, and posters. They also include new media such as CD-ROMs and DVDs. They are an important part of the emerging green sphere, but have not attracted scholarly attention. These media materials are alternative in the sense that they are not controlled by the government, but are edited and produced by NGO staff or volunteers and distributed through informal channels. Some of these publications do not look very different from official publications, yet they are unofficial because they do not have official ISBN numbers. For example, Friends of Nature (FON) has published a bimonthly newsletter since 1996. It looks identical to a regular magazine in its professional appearance, yet is not officially registered and has no ISBN number. The advantage of having no official registration is that environmental groups can largely publish what they want. The disadvantage is that without an official ISBN, the publications cannot be distributed publicly, which limits their reach (interview with FON staff July 2002).

Almost all the NGOs we encountered in our research have publicity materials. The number and frequency of publications depend on their professional and financial resources. Production costs are usually covered with funding raised from corporate sponsors, foundations, or foreign donors. Staff and volunteers are responsible for editorial work and distribution. In membership-based organizations like Friends of Nature, members automatically receive the organization's official newsletter. Many organizations distribute their publications for free while some sell them to defray production costs.

These publications cover a broad range of environmental issues in a variety of genres. There are many personal stories and perspectives, indicating an emphasis on personal experience and individual viewpoints commensurate with an ethics of participation and respect. The publications of Green Camp mentioned above illustrate the uses of alternative media, especially in promoting participation and voice.

Since 2000, Green Camp has published a collection of personal stories and investigative reports after each year's summer camp activities. The size of a magazine, of about 100 pages each, and interspersed with colorful pictures, these collections contain investigative reports, personal stories, commentaries, diaries, and letters. The title of a column called "Of Seven Mouths and Eight Tongues" (*qizui bashe*) in the 2003 volume most vividly captures the nature of these publications. "Of seven mouths and eight tongues" is a Chinese idiom for describing a conversational situation where everybody chips in. It is a fitting metaphor for the diverse voices featured in these publications. Titled *The Wetlands That Once Were*, the 2003 volume contains, among other things, four investigative reports respectively about the cultural, community, ecological, and wetlands conditions of Ruo'ergai county in Sichuan province, seven firsthand reports about similar Green Camp activities organized by college students in the provinces (Green Camp is based in Beijing), and 20 essays in the "Of Seven Mouths and Eight Tongues" column. These 20 essays are personal reflections about camp experiences – the beauty and mystery of nature, the sense of pain at seeing the poverty of China's rural areas, admiration for the simple lifestyles of farmers, the challenges of doing good teamwork, and so forth. In an essay titled "Step Forward – and Speak Up!" the author describes how she and a few fellow campers, after much hesitation, braced themselves to confront a tourist group that was trampling the protected grassland in a wild jeep ride. The essay ends by calling on citizens to stand up against those who have no regard for the environment.[34]

The alternative publications of the green sphere have historical predecessors. In production and distribution, they resemble the unofficial publications during the Democracy Wall movement in 1978 and 1979. Those earlier publications were also edited by self-organized groups and distributed through independent and informal channels.[35] They differ in contents and context, which partly explains why the Democracy Wall publications were quickly banned and the alternative green media today have thrived. In contents, the unofficial publications during the Democracy Wall movement were much more radical. They were full of direct denunciations of state policies and calls for democracy and political reform. At a time of great political uncertainty, the Democracy Wall movement helped to mobilize public support for Deng Xiaoping's political maneuvers against Mao's designated successor Hu Guofeng, but when its radicalism challenged Deng, it was suppressed. In contrast, today's alternative green media do not carry politically radical contents such as calls for democracy or political reform. Instead they focus on environmental education and discussion. Rather than challenging state legitimacy, they operate largely within the parameters of state policies. This approach provides some degree of legitimacy and explains the survival of alternative green media.

The internet

Compared with print media, the internet has the advantages of speed, broad reach, and interactivity. It favors open discussion, speedy communication, and wide dissemination. Chinese internet users have turned to the internet for public expression and political activism even as the government is stepping up control.[36] Have Chinese environmentalists embraced the internet? How and why?

Out of the many different types of network functions, environmentalists often use three. The first is web sites. A survey of the web presence of environmental NGOs conducted in March 2004 finds that of the 74 organizations surveyed, 40 (or 54 percent) have web sites. The second is mailing lists. Several environmental NGOs maintain active mailing lists for sending environmental information regularly to subscribers by email. The third is bulletin boards. Twenty-four (or 60 percent) of the 40 web sites mentioned above have bulletin boards for public discussion.[37] In addition, some commercial portal sites and large online communities such as Tianya Club run "green" web forums. Blogs are a new popular form, but we have yet to find any influential blogs on environmental issues.

Environmental mailing lists and web sites deserve special attention. Some environmental groups encourage users to sign up for their mailing lists and publicize subscription information on their web sites. The reasons for the mailing lists vary. Some mainly publicize activities. One mailing-list we are familiar with sends a 15–20 page collection of up-to-date environmental information five days a week. In a few cases, mailing lists were set up as a campaign tool to discuss strategies and send action alerts. Although not open to the general public, mailing lists are nevertheless linked to the public sphere by channeling information there. They have the advantage of fostering free discussions within bounded social circles of people scattered in different parts of the country, which may otherwise be hard to sustain due to both political and financial limits.

Environmental web sites are growing in number and influence. Not only do green NGOs have web sites, but official environmental protection agencies have also created many. In addition, there are many personal home pages on environmental topics. All contribute to environmental discussions in their own way. For example, the environmental laws and policies archived on the web site of the State Environmental Protection Administration (SEPA) provide useful information resources. Judging by the number of times a web site is linked to by another, however, those maintained by NGOs are more influential. The above-mentioned survey shows that on average, each of the 40 NGO web sites contains links to eight domestic NGOs, two international NGOs, and one governmental environmental agency. Friends of Nature is linked to 19 times, whereas SEPA is only linked to eight times.[38]

The networked nature of environmental web sites suggests that the circulation of discourse among environmental NGOs is vigorous, more so than between NGOs and government agencies. This implies that if just a few NGO web sites are actively engaged in publishing and discussing environmental issues, then the issues are likely to spread to other web sites (and hence other audiences). It is essential to bear these dynamics in mind in discussing the discourse in any single environmental web site.

The web site of Friends of Nature offers a good example of the discourse produced in environmental web sites. FON went online in December 1998 and launched its first web site in June 1999 (interview with FON staff, July 2002). Besides publishing activities and showcasing the organization's projects, the web site supports an active bulletin board system (BBS) and publishes the electronic version of FON's print newsletter and an electronic digest.

A slimmer and less formal publication than the print newsletter, the electronic digest was inaugurated on 25 July 2002. From then to 10 June 2004, 12 issues were released. They featured interesting debates about such topics as the relationship between traditional culture and sustainable development, the meaning of "development," the environmental lessons of the SARS epidemic, and animal rights. The animal rights debate in the 4 March 2003 issue contains two lengthy articles. One explicates the importance of promoting animal rights from the perspective of environmental philosophy. The other argues that animal rights is a Western discourse with hidden imperialist pretensions, because in this discourse non-Western societies with different attitudes toward animals are portrayed as primitive and uncivilized.[39]

The editors were concerned less with who was right or wrong than with using web forums to foster discussion. As the editorial accompanying the two essays explains, "we want to provoke your thinking. We believe that the independent thinking of ordinary people is no less significant and no less valuable than that of the experts."[40] To facilitate discussion, at the end of each article a hot link was set up to FON's BBS, where dozens of messages were posted in response to the debate. One message reads, "I haven't had time to read the articles … but I'd like to state my views first. In my personal view, the rights of animals are the rights to existence and to free activity, which are endowed by Great Nature and shared by all creatures."[41] It is personal voices like these that find channels of expression in the green web sites.

The green sphere in action: the campaign to stop dam-building on the Nu River

We started this article with the campaign to stop dam-building on the Nu River. We now return to the case to illustrate China's green public sphere in action. We highlight the interactive dynamics of civil society and different types of media in the campaign.

The hydropower project on the Nu River was approved by the National Development and Reform Commission on 14 August 2003. Its core components are a series of 13 dams on the lower reaches of the river, which fall within Yunnan province. According to the project design, the total installed capacity of the dams will be 21 million kilowatts, exceeding even that of the ongoing project at the Three Gorges. The project aroused immediate controversy.[42] Supporters of the project claimed that it would accelerate the economic development of the river valley regions and help alleviate poverty.

Environmentalists who campaigned against the project held that the ecological treasures of the Nu River – its breathtaking natural beauty and biodiversity – were unique in the world and that they belonged not only to China but were also a world heritage.[43] In framing the debate, they stressed that the Nu River was part of the Three Parallel Rivers of Yunnan Protected Areas, which had just been listed as a World Heritage Site by UNESCO on 3 July 2003. Damming the river, they argued, would threaten a world heritage. They also argued that the project would benefit the developers more than the local residents, citing the potential problems of population resettlement and potential destructive effects on indigenous cultures. Finally, although they did not frame the issues in explicitly political terms, legal references were common. For example, a petition letter signed by 62 scientists, journalists, writers, artists and environmentalists called for the enforcement of the Environmental Impact Assessment Law.

The public views about the hydropower project on the Nu River were thus pitted between two opposing visions: the protection of natural heritage vs economic development and poverty alleviation. Not a surprising conflict, this was only the most recent Chinese version of the tensions inherent in the global discourse of sustainable development.[44] What is remarkable in this case is that the arguments of both sides entered China's public sphere and influenced policy.

The role of environmental NGOs in publicizing the campaign

The campaign against dam-building on the Nu River has some peculiar features.

First, it enjoyed the support of SEPA officials. When the campaign started, China's first Environmental Impact Assessment Law had just gone into effect (on 1 September 2003). Perhaps to demonstrate its commitment to the new environmental law, SEPA organized forums to assess the environmental impact of the proposed project. The first forum took place on 3 September 2003. More than 30 scholars and researchers attended. The predominant voice at the forum, led by a professor from the Asian International Rivers Center of Yunnan University, was harshly critical of the project.[45] National media, especially the *China Youth Daily*, covered the forum with such dramatic titles as "13 Dams to Be Built on the Last Ecological River, Experts Vehemently Oppose the Development of the Nu River" (*China Youth Daily*, 5 September 2003). From 14 to 19 October, SEPA led a group of experts on a study tour of the Nu River valley and then held another forum on 20 and 21 October, this time in Kunming. This second forum invited representatives from relevant government agencies at the provincial and prefectural levels in Yunnan province, as well as scientists and other scholars. At the forum, scholars from Beijing opposed the project, whereas the local Yunnan scholars and government officials defended it. SEPA officials were on the opposing side, but the controversy did not seem to be resolvable between the parties directly involved.

Second, environmental NGOs played a central role in tipping the balance in favor of the opponents of the project. They were instrumental in producing the "high level of social concern" cited by Wen Jiabao. Environmental NGOs launched a campaign as soon as they learned that the National Development and Reform Commission had approved the project. The China Environmental Culture Promotion Society organized one of the first influential public events. At its second membership congress on 25 October 2003, the organization issued a public petition to protect the Nu River. On 17 November 2003, the Tian Xia Xi Education Institute organized a forum to educate the public about the Nu River. The forum featured a speaker from the Yunnan-based NGO Green Watershed.[46] In December 2003, an NGO in Chongqing City collected more than 15,000 petition signatures opposing the Nu River project.[47] On 8 and 9 January 2004, five research and environmental organizations, including Friends of Nature and

Green Watershed, organized a forum in Beijing to discuss the economic, social, and ecological impact of hydropower projects, again directing its criticisms at the Nu River project. From 16 to 24 February 2004, about 20 journalists, environmentalists, and researchers from Beijing and Yunnan conducted a study tour along the Nu River. They returned to Beijing to organize a photo exhibit. Indicating the international dimension of the campaign, they took the exhibit to the UNEP 5th Global Civil Society Forum (GCSF) held in Jeju, South Korea, in March 2004 to mobilize international support. Together, these efforts created the momentum of a public campaign.

Strategic use of the mass media and the internet

Environmental NGOs made effective use of the mass media and the internet to produce and disseminate opposition. The most active NGO in mobilizing the media was Green Earth Volunteers. Green Earth Volunteers organizes monthly environmental salons for journalists. The two main organizers of the salons, Wang Yongchen and Zhang Kejia, are influential journalists *and* environmentalists. Wang is a senior journalist in China's Central People's Radio Station and a co-founder of Green Earth Volunteers. Zhang is a journalist in *China Youth Daily* and a main force behind *China Youth Daily*'s "Green Net," an online section of the newspaper devoted to environmental issues. The environmental journalists' salon had already proved to be an important base for mobilizing the media. It played a crucial role, for example, in mobilizing media opposition to the Dujiangyan Dam incident in 2003.[48] In the Nu River case, both Wang and Zhang were signatories to the petition letter of 25 October 2003, the first major public action in the movement. Wang organized the study tour of the Nu River in February 2004 and the subsequent photo exhibit in Beijing. Besides publishing many news reports about the debates surrounding the hydropower project, Zhang uses the "Green Net" of *China Youth Daily* to cover the debates.

In addition, web sites were used to disseminate information and foster discussion. Integrating new media with the traditional print media, the "Green Net" of *China Youth Daily* set aside a special column on the Nu River campaign and collected nearly 200 articles on the topic. The Institute for Environment and Development set up a campaign web site, which featured an online version of the above-mentioned photo exhibit about the Nu River, beautiful scenery of the river valley, as well as an archive of essays debating the issues. A campaign leader reported that after the web site was set up, she received letters and telephone calls just about every day and people would tell her how excitedly they were browsing the web site and how they hoped that it could be updated more frequently.[49]

Debates about the case also appeared in the bulletin boards run by environmental NGOs and commercial web sites. For example, a keyword search on 19 August 2004 for "Nu River" in the BBS of the popular Tianyaclub.com yielded dozens of postings debating the "Nu River" project. The opinions in these postings were divided; some were expressed in very angry tones. One posting laments: "Population and economic growth are the natural enemy of environmental protection!" Again, what matters here is not who is right or wrong, but that people were debating the issues in the public arena. All this shows that there was indeed a high level of "social concern" about the project, which prompted the central government to temporarily halt it in April 2004. It shows that China's fledgling green public sphere was instrumental in producing this social concern.

Discussion and conclusion

We have shown that a fledgling green public sphere is emerging in China. Environmental NGOs are its primary discourse producers. The mass media, alternative media, and the internet provide the communicative spaces, but are used in differential ways because of different institutional and political constraints associated with them. With the rise of a green public sphere, new ways of talking about the environment have been introduced to the Chinese public. As much as the discourse itself is important, however, it is no less important to highlight the communicative spaces in which it appears. These spaces are prerequisites for citizen involvement and political participation. They are essential for sustained and ongoing public discussion.

A green public sphere is not a homogeneous entity, but consists of multiple actors, multiple media, and multiple discourses. Nor is it completely autonomous or equal. Civil society actors must heed the political context; the different types of media are subject to varying degrees of political control. In a sense, the green sphere is a product of what the editors of this volume refer to as "embedded environmentalism." It is embedded in politics, in civil society, and in communications technologies. While being embedded in Chinese politics is often a constraining factor, civil society and communications technologies provide fertile soil for its growth. Embeddedness can be both constraining and enabling.

To be more open to meaningful participation, the fledgling green public sphere must overcome many challenges. Political control remains the main obstacle, because there are still limits to what can and cannot be brought into the public sphere. While recognizing this condition, however, citizens and civil society organizations must not wait for further political opening to happen. It is important to be actively engaged in public discourse in the available spaces. It is through persistent participation that political boundaries can be pushed back.

Future research should explore further the complex dynamics between the green public sphere and the political realm dominated by China's Party-state. Neither the green public sphere nor the political realm is a homogeneous entity. Conflicts of interest, values, and political visions exist in both realms. These conflicts may result in unexpected alliances between actors in the green sphere and those in the political sphere, facilitating some degree of political "disembeddedness," so to speak. Future analysis should explore the conditions under which such alliances may develop or break down, and their implications for the shifting boundaries of the green public sphere.

A major concern in the scholarship on contemporary Chinese society is the relationship between the state and society. An influential perspective is state corporatism, which argues that the state permits the development of social organizations, such as NGOs, provided that they are licensed by the state and observe state controls on the selection of leaders and articulation of demands.[50] More than ten years ago, Unger and Chan articulated a strong case for a state corporatist perspective on Chinese society.[51] Based on an analysis of more recent trends, Howell argues, however, that the state corporatist perspective is no longer adequate for capturing some new directions in Chinese society, such as the emergence of new types of civil society organizations working on marginalized interests.[52] Our analysis lends support to Howell's conclusion. Although the Chinese government undoubtedly plays an important role in fostering the green public sphere, such as by sponsoring the "China Environment Centennial Journey" campaign, a state corporatist perspective does not give enough credit to the agency of non-state actors.[53] The evidence we presented shows that the constitution of a Chinese green public sphere depends crucially on citizens and citizen organizations and on their creative use of the internet, alternative media, and the mass media. We will need to develop a perspective that emphasizes the interpenetration and mutual shaping of state and society.

To some extent, the green public sphere is exemplary of the general development of the public sphere in China. There is an implicit politics to it quite beyond the environment, a politics of expanding general public discourse. This politics can also be discerned in other social arenas (such as rural poverty), where citizens and voluntary associations are similarly engaged in public discussion and in finding ways to engage policy makers. Public discussion in these other social arenas also depends on various types of media. A future research agenda therefore is to study the discourse, publics, and media in these other social arenas and explore the sources and consequences of potential synergies between different issue-specific public spheres.

Notes

1 D. Torgerson, "Farewell to the Green Movement? Political Action and the Green Public Sphere," *Environmental Politics*, Vol. 9, No. 4 (Winter 2000), pp. 1–19.

2 On new information technologies and transnational public spheres, see C. Calhoun, "Information Technology and the International Public Sphere," in D. Schuler and P. Day (eds) *Shaping the Network Society: The New Role of Civil Society in Cyberspace*, Cambridge, MA: MIT Press, 2004, pp. 229–51.

3 Peter Ho indicates that green activism is a kind of de-politicized politics; see P. Ho, "Self-imposed Censorship and De-politicized Politics in China: Green Activism or a Color Revolution?" Ch. 2 in this volume. Martens similarly argues, "from an environmental perspective, civic involvement does not have to be political in order to be significant. A focus on political participation – on citizens merely in their capacity as political producers and consumers – is thus too narrow"; see S. Martens, "Public Participation with Chinese Characteristics: Citizen Consumers in China's Environmental Management," *Environmental Politics*, Vol. 15, No. 21 (April 2006), p. 213.

4 J. Shapiro, *Mao's War Against Nature*, Cambridge: Cambridge University Press, 2001.
5 J. Habermas, "The Public Sphere," in S. Seidman (ed.) *Jürgen Habermas on Society and Politics: A Reader*, Boston: Beacon Press, 1989, p. 231.
6 See, among others, N. Fraser, "Rethinking the Public Sphere – A Contribution to the Critique of Actually Existing Democracy," in C. Calhoun (ed.) *Habermas and the Public Sphere*, Cambridge, MA: MIT Press, 1992, pp. 109–41.
7 J. Habermas, "Further Reflections on the Public Sphere," in C. Calhoun (ed.) *Habermas and the Public Sphere*, Cambridge, MA: MIT Press, 1992.
8 C. Calhoun, "Tiananmen, Television and the Public Sphere: Internationalization of Culture and the Beijing Spring of 1989," *Public Culture*, Vol. 2, No. 1 (1989), pp. 54–71.
9 M.B. Rankin, "Some Observations on a Chinese Public Sphere," *Modern China*, Vol. 19, No. 2 (1993), pp. 158–82. W.T. Rowe, "The problem of 'civil society' in Late Imperial China," *Modern China*, Vol. 19, No. 2 (1993), pp. 143–8.
10 F. Wakeman, Jr, "The Civil Society and Public Sphere Debate: Western Reflections on Chinese Political Culture," *Modern China*, Vol. 19, No. 2 (1993), pp. 108–38.
11 M. M.-H. Yang, "Spatial Struggles: Postcolonial Complex, State Disenchantment, and Popular Reappropriation of Space in Rural Southeast China," *Journal of Asian Studies*, Vol. 63, No. 3 (August 2004), pp. 719–55.
12 G. Yang, "Civil Society in China: A Dynamic Field of Study," *China Review International*, Vol. 9, No. 1 (Spring 2002), pp. 1–16. E. Lean, "The Making of a Public: Emotions and Media Sensation in 1930s China," *Twentieth-Century China*, Vol. 29, No. 2 (April 2004), pp. 39–62.
13 Wang Lili, *Green Media: Environmental Communication in China* [*Lü meiti: Zhongguo huanbao chuanmei yanjiu*], Beijing: Tsinghua University Press, 2005. A Chinese discourse on civil society already began to emerge in the 1980s. See S.Y. Ma, "The Chinese Discourse on Civil Society," *The China Quarterly*, No. 137 (1994), pp. 180–93.
14 See H. Rom, J. Brockmeier and P. Muhlhausler, *Greenspeak: A Study of Environmental Discourse*, Thousand Oaks, CA: Sage, 1999, p. 2.
15 J. Greer and K. Bruno, *Greenwash: The Reality Behind Corporate Environmentalism*, Penang: Third World Network, 1996.
16 On Western new social movements, see C. Calhoun, "'New Social Movements' of the Early Nineteenth Century," in M. Traugott (ed.) *Repertoires and Cycles of Collective Action*, Durham, NC: Duke University Press, 1995, pp. 173–215.
17 Environmentalists articulated these visions forcefully in a public debate in 2005 that has come to be known as the "respect/fear nature" (*jingwei ziran*) debate. See Bao Hongmei and Liu Bing, "*Zhongsheng xuanhua: 'jingwei ziran' da taolun*" [The big debate regarding "respect or fear nature"] in Liang Congjie (ed.) *Huanjing lü pishu 2005 nian: Zhongguo de huanjing weiju yu tuwei* [Green Book of Environment 2005: Crisis and Breakthrough of China's Environment], Beijing: Shehui kexue wenxian chubanshe, 2006, pp. 119–30.
18 The original Chinese is *toucai shanhuo*. *Shanhuo*, here translated as "mountain stuff," refers to profitable products from the mountains, such as expensive mushrooms and ginseng.
19 Zhou Wei, "*Zoujin xianshi, zaisi huanbao*" [Up close to reality, rethinking environmental protection], in Green Camp 2000 (ed.) *Ba qian li lu yun he yue: 2000 Zhongguo daxuesheng luseying wenji* [Eight Thousand Li of Road, Clouds, and Moonlight: Collected Works of Chinese University Students' Green Camp in 2000], p. 31.
20 Haxi Zhaxiduojie, "*Sanjiangyuan de huhuan*" [The call of the three-river source], *Friends of Nature Newsletter*, Issue 4 (2002), online, available at www.fon.org.cn/index.php?id=3009 (accessed 2 April 2003).
21 Ho, "Self-imposed Censorship and De-politicized Politics in China".

22 M. Warner, "Publics and Counterpublics," *Public Culture*, Vol. 14, No. 1 (2002), p. 50.
23 P. Ho, "Greening without Conflict? Environmentalism, NGOs and Civil Society in China," *Development and Change*, Vol. 32, No. 5 (2001), pp. 893–921.
24 Lu Hongyan, "Bamboo Sprouts after the Rain: The History of University Student Environmental Associations in China," *China Environment Series*, Issue 6 (2003), pp. 55–65.
25 G. Yang, "Environmental NGOs and Institutional Dynamics in China," *The China Quarterly*, No. 181 (March 2005), pp. 46–66.
26 Xinhuanet, "Number and Staff of Environmental NGOs to Increase by 10 to 15%" [*Huanbao minjian zuzhi shuliang he renshu jiang yi 10% zhi 15% sudu dizeng*], 28 October 2006, online, available at http://news.xinhuanet.com/environment/2006–10/28/content_5261309.htm (accessed 29 October 2006).
27 S. Wang and J. He, "Associational Revolution in China: Mapping the Landscapes," *Korea Observer*, Vol. 35, No. 3 (Autumn 2004), pp. 485–534.
28 P. Ho, "Greening without Conflict?" On environmental regulation in China, see R.L. Edmonds, *Managing the Chinese Environment*, Oxford: Oxford University Press, 2000.
29 On the international dimension of Chinese NGO development, see J.G. Bentley, "The Role of International Support for Civil Society Organizations in China," *Harvard Asia Quarterly*, Vol. 7, No. 1 (Winter 2003), pp. 11–20, and K. Morton, "The Emergence of NGOs in China and their Transnational Linkages: Implications for Domestic Reform," *Australian Journal of International Affairs*, Vol. 59, No. 4 (December 2005), pp. 519–32.
30 Friends of Nature, "*Zhongguo baozhi de huanjing yishi*" [Survey on environmental reporting in Chinese newspapers], 2000.
31 "*Cong hongda maque dao shouhu dayan*" [From chasing and killing sparrows to guarding wild geese], *Beijing Youth Daily*, 7 September 1999, online, available at www.bjyouth.com.cn/Bqb/19990907/GB/3998^0907B09918.htm (accessed 1 September 2004).
32 R. Akhavan-Majid, "Mass Media Reform in China: Toward a New Analytical Framework," *Gazette*, Vol. 66, No. 6 (December 2004), pp. 553–65. Y. Zhao, "From Commercialization to Conglomeration: The Transformation of the Chinese Press within the Orbit of the Party State," *Journal of Communication*, Vol. 50, No. 2 (2000), pp. 3–26. G. Wu, "One Head, Many Mouths: Diversifying Press Structures in Reform China," in Chin-Chuan Lee (ed.) *Power, Money, and Media: Communication Patterns and Bureaucratic Control in Cultural China*, Evanston, IL: Northwestern University Press, 2000, pp. 45–66.
33 Wang Lili, p. 114.
34 Jin Cuihong, "*Zhan chulai – shuohua*" [Step forward – and speak up!], in Green Camp 2003 (ed.) *Ceng jing shi di* [*The Wetlands That Once Were*], p. 44.
35 On the unofficial publications in the Democracy Wall movement, see D.S.G. Goodman, *Beijing Street Voices: The Poetry and Politics of China's Democracy Movement*, London: Marion Boyars, 1981.
36 G.R. Barme and G. Davies, "Have We Been Noticed Yet? Intellectual Contestation and the Chinese Web," in E. Gu and M. Goldman (eds) *Chinese Intellectuals Between State and Market*, London: RoutledgeCurzon, 2004, pp. 75–108. Y. Zhou, "Living on the Cyber Border," *Current Anthropology*, Vol. 46 (2005), pp. 779–803. G. Yang, "Between Control and Contention: China's New Internet Politics," *Washington Journal of Modern China*, Vol. 8, No. 1 (Spring/Summer 2006), pp. 30–47.
37 G. Yang, "Activists Beyond Virtual Borders: Internet–mediated Networks and Informational Politics in China," *First Monday*, Vol. 11, No. 9 (September 2006), online, available at http://firstmonday.org/issues/special11_9/yang/.
38 G. Yang, "Activists Beyond Virtual Borders".

39 The two sides of the debate are represented in a recent issue of the electronic newsletter of Friends of Nature. See *Friends of Nature Electronic Newsletter*, No. 5 (4 March 2003), online, available at www.fon.org.cn/enl/view.php?id=49.
40 Available at www.fon.org.cn/enl/view.php?id=49#pinglun.
41 Available at www.fon.org.cn/forum/showthread.php?threadid=2527.
42 The campaign is still under way as of October 2006. The controversy analyzed here covered the period from August 2003 to April 2004.
43 The controversy is widely covered by the media. Transcripts of detailed arguments against the hydropower project made by scientists and environmentalists at a forum organized by the State Environmental Protection Administration are available on the "Green Net" web site of the *China Youth Daily*. See http://202.99.23.201/cydgn/gb/cydgn/content.759348.htm (accessed 1 May 2004).
44 According to one study, the meaning of "sustainable development" is so ambiguous and so often contested that by 1992, five years after it was first introduced by the World Commission on Environment and Development, about 40 different definitions had appeared. See D. Torgerson, "Strategy and Ideology in Environmentalism: A Decentered Approach to Sustainability," *Industrial and Environmental Crisis Quarterly*, No. 8 (1994), pp. 295–321.
45 See transcripts of the forum published on the "Green Net" web site of the *China Youth Daily*, online available at http://202.99.23.201/cydgn/gb/cydgn/content.759348.htm (accessed 1 May 2004).
46 Announcement to Tian Xia Xi's mailing list, 4 November 2003.
47 J. Yardley, "Dam Building Threatens China," *New York Times*, 10 March 2004.
48 "Insiders' Reflections on the Fight to Protect the Dujiangyan Dam," document circulated for discussion on a private environmental mailing list.
49 Discussions at an environmental salon held in Beijing on 17 May 2004. Source: Nu River mailing list, 31 May 2004.
50 P.C. Schmitter, "Still a Century of Corporatism?" in F.B. Pike and T. Stritch (eds) *Social-Political Structures in the Iberian World*, Notre Dame, IN: University of Notre Dame Press, 1974, pp. 85–130.
51 J. Unger and A. Chan, "China, Corporatism, and the East Asian Model," *Australian Journal of Chinese Affairs*, No. 33 (January 1995), pp. 29–53.
52 J. Howell, "New Directions in Civil Society: Organizing around Marginalized Interests," in J. Howell (ed.) *Governance in China*, Lanham, MD: Rowman and Littlefield, 2004, pp. 143–71.
53 On the role of non-state actors in China's media transformation, see Akhavan-Majid, "Mass Media Reform in China".

5 Grassland campaigns during the collective era

Socialist politics and local strategies in Uxin Ju[1]

Hong Jiang

In comparison with the Qing dynasty, the Maoist regime has accrued much more administrative, economic and political resources and has exercised tighter control over its vast territories. During the collective era beginning in 1958 and lasting until the late 1970s, such control virtually eliminated any possibility of grassroots resistance, and those who engaged in overt protests to the state, whether peaceful or violent, were met with severe persecution or execution. During the Great Leap famine (1958–61), for example, the starving farmers who attacked state granaries or robbed grain from cargo trains received gruesome physical torture and capital punishment;[2] dissenting from state control even just slightly could invite severe punishments.[3] The Maoist government forced its citizens to destroy forests to fire backyard furnaces and to fill lakes in order to make crop fields; grave environmental disasters ensued.[4] Since the dismantling of the collective institution in the late 1970s, China has experienced greater political openness, and grassroots environmental movements have started to emerge. Environmental NGOs have grown rapidly, and channels for citizens' grievances and protests have increased significantly.[5] The semi-authoritarian nature of Party-state control today has enabled increased participation in environmental affairs in recent years, albeit in an embedded fashion. This represents a departure from the repressive governmental control during the collective era.

Does this mean that the collective era can only be viewed through the lens of state politics? This contribution uses this question as a point of departure and explores local strategies employed in response to state grassland policies, arguing for the important role of local agency in such endeavors. Such agency was not expressed in the form of organized protests but through subtle resistance to and creative use of state grassland policies. In this sense, there are certain similarities with contemporary "embedded social activism" even though the post-Mao Party-state has exercised a different form of control. At the basis of this study are an affirmation of local agency and a non-dichotomous approach to the study of state–society relationships. Uxin Ju (Wushenzhao), a Mongolian community in Inner Mongolia, provides a case study. This exploration not only places the post-Mao environmental actions in a historical context, but also helps inform our understanding of current grassroots environmental strategies, as I will discuss in the concluding section.

Scott's notion of resistance[6] and de Certeau's theory of everyday practice[7] provide the framework for my exploration of local agency during the collective era. In power-laden settings of an oppressive system, instead of organized protests and movements, subordinate groups utilize subtle and unorganized methods of behavioral and symbolic resistance, such as "foot dragging, dissimulation, desertion, false compliance, pilfering, feigned ignorance, slander, arson, sabotage"[8] and "slander, gossip, and character assassination."[9] Hidden dissent behind conforming public transcripts is common.[10] Subtlety and disguise are characteristic of these strategies, enabling the resisters to escape the severe lash of punishment by the oppressive system. Although not aiming at revolutionary social changes, such resistance slowly erodes the effectiveness and legitimacy of the dominant system. De Certeau takes this empowerment of subaltern groups a step further by explicating their creative use of dominant structures. The subaltern uses various flexible "tactics" to counter the rigid rules or strategies of the dominant system, and in the process "poaches" from the system to its own advantage.[11] Such poaching does not necessarily create direct opposition to the system, but instead exploits the space between the system's production and its use. Employing the example of the Indian's use of the Spanish rule, de Certeau states,

> [T]he ambiguity that subverted from within the Spanish colonizers' "success" in imposing their own culture on the indigenous Indians is well known. Submissive, and even consenting to their subjection, the Indians nevertheless often made of the rituals, representations, and laws imposed on them something quite different from what their conquerors had in mind.... [T]heir use of the dominant social order deflected its power, which they lacked the means to challenge; they escaped it without leaving it.[12]

Both resistance and creative use have been practiced under the socialist regime, by groups to give meaning to places and by individuals to render personal life bearable. In a monumental work on nature protection in the Soviet Union, Weiner explores the discursive space carved out by independent social and professional identities within the repressive authoritarian system, observing, "Nature protection serves as a surrogate for politics, as actual political discourse was prohibited and punished."[13] Space for freedom was found in inconspicuous fissures left by the authoritarian system. In China's collective era, similar spaces were found between "official ideology" and "lived experience."[14] By uncovering the innovative ways in which women were able to subvert state control, Zhong and her colleagues affirm the power possessed, however constrained, by people in their daily lives, thus challenging the dominant "dark age narrative" of the Mao era. Friedman, even when examining the tight grip of central politics on model communities such as Dazhai, also emphasizes the importance of local independence and strategies.[15] While local agency did not prevent millions upon millions from falling victims to socialist policies and political campaigns,[16] it did offer some localities, groups and individuals opportunities to transcend the regime's oppression. Uxin Ju offers one such example.

My discussion centers on Uxin Ju's experience in grassland use. From 1958 to 1965, the Mongols participated actively in the campaign to improve the grassland and control sandy land. They removed a toxic grass from the pasture, converted moving sand dunes into usable pastures, and constructed pasture enclosures. These achievements were recognized by the state in 1965. After the Shanxi village Dazhai was promoted by the state as a national model in agriculture in 1964, Uxin Ju was named a "pastoral Dazhai."[17] As a state-promoted model, Uxin Ju's experience might be judged immediately as an outcome of political obedience. However, my research suggests that such an assessment is overly simplistic. While the Mongols were pressured into complying with the politics of the campaign, they utilized them creatively to fit their locally based goals, thus transforming them into local meanings, actions, and history centered around their lived experiences on the grassland. In the beginning, subtle resistance was plentiful, but over time it became weakened, not by political pressures but by the Mongols' changing attitudes toward nature after seeing the increasing reformability of the grassland. The Mongols' active participation in the grassland campaign challenges the simple dichotomy of state oppression and local action.

Anti-grassland backlash accompanied and then overwhelmed the grassland campaign, not for lack of local participation but because of state policies that reversed it. In the beginning of the Cultural Revolution (1966–76), Inner Mongolia was taken over by a leftist military leadership that ordered the opening of grassland to cultivation. Grassland improvement stopped. Local strategies during this anti-grassland campaign differed from the previous period. Rather than active participation and creative use, the Mongols now protested both overtly and subtly, and their public compliance was laden with hidden transcripts of dissent and self-protection. While the Mongols' strategies during the grassland campaign are more informed by de Certeau's idea of "poaching" or creative use, their resistance to the anti-grassland campaign, which so fundamentally contradicted with their own interests, can be better understood through Scott's notion of resistance.

The rest of this contribution proceeds as follows. I will first describe Uxin Ju's physical environment, which provides the background for understanding the impacts of state policies on the grassland. Since all socialist models were intimately connected with state politics at the national and regional levels,[18] I will then discuss political struggle involving grassland use and Uxin Ju's elevation by the state as a model community. My emphasis, however, is on local responses to state power. Next I will chronicle Uxin Ju's participation in the grassland campaign from 1958 to 1965, separating official depictions from local realities and analyzing local motivations. Discussion of the anti-grassland campaign during the Cultural Revolution will follow, where I will trace the Mongols' resistance in its historical context. This contribution will conclude with a summary of local agency during the collective era and its implications for post-Mao environmental actions. This contribution is based on field visits and interviews with over 80 Mongols conducted from 1998 to 2001, as well as extensive collections of newspaper articles, government documents, and unpublished local archival materials.

Uxin Ju: a land of sand and bad water

Uxin Ju is a Mongolian community situated in Uxin banner, Ih-Ju league in western Inner Mongolia (Figure 5.1). Its total area is 1,744 sq km. Uxin Ju is located at the center of Mu Us Sandy Land, *Mu Us* being Mongolian for "bad water." A dry land covered mostly by sand, Uxin Ju has a low amount of precipitation, averaging only 321 mm per year. With high rainfall variability, drought is frequent. With no water outlets, precipitation either filters quickly into the ground or accumulates on lowland to form stagnant bodies of water, which can

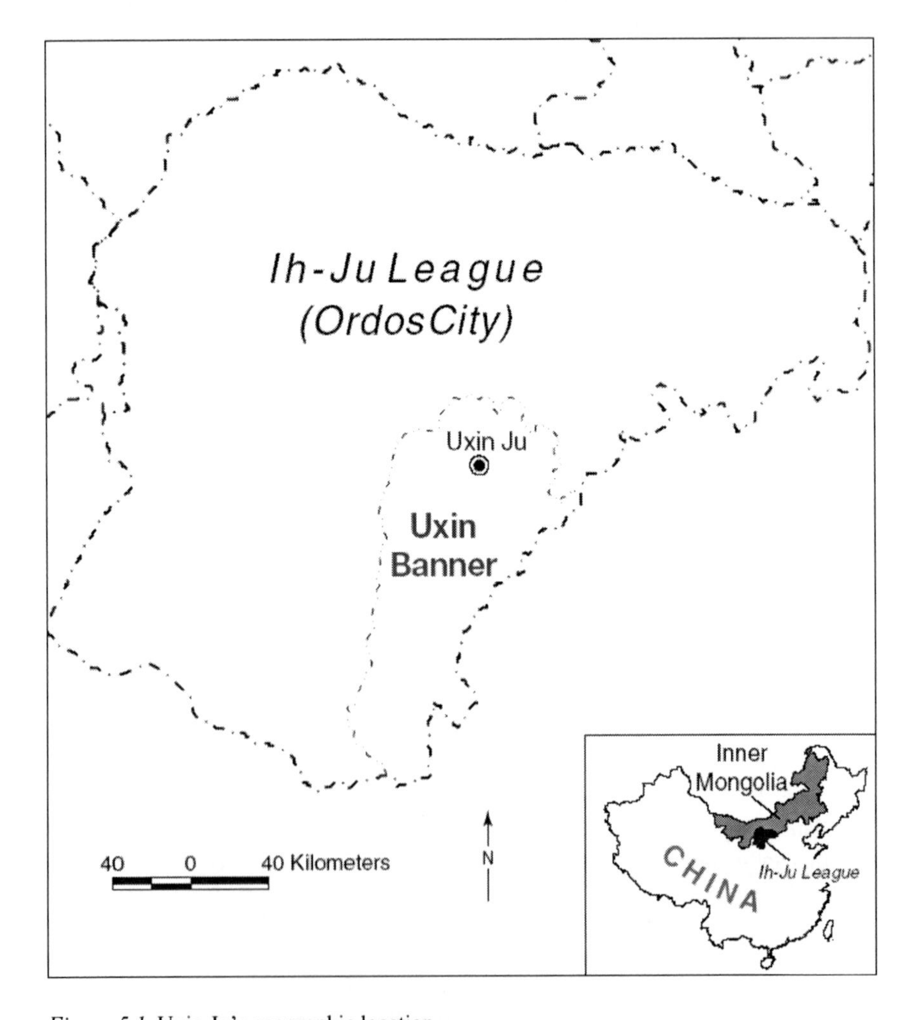

Figure 5.1 Uxin Ju's geographic location.

become a breeding ground for diseases – thus the name "bad water."[19] These water bodies cover less than 3 percent of the area, however.

Most of the land is sandy, either exposed or vegetated. An unfortunate accompaniment of the dry and sandy environment is strong wind, especially in the spring (average wind speed 4.4 m/second), as the Pacific warm air mass contends with the Siberian cold air mass. Vegetation cover also happens to be the lowest in the spring, making the wind more detrimental for the establishment of newly planted or vulnerable vegetation. The sand, once exposed, is easily displaced by the wind, which, on average, shifts an entire sand dune in the direction of the prevailing wind by up to 6–10 m each season. Exposed sand and sand dunes occupy slightly more than half of the area; one can imagine the harshness and high variability of the environment. The rest of the landscape can be called grassland, either shrubs on sandy (and occasionally rocky) surface, or grass in earthy lowlands, which is the best pasture in Uxin Ju.

As an administrative unit, Uxin Ju is a *sum*, with four smaller communities, each called a *gacha*.[20] Located north of the Great Wall, the area has historically been a Mongolian homeland. Uxin Ju's harsh environment makes agriculture impossible without irrigation techniques, which have only been developed significantly since the 1980s. Uxin Ju remains a Mongolian place, with animal grazing as the traditional basis of its economy. At the beginning of the grassland campaign in 1958, Uxin Ju had 2,300 residents, 96 percent of whom were Mongols, the rest Chinese. Among its 50,000 livestock, 62 percent were goats, and the rest included sheep, horses, and other animals. Elements of nomadic practices still remained, as the Mongols rotated their livestock on the grassland, a grazing method established after a long adaptation to the environment. Sedentarization had started and many had built semi-permanent homes. While the pastoral conditions have changed, e.g. goats gave way to sheep and agropastoral practices replaced pastoral grazing,[21] grassland change has been an important part of the transformation that started with the grassland campaign in 1958.

The politics of the grassland campaign

The grassland campaign in Uxin Ju was initiated at the onset of the Great Leap Forward, which hastened the integration of Inner Mongolia into socialist politics. Prior to that, Inner Mongolia, being an autonomous region established in 1947, was allowed some unique regional policies that fit the regional situations. In land reform, for example, while landlords in most of China were identified as hostile elements and their properties seized and distributed to poor farmers, Inner Mongolia, under the leadership of Ulanhu, implemented a unique regional policy for its pastoral areas that was referred to as "Three-Nos": no execution of herd-owners, no distribution of livestock, and no class differentiation among herders. In the Tumed agricultural area, due to their unfamiliarity with farming, Mongols received double the cropland of the Chinese.[22] These regional policies allowed for a much more gradual transition to socialism in Inner Mongolia than much of the rest of China. In addition, grassland protection regained its central importance. The region's "1951 method of land reform in Mongol areas" mandated that "land reform must take into consideration the development of animal husbandry, and firmly protect grassland and livestock, and absolutely ban grassland opening."[23] These regional policies served to energize the grassland economy and the Mongolian way of life.[24]

The level of autonomy that Inner Mongolia gained was due to Ulanhu's careful negotiation with the central government. While details of Ulanhu's life and his effort to serve both the center and the Mongols' interests have been examined by Bulag,[25] here I will discuss briefly his maneuvers in protecting the grassland. Using the socialist discourse of ethnic equality and development, Ulanhu maintained that grassland grazing was a "Mongolian economy," the development of which would promote the socialist goal of unity between the Chinese and Mongols. To counter the age-old Chinese bias that considers any unfarmed land as wasteland (*huangdi*), Ulanhu invited China's best grassland biologist Li Jisheng in the late 1950s to be the vice president of the region's flagship university, Inner Mongolia University, and promoted grassland science and research.[26] By linking grassland policies with the Mongolian ethnicity, Ulanhu was able to withstand the Chinese encroachment until 1958.

In the Great Leap Forward movement, Mao's call to build socialism "more, faster, better, and more efficiently" (*duokuai haosheng*) pressured Inner Mongolia to join in. It was in this context that Inner Mongolia advocated a grassland campaign. In May 1958, Ulanhu published an article in *Renmin Ribao* (*People's Daily*) entitled "Liberate Thoughts, Unite Nationalities, and Leap Forward Production," in which he asked the Mongols to plant shrubs on degraded grassland and to control the ever expanding desert.[27] As he announced that "transforming the desert is a historical task in our autonomous region," he added that "transforming the desert must be combined with the development of pastoral economy,"[28] implying that the campaign was not about converting to farming. The grassland campaign, by using the aggressive socialist approach to nature but without jeopardizing the grassland economy, embodied Ulanhu's double role of serving both the Mongols and socialism. Uxin Ju's participation in the grassland campaign was set in this juncture of central and regional politics.

The later elevation of Uxin Ju as the "pastoral Dazhai" represents yet another attempt by Ulanhu to protect the Inner Mongolian grassland. In 1964, after the Dazhai village of Shanxi province was promoted by Mao as a national model, the "Learn from Dazhai" movement soon swept the entire country. As a farming village, some of Dazhai's main achievements were to build terraces, improve soils, and achieve high yields. In socialist orthodoxy and uniformity, the Dazhai movement forced communities to cut down trees in order to build terraces and fill lakes to make croplands.[29] These efforts failed miserably, causing disastrous environmental damage. If Inner Mongolia were to be swept into similar frenzy, its grassland would soon turn into "Dazhai crop fields." In 1965, to counter the pressure to follow a farming model, Ulanhu praised Uxin Ju's grassland campaign and elevated it as the "pastoral Dazhai" and promoted Uxin Ju's leader Baoriledai, the Mongols' counterpart of Dazhai's Chen Yonggui.[30] Ulanhu's strategy is referred to by Bulag as "resistance-within-collaboration."[31] To use Scott, the hidden transcript of grassland protection was disguised in a public endorsement of Dazhai; or to follow de Certeau, Ulanhu utilized the socialist discourse creatively to serve the Mongols' interests.

Transforming the grassland

Removing a toxic grass

In 1958, in response to the Great Leap Forward campaign, Uxin banner conducted a "socialist competition," ranking all its 16 *sum*s/townships according to their economic achievements. Uxin Ju was ranked the lowest. The previous year, during a severe drought, 11 percent of its livestock died. On the competition chart, Uxin Ju was portrayed as riding backward on a black pig. Coming back from the meeting in the banner, the then-party secretary Han Fuchen immediately organized a meeting with the *sum* leaders to devise ways to catch up to the other *sum*s. They decided to follow the government's call to remove horse-poisoning grass.[32]

Horse-poisoning grass (*Oxytropis glabra*, in the legume family) was the foremost direct cause of livestock death in 1957. It is a toxic grass that flourishes during drought, and livestock will weaken and die after grazing on it for any extended period of time. The decision to remove it, however, was met by strong opposition, particularly from the elders, and from Uxin Ju temple lamas as well.[33] According to traditional beliefs, horse-poisoning grass is a sacred plant which must be appeased. Should the grass be removed, god would be offended and livestock would suffer even more. As Baoriledai, a campaign leader, recalled her mother's devout belief in the grass,

> On May 5th [of the lunar calendar], my mother would take me [to worship]. I was told to be quiet. She sprayed fresh milk on horse-poisoning grass, praying that it would not harm us … We were really up against traditional forces – we decided to launch the campaign [to remove this sacred grass] on May 5th.[34]

The "we" Baoriledai referred to included *sum* leaders, party members, and many young Mongols who were inspired by the hope of socialism. Mobilization meetings were held, and pastoralists debated whether horse-poisoning grass was sacred or harmful. Some contended that well-digging and cutting of the *Salix psammophila* shrub around the lama temple during the previous year were offensive, so "the sacred grass was angry, and therefore livestock died." But others reasoned that if the grass were sacred, it should give the livestock longevity; therefore, the grass must be harmful – as harmful as the herdlords.[35] In the end, although fearful, many endorsed the removal decision.

For the first time in Uxin Ju's history, the Mongols adopted a confrontational approach to the grassland. Altogether about 1,300 people, more than half of its total population, were mobilized. They formed four teams, each of which worked, ate, and slept on the grassland, removing the poisonous grass by hand and shovel during the day and attending political meetings to strengthen socialist ideology in the evening. According to the official report, they acted like soldiers on the battlefield.[36] The campaign lasted for 24 days, and they removed the toxic grass from 420,000 *mu* (28,000 ha) of the grassland.

The removal did not begin with enthusiastic support by the masses, as the official claim has it. For example, not all teams were mobilized on 5 May, and not all participants were convinced that the grass should be removed. Gongbu, a Mongol in Buridu *gacha*, recalled in an interview:

> I was not enthusiastic about the removal campaign. The elders said it was sacred, and I was afraid ... Then, very soon, I changed my mind about the grass: reality proved the efficacy. After the removal, livestock did not die!

Indeed, livestock death rate decreased to 3 percent and total number increased by 32 percent the following year.[37]

During the removal, the *sum* organized competitions: the team that placed first would receive a red banner on their "battle ground," and extraordinary deeds, such as that of eight mothers who participated in the campaign while carrying their infant children, received praise on the blackboards that were erected on the grassland. Peer pressure from team members turned the campaign into a game to be won. For example, to strengthen their chance of winning, a group of young Mongols in Buridu decided to spend two hours at night digging irrigation wells. As political pressure turned into cultural prestige, volunteerism took over. It is no wonder that most of the participants whom I interviewed recalled fun memories from the campaign, despite the fact that the campaign originated from political pressure.[38]

This removal campaign represented a landmark transition in Uxin Ju, ending the long tradition of adaptation to the grassland and initiating a human-centered approach to pastoral economy. During the next seven years, similar removal campaigns were mobilized three more times, and fear of the grass was largely dispelled.[39]

Turning the sand dunes green

In the autumn of the following year, 1959, spurred by the success of toxic grass removal and urged on by government's call to transform the desert, a campaign to control the sandy land was launched. Prior to that, a mutual help team in Buridu *gacha* achieved some success in planting *Artemisia ordosica* on the moving sand. Now, Baorong, a leader from Uxin banner who was sent down to Uxin Ju to assist in the local work, believed that the success could be expanded, so he helped launch the sandy land conversion campaign. A Mongol cadre intellectual who had invented a method of digging *Salix*-walled irrigation wells,

Baorong was well respected. He was to play a key role in the creation of methods for sand dune conversion.

The first sandy land campaign lasted for 20 days. Similar to the campaign in 1958, the Mongols worked and slept on the sandy land, carrying *Artemisia* shrubs they dug out of the grassland and planting them on the moving sand (Figure 5.2). However, their efforts were met with limited success. Siqingerile, an active participant in Chahanmiao *gacha*, recalled, "We planted on an entire sand dune in 1959, but in the following spring, only three stems survived."

The pastoralists had no experience in converting the sand dunes. They indiscriminately planted on the dune sand, which provided little moisture to sustain the planted shrubs, and strong winds in the spring helped blow out their roots. The failure encouraged the belief that planting on the sand dunes was against god's design. In Buridu, a Mongol lost quite a few livestock to miscarriages, and when he went to the lama temple for divination, a lama told him, "In front of your house there used to be pure white [sand], now black dots [survived *Artemisia*] cover it. That is why your livestock died."[40] This incident was quickly seized by the *sum* as a negative example in thought education. Baorong led a meeting on the sand dunes to assess their plantings, and discovered that those that survived were all located on the lower leeward side. This discovery gave them hope, and during the next few years they first planted on lower leeward locations, and the survival rate increased.

Figure 5.2 Planting on sandy land in the 1960s (source: photo courtesy of Arbijihu).

After several years of trial and error, a system of successful steps for sand dune improvement was established. It was summarized by the pastoralists as "block the front and pull the back" and "boot first, robe second, and then hat." Planting of trees and shrubs started from the bottom of the sand dunes on both windward (front) and leeward (back) sides. As the vegetation grew to wrap around the sand dunes, the "boots" vegetation was established, and planting continued upward to the tops of the sand dunes, covering the dunes with "robes" and eventually "hats," thereby turning the entirety of each sand dune green. Wind blocks, made of dry shrubs, were erected between the planted rows to reduce wind damage and to increase the survival rate. These methods were later summarized and spread to other areas in Inner Mongolia by the scientists at the Uxin Ju Sand Control Station, which was established in 1959 to facilitate the desert campaign.

According to official statistics, by 1965, Uxin Ju succeeded in turning 60,000 mu (4,000 ha) of sand dunes into green pastures.[41] This was less than 5 percent of Uxin Ju's sandy land. However, the success gave Uxin Ju great hope, and the *sum* leaders proposed to continue with sandy land conversion in order to increase "grass, water, trees, livestock, animal feed, and grain" – the so-called "six demands" from the sandy land.[42] In 2001, during my interviews, Erdun, an active campaigner in Bayintaolegai *gacha*, led me to an area of *Artemisia* and *Salix* shrub that had been converted from sand dunes during the 1960s (see Figure 5.3), and recounted proudly how they had achieved such improvement through persistence and hard work. Indeed, much of the success was the result of repeated planting, sometimes up to nine times.[43]

Figure 5.3 Sandy land converted during the 1960s (source: photo taken by author in 2001).

Enclosing the grassland

Grassland enclosure, called *kulumn* in Mongolian, was developed in Uxin Ju as a result of highly managed grassland. Following sandy land conversion, both cropland and tree planting expanded. Dryland farming of broomcorn millet had been practiced since the late 1800s when Chinese immigrants moved into the area, but irrigation only started in 1957 when Baorong led the digging of the first irrigation well. During the ensuing decades, irrigation expanded to 54 ha.[44] Tree planting started in the early 1960s. Traditionally, pastoralists did not like planting trees since grass does not grow well underneath them.[45] In 1949, Uxin Ju had 94 willows around the lama temple. By 1965, however, the number of planted trees accrued to 60,000, mostly willows (*Salix matsudana*) and poplars (*Populus spp.*), whose leaves could be used to feed livestock.

To protect the cropland and trees from destruction by livestock, pastoralists in Chahanmiao *gacha* built the first enclosure. It measured 90 ha in area and 5 km in circumference. Building the enclosure was time-consuming, since the 2 m-high wall was made with sod dug out of and carried from nearby lowlands. Each person could build only 1 m of wall a day, making the labor requirement of the enclosure total 5,000 labor days. Additional work was done to flatten the sandy land inside the enclosure in order to create cropland, and, in the process, the Mongols removed numerous sand dunes.[46] The success of the first enclosure was inspiring, and over time more enclosures were built throughout Uxin Ju. In addition to protecting planted crops and trees, these enclosures also helped in the production of high-quality hay as well as the grazing of weak and young livestock in cold seasons. By 1972, there were over 50 large enclosures. Smaller enclosures of several hectares numbered over 100. All together, over 2,700 ha, or 4 percent of usable grassland, was enclosed by more than 250 km of walls. Most of the hay was cut from these enclosures. In cold seasons, over 50,000 weak and young livestock, about half of the area's total livestock, grazed in the enclosures.[47]

Uxin Ju's recognition

In 1965, in the central-regional political context that was discussed earlier, Uxin Ju gained nationwide recognition. On 2 December, *Renmin Ribao*, the key newspaper of the national government, published a report and an editorial about Uxin Ju on its front page, praising its spirit and achievements in grassland construction and naming it the "pastoral Dazhai." *Neimenggu Ribao* (*Inner Mongolia Daily*) syndicated the same articles on 3 December, with Ulanhu's praises. Uxin Ju's Party secretary Baoriledai was soon promoted to the Inner Mongolia regional government. In Inner Mongolia, "learn from Dazhai and catch up with Uxin Ju" became a popular slogan.[48] In the first half of 1966, delegations from China's pastoral areas, as well as international visitors, came to visit Uxin Ju to learn from its spirit. Chen Yi, a central government official, visited the area and gave praises.

Who created the "pastoral Dazhai" model?

The question of who created the "pastoral Dazhai" model can be answered in one of two ways: the regional government led by Ulanhu, or the local people. Of course, model creation in socialist China was a political process.[49] Given my earlier discussion on the politics of the grassland campaign, the first answer is surely correct, in part. But I will also explore this question from the perspective of the Mongols, most of whom were outside the regional political process and for whom the "pastoral Dazhai" experience only had local meaning.

In 2001, to my question of the creation of the "pastoral Dazhai" model, my informants answered overwhelmingly that "we, the pastoralists in Uxin Ju, created it." Some did mention the leadership of Baorong and Baoriledai, but most insisted that they worked hard in the campaign and invested their energy and ingenuity in it. All of my informants believed that the direction of the campaign was correct, although some pointed out that the methods of sand dune conversion were not as well developed as they are today; especially, wind-blocking strategies were not sufficient in the past. Some informants offered sentiments similar to those of Erdun, who stated, "If we had not planted trees then, it would be all sand now."

What motivated the Mongols to participate in the campaign? Of course, in Mao's China, political pressure was always an important factor. But since the Mongols themselves claim ownership of the campaign, we must examine other, more immediate reasons that motivated their participation. Some of these reasons were alluded to in the last section. Indeed, the sandy environment was detrimental to the sedentarized lifestyle, as moving sand could easily destroy human and livestock shelters. One slogan often used in referring to the sandy land before the campaign was "as sand advanced, people retreated." Humans were on the losing end of the contest with sand; some propaganda material even called sand an "enemy."[50] In addition, the shortage of pastureland became severe as the number of livestock increased rapidly. From 1949 to 1958, for example, the number of livestock more than doubled, and it doubled again in the decade after 1958.[51] Only by improving the sandy land could pasture increase. With this reasoning, an informant, Dongrebu, commented:

> Pastoral development, which started in 1958, was forced by natural and economic conditions; removing horse-poisoning grass was a good deed. At a time when grazing relied on nature [kaotian yangxu], Uxin-Ju *sum* adopted "grassland construction" for the first time. This was a great initiative. Only through grassland construction can we increase grassland and pastoral economy.

Sedantarization and grassland shortage have deep historical roots. The Qing Dynasty organized the Mongols into mutually exclusive banners, each fixed to a specific piece of land called *nutag* homeland.[52] This was the beginning of the end of nomadic mobility. Chinese immigration and grassland opening after the late Qing took away some of the best pastures from the Mongols, and land shortage became keenly felt. Land disputes between neighboring banners, such as the one between Uxin and Otog, became common.[53] The socialist regime further fixated the Mongols to specific land by promoting permanent settlements. As the pasture-to-livestock ratio decreased, the only way that the Mongols could survive was to "construct" so as to expand their grassland. Seen in this context, grassland construction, just as Dongrebu indicated, was both forced upon and chosen by the Mongols as a way to adapt to increasingly unfavorable human–environmental conditions.

"Grassland construction" was proposed by Ih-Ju league prior to 1958, but the Mongols in Uxin Ju believed that they were the first to apply it in the grassland campaign. My research in the surrounding communities confirms that Uxin Ju was the first to remove toxic grass. Their claim of the invention of sand dune control methods can be supported by the metaphoric descriptions using robes, boots and hats of traditional Mongolian dress. Unlike some other Mao-era models that exaggerated or even falsified their achievements, the Mongols' effort created real grassland improvement in Uxin Ju. This judgment comes not only from my interviews and field visits, but also from a report from the *Siqing* (four-cleanup) team that the reported figures of grassland improvement were accurate.[54] The *Siqing* team came to Uxin Ju in 1966 to investigate cadre corruption. Experiences with *Siqing* teams elsewhere show their ruthless criticism of cadres;[55] therefore I take the team's validating report as an additional confirmation of Uxin Ju's significant grassland improvement.

The Mongols' changing attitude toward nature indicates that the grassland campaign has been internalized. Not only did they see the disappearance of gods from the landscape, they also came to see the necessity of reformation of the grassland. "Nature is not reliable," said one informant. The traditional value of nature–culture unity[56] was replaced with a trust in human ability to improve on nature that was encouraged by the campaign's ideology and practice. Such changes have extended beyond the campaign itself and are influencing today's household-based land use in the post-reform era after the 1980s. My interviews concerning household economic planning suggest that most households planned to improve their sandy pastureland with planting or seeding, and many also hoped to expand irrigation in order to produce more fodder and feed. Human improvement has become an integral part of agropastoral economy in today's Uxin Ju. In other words, the early grassland campaign has laid foundations in both attitudes and practices for today's agropastoral development.[57]

Since the 1980s, Mao-era models have fallen out of favor, and some (such as Dazhai) have been discredited; but my informants still consider their campaign a worthy effort and the "pastoral Dazhai" model is still granted prestigious status. In the courtyard of the Uxin Ju *sum* government, two tablets have been proudly erected displaying laudatory remarks from Ulanhu and Chen Yi. Uxin Ju's grassland campaign represents an important chapter in the experiences of rural China during the collective era, one that testifies that landscape improvement did occur and that local people's efforts did make a difference.

Some may argue that the Mongols' motivations (i.e. building a stronger pastoral economy) and changes in attitude (i.e. their increasing trust in human power) testify to the very success of state policies. That may be true; yet I want to emphasize the Mongols' local goals, that give meaning to local life that is deeply rooted in the Mongols' experience living on the grassland. Using de Certeau's framework, the Mongols "poached" state policies and created their own life and history through their participation in the grassland campaign. That campaign is still being reinvented today, as the Mongols trace current grassland improvements to the campaign, and as Bao Qingwu, a reporter of Uxin Ju during the grassland campaign and later a grassland scientist, claimed in an interview that Uxin Ju's campaign was a start of an ecological-based pastoral economy.

Anti-grassland campaign during the Cultural Revolution

It would be false to assert that people in Uxin Ju were completely free to reinvent their land and life in the collective era. In the grassland campaign, while the methods were against the traditional approach, the end goal could be made to fit with local needs, thus allowing room for local appropriation. In the on-going political contention between the regional effort for grassland protection and the center's push for grassland opening, the grassland campaign represented the conditional victory of the former. As I will show in this section, that victory was both incomplete and unsustained. During the grassland campaign, a wave of central pressure hindered the regional effort to protect the grassland; at the height of the Cultural Revolution, the center (the left faction) completely won out, effectively replacing the grassland campaign with an anti-grassland campaign to convert more grassland to cropland. While the Mongols mobilized the space available in the grassland campaign to their own advantage, their interests were so completely disregarded in the anti-grassland campaign, and the politics so oppressive, that the only room left to them was subtle resistance. The previous discussions show the existence of creative local strategies; now it becomes clear that state policy structures different possibilities for local responses.

The anti-grassland campaign severely damaged Uxin Ju's grassland. In 1966, at the beginning of the Cultural Revolution, Ulanhu was removed from his position. The political struggle that ensued turned Inner Mongolia into turmoil, in which a military government appointed by the center took over control of the region in 1969.[58] This new government, closely in line with the leftist faction at the center, issued various policies that were detrimental to the Mongolian population.[59] Taking the slogan "Grain First," it issued orders to open the Inner Mongolian grassland. Cropland in Uxin Ju increased to 1,625 ha in 1970, more than double the area in 1968 (see Figure 5.4).

In terms of cropland area, 1970 was not the highest. In the aftermath of the disastrous Great Leap famine,[60] the central government pressured Inner Mongolia to increase grain production, and Uxin Ju's cropland reached its historical record in 1961 at 2,000 ha (Figure 5.4). However, grassland opening during the Cultural Revolution was more detrimental environmentally. If the previous opening had chosen more flat areas that could be better protected, this round of

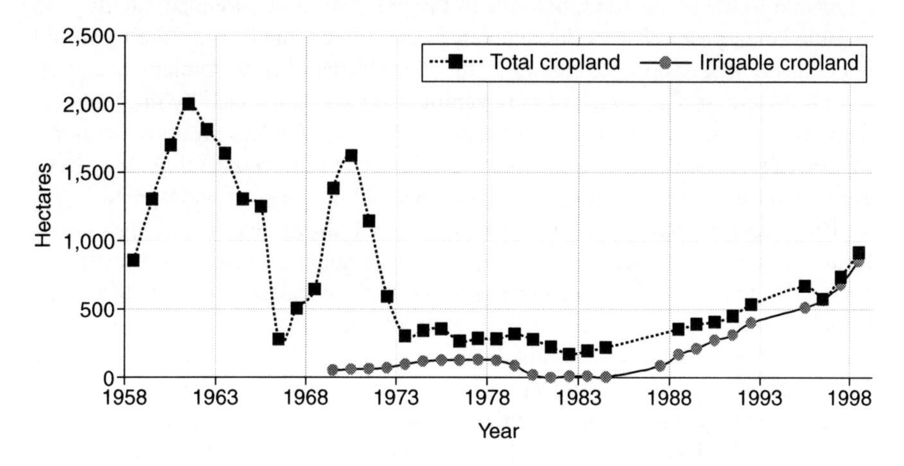

Figure 5.4 Cropland change since 1958. Since the 1980s, cropland has again increased, but most areas are nor irrigated. Unlike dryland farming, irrigable croplands do not cause on-site sandification (source; data from unpublished Uxin Ju statistics, 2001).

opening destroyed more vulnerable sandy land. My informants described it as "indiscriminate opening" (*lankai*). Compounded by overgrazing, grassland opening degraded the sandy land and cancelled the positive effects of the previous grassland improvement. A 1973 report from the Inner Mongolia Department of Animal Husbandry assessed that Uxin Ju's speed of improvement lagged far behind that of degradation.[61] My earlier description of the sandy and windy environment helps to explain that, once opened, not only does the grassland become quickly sandified, but the adjacent pastures are also covered by sand.[62] Furthermore, degradation takes years or decades to recover.

Grassland opening in Inner Mongolia has historical roots. The opening of Mongolian grassland by the Chinese started from as early as the second century BC. In recent history, the late Qing Dynasty and the Nationalist government undertook large-scale grassland openings after 1902,[63] and Mongolian resistance against Chinese colonization and grassland opening was widespread.[64] Many organized rebellions occurred in Ih-Ju league,[65] among which the 1943 "Ih-Ju Incident" was the most prominent. Provoked by the Nationalist general Cheng Changjie's attempt to open the grassland around the Mongols' sacred locations of lama temples and even Chinggis Khan's shrine, the Mongols in Ih-Ju league rebelled.[66] The Mongols' historical loss of grassland to openings by the Chinese made grassland one of the most important ethnic issues.

While historical experience of protest against grassland opening is rich in Uxin Ju, the Mao era's oppressive control left little room for organized rebellion, and open resistance was limited. In one incident during the Cultural Revolution, some Mongols fought the Chinese farmers who were invited by the local government to come and help in cultivation.[67] But this fight should not be compared to the organized rebellions of the pre-1949 time. It ended quietly, and as Huhelao, a pastoralist in Bayintaolegai *gacha*, recounted, "that [the fight] did not stop the grassland opening, since the local leaders had to implement upper-level policies." In fact, much of the opening had to be conducted by the Mongols themselves. The socialist regime's control of its local leaders through hierarchical structure, as well as Mao's ruthless machine of political discipline, may have accounted for the reasons why open resistance did not become widespread.

The cropland increase as shown in Figure 5.4 shows that the Mongols conformed to grassland opening, but behind their compliance were subtle resistance and hidden transcripts, which can be explored in three aspects. First, the Mongols' compliance has to be seen as Scott's notion of performance[68] for the purpose of self-protection. Bao Qingwu recalled that although the Mongols did not know why Confucius was criticized, they were very vocal at the criticism meetings; yet at the same time, "they did not care in their hearts." This "hearts not caring" is at the foundation of hidden transcripts. Although Uxin Ju was elevated by Ulanhu, Baoriledai criticized Ulanhu in 1968 and claimed that Uxin Ju was a model of following Mao's instructions.[69] This distancing from Ulanhu may have prevented the Mongols in Uxin Ju from falling too precipitately with Ulanhu.[70] Second, the Mongols used various safe "excuses" to resist grassland opening. Instead of questioning the Chinese imposition, they insisted that "Mongols do not know how to farm" since "our ancestors did not farm."[71] These attitudes were criticized as pessimism and self-doubts, instead of political threat to the state.

Third, passive resistance has always been the last resort of the subordinate groups, both elsewhere[72] and here in Uxin Ju. Low energy and foot-dragging were widespread in Uxin Ju, and cropland productivity was extremely low. Yield remained 10–25 kg/mu from the 1950s to the early 1970s, as compared to 60–85 kg/mu in the 1980s as irrigation increased.[73] The Mongols "bought time" through their endurance, and their strategies worked, since in time, grassland opening was stopped. In the heightened political pressure of the collective era, much of the subtle resistance and hidden transcripts have to be understood not from people's behaviors but from their ideologies. Even though the state controlled their landscape, the Mongols' resentment and dissent for grassland opening belied any true existence of the socialist hegemony.[74]

It was during the grassland opening of 1969–71 that Uxin Ju's model status was diminished. Local work focused on political mobilization, and the propaganda material praised the Mongols who were active in studying Mao's thoughts.[75] Newspapers stopped advocating the Uxin Ju model, which represented the "grassland focus" of Ulanhu's "counter-revolutionary" path. My archival research uncovered few reports of Uxin Ju during this time. Those I did find applauded the Mongols' participation in political studies but made no mention of grassland improvements.[76] During the time the grassland was being opened, persecution of Mongols led to thousands of deaths in Inner Mongolia.[77] Uxin Ju's campaign leader Baorong committed suicide because of alleged past membership in the Nationalist Party. Huhelao, an ordinary Mongol, had to write self-criticism 20 times since he had been friendly with a herdlord's daughter.[78]

As the Mongols' suffering and outcry of injustice mounted, the central government halted the policies of the military government in 1971. What followed was a period of slow rectification with a refocus on pastoral economy. Cropland acreage was reduced, and grassland improvement became "correct" again. Uxin Ju regained its model status as the "pastoral Dazhai,"[79] newspaper reports resumed,[80] and the reception of visiting delegations from China's pastoral areas resumed. But the grassland campaign did not really return. Local energy was tied up in the "criticizing Lin Biao and Confucius" political movement,[81] and the main achievement on the landscape was the planting of trees and the maintenance of 17 sites for visitors.[82] My informants used the following image to comment on this period: "the thunder [propaganda] was loud but raindrops [actual improvements] were small." As part of this loud "thunder," the Inner Mongolian regional government sent a Mongolian reporter, Chaogetu, to Uxin Ju to produce regular reports. It was at this time that Bao Qingwu, a young educated Mongol, was sent to assist the local government in producing propaganda material. In my interviews in 1999 and 2001, both writers expressed considerable respect for the Mongols in Uxin Ju.

Conclusion

Maoist China cannot be understood through state politics alone, but has to be explored in local actions. Even though the oppressive state largely suppressed overt protests and resistance, local people were able to find creative and subtle ways to express their agency, similar to what green activists do in China's embedded, semi-authoritarian context today. This contribution has examined how the Mongols in Uxin Ju adopted various strategies in response to state grassland policies during the collective era. In the grassland campaign, they utilized the state's aggressive approach to grassland improvement to their own advantage, and in the process created local-based grassland practices and brought about grassland improvement. During the state's anti-grassland campaign, especially during the Cultural Revolution, however, the policies of grassland opening ran contrary to the Mongols' cultural and economic interests, making it impossible for the Mongols to appropriate the campaign. Instead of open protest to state power, the Mongols resisted subtly, mostly through hidden ideological dissent.

This study highlights four aspects of state–society relationships in China. First, the state is not a uniform entity, and state politics are structured in more complicated layers than simple delineation of state and society.[83] Inner Mongolia's grassland policies were part of a central–regional political contention between grassland opening (center) and grassland protection (regional). During the grassland campaign, the regional Mongol elites led by Ulanhu mobilized the center's political discourse to protect the grassland, just like the Mongols in Uxin Ju who turned the campaign politics into local pastoral goals. The anti-grassland campaign came as a result of Ulanhu's loss of power, which subjected both Uxin Ju's grassland and its model status to trial. Second, the close connection between local experience and state political processes indicates that state policies are important in structuring local actions, but they do not determine them. The different strategies the Mongols adopted during the two campaigns demonstrate this point. When state policies were more in line with local needs, local energies and ingenuities were more easily mobilized; but when state policies were both oppressive and detrimental to local interests, limited avenues were left to the local people except for subtle resistance. Third, even with strict state control, local agency cannot be drowned out. While the Mongols in Uxin Ju were not able to effect direct changes in state policies, they assisted indirectly in grassland protection through the use of Uxin Ju by the regional government to counter anti-grassland political forces. Four, the Mongols' actions served to give meaning to local life and were vital to the functioning of local society and culture.

Uxin Ju's story has important implications for our understanding of grassroots environmental actions in China today. Let us look at several such implications. First, during the post-reform era, although the state has lessened its stiff control over society, many of the environmental programs are still channeled through the government.[84] It is incorrect to assume, however, that environmental movements cannot occur in this semi-authoritarian context. Uxin Ju's experience indicates that even in fully state-led programs and campaigns, considerable local initiative and change can occur. Compared to the collective era, post-reform policies have been more attuned to local needs and concerned about the environment. Thus they leave spaces for creative grassroots appropriation. Many environmental NGOs have been effective in utilizing state policies to achieve their own goals. Compared with open resistance, creative use of political structures, networks, and policies can be more difficult to formulate, requiring more skillful maneuvers; also required is familiarity with the "terrains" involved, just as the Mongols' ability to utilize state politics depended upon their intimate experience of living on the grassland. Furthermore, we need to pay attention to subtle ideological resistance, which may not create open confrontation to the state but over time can erode hegemonic state control.[85] As we can see throughout this volume, in China's current political environment, indirect resistance, non-confrontational protest, and careful maneuver may provide vital alternatives to organized resistance, enabling grassroots groups to achieve their environmental goals and political aims. Appreciating subtle resistance requires in-depth understanding of attitudes, intent, and goals of local actors behind public environmental actions. Finally, while organized environmental actions (such as NGOs) have sprung up in China, opportunities for such participation in China's vast rural areas are still limited. Therefore, we are called to pay attention to unorganized environmental actions that continue to be the principal means of local participation. These local responses are often closely related to the effectiveness of broad-based policies, since rural environmental issues often have their origin in economic policies and resource use. Given the persistent rural ecological degradation,[86] studying rural local responses becomes even more urgent.

Notes

1 Field research for this contribution would not have been possible without the help and participation of many Mongols in Uxin Ju. I am grateful for the encouragement and suggestions made by the volume editors, Peter Ho and Richard Edmonds. My thanks also go to Mark Selden, Mario Rutten, and Uradyn Bulag for their insightful comments on earlier drafts. Richard Hennessey provided editorial assistance.
2 J. Becker, *Hungry Ghosts: Mao's Secret Famine*, New York: The Free Press, 1996.
3 Tumen and Dongli Zhu, *Kangsheng yu "neirendang" yuanan* [Kansheng and the "Inner Mongolia People's Party" Tragedy], Beijing: Zhonggong Zhongyang Dangxiao Press, 1995.
4 J. Shapiro, *Mao's War Against Nature: Politics and the Environment in Revolutionary China*, Cambridge: Cambridge University Press, 2001.
5 See Brettell, contribution in this volume; P. Ho, "Greening without Conflict? Environmentalism, Green NGOs and Civil Society in China," *Development and Change*, Vol. 32 (2001), pp. 893–921; K. Wong, "The Environmental Awareness of University Students in Beijing, China," *Journal of Contemporary China*, Vol. 12, No. 36 (2003), pp. 519–36.
6 J.C. Scott, *Weapons of the Weak: The Everyday Forms of Peasant Resistance*, New Haven: Yale University Press, 1985; J.C. Scott, *Domination and the Arts of Resistance: Hidden Transcripts*, New Haven: Yale University Press, 1990.
7 M. de Certeau (translated by S. Rendall), *The Practice of Everyday Life*, Berkeley: University of California Press, 1984; J. Ahearne, *Michel de Certeau: Interpretation and its Other*, Stanford: Stanford University Press, 1995.
8 Scott, *Weapons of the Weak*, p. xvi.
9 Ibid., p. 25.
10 Scott, *Domination.*
11 Certeau, *The Practice of Everyday Life*; see also H. Jiang (forthcoming), "Poaching State Politics in Socialist China: Uxin Ju's Grassland Campaign during 1958–1966," *Geographical Review*.
12 Certeau, *The Practice of Everyday Life*, p. xiii.
13 D.R. Weiner, *A Little Corner of Freedom: Russian Nature Protection from Stalin to Gorbachev*, Berkeley: University of California Press, 1999, p. 444.
14 X. Zhong, Wang Zheng, and Bai Di, *Some of Us: Chinese Women Growing Up in the Mao Era*, New Brunswick, NJ: Rutgers University Press, 2001, p. xvii.
15 E. Friedman, "The Politics of Local Models, Social Transformation and State Power Struggles in the People's Republic of China: Tachai and Teng Hsiao-P'ing," *The China Quarterly*, Vol. 76 (1978), pp. 873–90.
16 See A.G. Walder. and Yang Su, "The Cultural Revolution in the Countryside: Scope, Timing and Human Impact," *The China Quarterly*, Vol. 173 (2003), pp. 74–99; Becker, *Hungry Ghosts*.
17 *Renmin Ribao* (*People's Daily*), 2 December 1965.
18 Friedman, "The Politics of Local Models"; U.E. Bulag, *The Mongols at China's Edge: History and Politics of National Unity*, Lanham, MD: Rowman and Littlefield, 2002.
19 Wushen Qi (Uxin banner), "*Wushenqi zhi*" [Gazetteer of Uxin banner], unpublished material, 1998.
20 *Sum* is a Mongolian word for "township," and *gacha* is the equivalent of "village."
21 For details, see H. Jiang, "Cooperation, Land Use, and the Environment in Uxin Ju: A Changing Landscape of a Mongolian–Chinese Borderland in China," *Annals of Association of American Geographers*, Vol. 94, No. 1 (2004), pp. 117–39.
22 U.E. Bulag, "From Inequality to Difference: Colonial Contradictions of Class and Ethnicity in 'Socialist' China," *Cultural Studies*, Vol. 14, No. 3–4 (2000), pp. 531–61.

23 YMDBW (Yikezhao Meng Difangzhi Bianzuan Weiyuanhui) (Ih-Ju Gazetteer Compilation Committee), *Yikezhao mengzhi* [Ih-Ju Gazetteer], Vol. 2, Beijing: Xiandai Press, 1994, p. 72.
24 Although Uxin Ju, as part of Suiyuan province, did not join Inner Mongolia Autonomous Region until 1954, its grassland policies followed that of Inner Mongolia after 1949. See YMDBW, *Yikezhao mengzhi*, p. 388.
25 Bulag, "From Inequality to Difference"; Bulag, *The Mongols at China's Edge.*
26 U.E. Bulag, personal communication, 2004.
27 Ulanhu (compiled by Temuer *et al.*), *Wulanfu lun muqu gongzuo* [Ulanhu's Talks on Pastoral Work], Hohhot, China: Neimenggu Renmin Press, 1990, p. 160.
28 Ibid., p. 162.
29 Shapiro, *Mao's War.*
30 Bulag, *The Mongols at China's Edge*, p. 160; see also *Neimenggu Ribao* (*Inner Mongolia Daily*), 3 February 1966.
31 Bulag, *The Mongols at China's Edge*, p. 191.
32 Interviews, 2001.
33 Uxin Ju was named after the lama temple: Uxin refers to a Mongol tribe in the area and *Ju* means temple.
34 Interviews, 2001.
35 Baoriledai, "*Baoriledai Jianghua*" [Baoriledai speech], unpublished, 1965.
36 *Renmin Ribao*, 2 December 1965.
37 "*Wushenzhao tongji*" (Uxin Ju statistics), unpublished material, 2001.
38 The reward for hard work, however, was political: at the end of the campaign, 20 young people were admitted to the Youth League, and 11, including Baoriledai, became Party members.
39 Collection 4, "*Wushenzhao tuchu zhengzhi jiangshe caoyuan*" (Uxin Ju emphasizes politics in grassland construction), unpublished material, 1966.
40 Interviews, 2001.
41 *Renmin Ribao*, 2 December 1965.
42 Wushenzhao Jiedaizhan (Uxin Ju Reception Station), "*Wushenzhao gaikuang jieshao cailiao*" [General introduction to Uxin Ju], in Collection 20, unpublished material, 1974.
43 Interviews, 2001.
44 "*Wushenzhao tongji*" [Uxin Ju statistics].
45 Interviews, 2001.
46 *Neimenggu Ribao*, 14 February 1966.
47 Wushenzhao Dangwei (Uxin Ju Party Branch), untitled unpublished material, 1972.
48 *Neimenggu Ribao*, 7 December 1965.
49 Friedman, "The Politics of Local Models"; E. Friedman, P.G. Pickowicz, M. Selden, and K.A. Johnson, *Chinese Village, Socialist State*, New Haven: Yale University Press, 1991; Bulag, *The Mongols at China's Edge.*
50 *Renmin Ribao*, 2 December 1965.
51 Wushen Qi, "*Wushenqi zhi,*" p. 45.
52 T.J. Barfield, *The Perilous Frontier. Nomadic Empires and China, 221 BC to AD 1757*, Cambridge, MA: Blackwell, 1989.
53 C. Atwood, "National Party and Local Politics in Ordos, Inner Mongolia (1926–1935)," *Journal of Asian History*, Vol. 26, No. 1 (1992), pp. 1–30.
54 Siqing Gongzuozu [Four-cleanup work team], "*Wushenzhao Gongshe Siqing yundong zongjie*" [Uxin Ju Commune's Four-cleanup summary] in "Collected material of Four-cleanup," unpublished material, 1966.
55 For example, A. Chan, R. Madsen, and J. Unger, *Chen Village under Mao and Deng*, 2nd edition, Berkeley: University of California Press, 1992.
56 See P.B. Tseren, "Traditional Pastoral Practice of the Oirat Mongols and their Relationship with the Environment," in C. Humphrey and D. Sneath (eds) *Culture and Environment in Inner Asia, Vol. 2*, Cambridge: White Horse Press, 1996, pp. 147–59.

57 The environmental consequence of such agropastoral development has been contradictory, however. This is to say that local agency does not automatically lead to environmental sustainability. See Jiang, "Cooperation, Land Use, and the Environment in Uxin Ju."
58 For details, see D. Sneath, "The Impact of the Cultural Revolution in China on the Mongolians of Inner Mongolia," *Modern Asian Studies*, Vol. 28, No. 2 (1994), pp. 409–30; Tumen and Dongli Zhu, *Kangsheng yu "neirendang" yuanan*; W. Woody; *The Cultural Revolution in Inner Mongolia*, Stockholm: Center for Pacific Asia Studies at Stockholm University Occasional Paper 20, 1993.
59 XBW (Xiuzhi Bianshi Weiyuanhui), *Neimenggu xumuye fazhangshi* [Development History of Animal Husbandry in Inner Mongolia], Hohhot: Neimenggu Renmin Press, 2000.
60 There were reports of hunger-induced symptoms such as water retention in Ih-Ju league, although Inner Mongolia did not suffer from hunger as severely as regions in central and southern China. See G.H. Chang and G.J. Wen, "Communal Dining and the Chinese Famine of 1958–1961," *Economic Development and Cultural Change*, Vol. 46, No. 1 (1997), pp. 1–34.
61 Neimeng Xumuju (Inner Mongolia Department of Animal Husbandry), "*Wushenzhao Renmingongshe jiben caomuchang de jianshe qingkuang*" (Basic situation of Uxin Ju commune's pastureland construction), in Collection 4, unpublished material, 1973.
62 For physical processes of sandification, see Z. Huang and B. Song, "*Neimenggu yikezhao tudi shamohua jiqi fangzhi*" (Desertification and its control in Ih-Ju league, Inner Mongolia), *Zhongguo Kexueyuan Lanzhou Shamosuo Jikan*, No. 3 (1986), pp. 35–47; J. Sun, "*Eerduosi gaoyuan shengtai huanjing zhengzhi de zhanlue yanjiu*" (Strategic research on the improvement of ecological environment in the Ordos Plateau), *Ganhanqu Ziyuan Yu Huanjing* [Journal of Arid Land Resources and Environment], Vol. 4, No, 4 (1990), pp. 45–51.
63 During 1902–11, under a policy called "Open the Mongolia wasteland" (*kaifang menghuang*), the Qing state, led by Yigu, opened over 155,800 ha of grassland in Ih-Ju league (XBW, *Neimenggu xumuye fazhangshi*, p. 56). The Chinese nationalist government accelerated this opening, and between 1912 and 1928 another 1.32 million ha was opened in Suiyuan (including Ih-Ju league and the surrounding area to the north east; ibid., p. 57).
64 Bayin, *Menggu youmu shehui de bianqian* [Change in Mongolian Nomadic Society], Hohhot: Neimenggu Renmin Press, 1998, pp. 80–5.
65 Since 1900, many *duguilang* movements were ignited by the grassland opening.
66 B. Liang, *Yikezhao meng de tudi kaiken* [Land Opening in Ih-Ju League], Hohhot: Neimenggu Daxue Press, 1991.
67 Interviews, 1999.
68 Scott, *Domination*, p. 3.
69 Baoriledai, "*Zhongyu maozhuxi shi zuida zuidai de gong*" [Loyal to Chairman Mao is the utmost highest collective], in Collection 3, unpublished material, 1968.
70 Baoriledai stood on the side of the radical faction that was against Ulanhu.
71 Interviews, 2001.
72 Scott, *Weapons of the Weak.*
73 *Wushenzhao tongji.*
74 See also Scott, *Weapons of the Weak.*
75 Wushenzhao Geweihui (Uxin Ju Revolutionary Committee), "*Women shi zenyang daban xuexiban, kaizhan geming de dapipan, zhuageming, cushengchan de*" [How we promote Mao study meetings, conduct revolutionary criticisms, and use revolution to lead production], unpublished material, 1968.
76 *Neimenggu Ribao*, 19 August 1970; 4 September 1970.
77 The fabricated resurrection of the separatist Inner Mongolian People's Revolutionary Party (Neirendang) led to the large-scale jailing and torture of Mongols, both leaders

and ordinary herders. Tumen and Zhu reported that at least 4,759 people died, and 15,000 people were injured or disabled. Altogether over 346,000 people were persecuted. Jankowiak uses a death figure of 16,222. See Tumen and Zhu, *Kangsheng yu "neirendang" yuanan*; W.R. Jankowiak, "The Last Hurrah? Political Protest in Inner Mongolia," *Australian Journal of Chinese Affairs*, No. 19/20 (1988), pp. 369–88.

78 Interviews, 1999. In Uxin Ju, no deaths of ordinary Mongols were reported during the Neirendang persecution. In Tuke, a neighboring *sum*, however, severe torture and persecution occurred (Q. Bao, personal communication, 2001).

79 Baoriledai's inclusion in the post-1971 Inner Mongolia leadership may have been a factor in Uxin Ju's revival. Other Inner Mongolia leaders, including Zhao Ziyang, visited Uxin Ju and confirmed its direction of grassland improvement (Bao, personal communication). Zhao was removed from the top Party post of general secretary during the 1989 Tiananmen student movement for his sympathy for the students' demand for democracy.

80 *Neimenggu Ribao*, 16 March 1973; 16 April 1973. A documentary movie about Uxin Ju's grassland campaign, called *Zhengfu shamo jian caoyuan* [Overcome Desert and Construct Grassland], was made in 1973 by the Beijing Film Studio for Scientific Education. The documentary was originally planned in 1966.

81 Lin Biao, Mao's trusted future successor, failed in a plot to seize power in 1971. The subsequent "criticizing Lin Biao and Confucius" movement absorbed much energy of the Mongols in Uxin Ju. In one report, within a month and half, 82 study meetings and 74 criticism sessions were held. See Wushenzhao Dangwei (Uxin Ju Party Committee), "*Women shi zenyang ba pilin pikong douzheng buduan yinxiang shenru de*" (How did we keep deepening the "Criticizing Lin and Confucius" movement), in Collection 7, unpublished material, 1974.

82 These sites were showpieces of the year effort in improving the grassland; most were established before 1965. The improvements on some of these sites have been maintained to this day.

83 See also V. Shue, *The Reach of the State: Sketches of the Chinese Body Politics*, Stanford: Stanford University Press, 1988; H. Jia and Z. Lin, *Changing Central–Local Relations in China*, Boulder: Westview, 1994.

84 Brettell, contribution in this volume.

85 P.M. Thornton, "Framing Dissent in Contemporary China: Irony, Ambiguity and Metonymy," *The China Quarterly*, Vol. 172 (2002), pp. 661–81.

86 R.L. Edmonds, *Patterns of China's Lost Harmony: A Survey of the Country's Environmental Degradation and Protection*, London: Routledge, 1994; E.C. Economy, *The River Runs Black: Environmental Challenges to China's Future*, Ithaca: Cornell University Press, 2004.

6 Channeling dissent

The institutionalization of environmental complaint resolution

Anna Brettell

Introduction

Environmental grievances are common in China today. Such grievances are not new and have given rise to formal complaints and disputes since imperial times.[1] In modern times, citizens send letters or visit officials (*xinfang*)[2] for simple problems like a neighbor's smoky coal furnace, as well as for the most intractable of problems, some of which have defied resolution in the courts and may involve death or destruction of livelihoods for whole communities. Since 1990, when environmental protection officials at the county level and above were required to keep records of complaints, the numbers of environmental grievances (*xinfang*) have increased dramatically. Between 1990 and 2004 the number of incidents about which people sent letters or visited Environment Protection Bureaus (EPBs) increased by 513 percent from approximately 111,359 in 1991 to approximately 682,744 in 2004.[3]

The increase in citizen environmental grievances could have political consequences for China.[4] How authorities deal with these *xinfang* (complaints) could impact state legitimacy and social stability as well as environmental quality. Similar grievances formed the supporting pillars of influential environmental movements that challenged authorities in Russia and Eastern Europe,[5] Japan,[6] Taiwan,[7] and South Korea.[8] Chinese authorities have reason to be concerned about rising citizen anti-pollution complaints and protests. Chinese leaders recognize that left unmanaged this rising tide of contention represents, at worst, a threat to social stability.[9]

However, complaints also represent citizen demands for a cleaner environment. If authorities are able to respond to citizen demands and improve the environment as other countries have done, such as Japan, the US, and South Korea, then confidence in and the legitimacy of the Chinese Communist Party (CCP) could be enhanced.[10]

On the other hand, if the *xinfang* system is inadequate and authorities are unable to respond satisfactorily at all levels of government by improving the environment, increasing governmental accountability, and including public input in policy processes, then distrust, dissent and the threat of social instability will grow.

Thus, it is crucial to understand why complaints have increased and how authorities have responded to them. The research reflected in this chapter is based on quantitative and qualitative methodology and seeks to assess explanations for variations in pollution complaints in China between 1990 and 2004. The quantitative survey tests hypothesized correlations among levels of SO_2 and wastewater effluent, general education levels, and per capita GDP, with the numbers of water and air pollution complaints to environmental authorities aggregated at the provincial level in a pooled, panel dataset.[11] The quantitative analysis also reveals several puzzles related to environmental protection data that are worthy of future research.

The results of the quantitative analysis do not fully explain variations in environmental complaints because certain omitted institutional, legal, and political variables cannot be meaningfully quantified; therefore, this chapter also includes a qualitative component that explores the relevance of legal and political variables, institutional development and the roles of environmental protection officials over time on variations in complaint numbers.[12] The qualitative analysis complements the quantitative analysis and is crucial for understanding why complaints have increased. Further, this chapter examines the imperfections in the environmental complaint system and discusses the relationship between increasing environmental complaints and the rise of an environmental movement in China.

Theoretical explanations for increasing complaints

This research will test the relevance of the levels of education, pollution, economic well-being, as well as political and legal contexts to variations in environmental complaints over time. Each of these factors is examined below. However, none of them alone are sufficient to explain variations in complaint levels.

First, Chinese officials often state that the reason for surging complaints has been the rise in environmental awareness brought about by environmental education programs.[13] Environmental awareness is difficult to measure systematically across China and across time, but since environmental education has become commonplace in many schools, one can look to more general measures of education to find quantifiable variables. In addition, other studies have found that education levels are correlated with variations in citizen complaints.[14] While there is some logic to the reasoning that citizens with higher education levels and greater environmental awareness are more likely to file environmental complaints, the correlation between changes in the levels of environmental awareness and changes in citizen action, including complaining,[15] is far from clear.[16]

Second, it is conceivable that a measurable rise in pollution will lead to more citizen environmental complaints. There is evidence in other countries that mounting pollution levels trigger citizen action to protect the environment, namely complaints.[17] However, previous research in China showed that there was no relationship between SO_2 density and complaints.[18]

Third, another theory suggests that citizen complaints have increased because citizens are economically more well off than before. This assertion implies the demand for environmental quality rises as income rises. Those who assert the validity of this theory argue environmental quality is a luxury good and will not be of concern to citizens who are worried about food, shelter, and economic survival.[19] In contrast, some research has found that citizens of both poor and rich nations are concerned about environmental quality.[20] Therefore, the evidence is mixed as to whether environmental quality is a luxury good and only sought after by those who reach a certain level of economic development.[21] A related hypothesis suggests that as income levels rise, citizens will be more inclined to participate in the political system.[22] However, some argue this may not be true for certain types of participation.[23] Some scholars argue lower-income citizens will turn to contacting officials because it requires fewer resources and because it represents the "most clear, direct, and immediate link between action and results."[24]

Fourth, the national and local level legal and political contexts are thought to influence citizen participation in China.[25] Kevin O'Brien and others have shown that an improved legal context has been conducive to increased complaints in rural China.[26] Elizabeth Perry and Mark Selden also discuss the importance of greater acceptance of some forms of protest, at least during some periods in China, on the part of authorities as an explanation for increasing protests.[27] However, these legal and political explanations by themselves pay too little attention to the presence of grievances brought about by changes objective conditions, such as industrial pollution.

Each of the four explanations alone is insufficient to explain in general the rise in environmental complaints in China and in addition, unlike in more developed countries, the development of China's institutional capacity for handling complaints should be given due consideration. Therefore, a more textured explanation is necessary; one that includes more than just a single pertinent variable. To build a better explanatory model, this chapter tests the importance of each of the above variables and examines how they interact. The goal is not to discover one variable that offers a parsimonious explanation, but to see how changes in objective environmental and economic conditions, education levels, institutional capacity, and political changes interact to lead to mounting complaints at the aggregate level.

Explaining variations in complaint levels: the quantitative perspective

The analysis of the variance in air and water pollution complaints across China between 1992 and 2004 revealed some very interesting patterns and relationships. This section will explore these patterns. First, it outlines the variables used in the statistical analysis and describes overall fluctuations in the national and provincial level complaint data. Then it discusses the results of linear regressions on the provincial level pooled panel data.

The independent variables used in the statistical analysis include general education level, industrial wastewater effluent, industrial SO_2 emissions, and economic well-being. General education level is operationalized as the number of students that enter regular secondary school as a percentage of those entering primary school.[28] The indicators for pollution levels are emissions of industrial sulfur dioxide (SO_2) and industrial wastewater effluent.[29] This research will test the relevance of pollution levels measured in three different ways: as a measure of absolute annual pollution discharges, a measure of natural density, and a measure of social density. The natural density of air pollution is calculated by dividing the total annual amount of a specific pollutant by the land area of the relevant administrative area. The social density of pollution is calculated by multiplying the natural density of pollution by the population density.[30] Economic well-being is measured by per capita gross domestic product (GDP).

Descriptive results

At the national level, the number of complaints increased over time for all types of pollution.[31] Both the total number of people complaining and the number of incidences rose. Between 1990 and 2004 the number of people sending letters and visiting EPBs increased by 410 percent, rising from around 140,681 in 1990 to around 726,192 in 2004. The number of incidents about which people sent letters or visited EPBs increased by 513 percent from around 111,359 in 1991 to around 682,744 in 2004.[32] The growth in complaints is not simply the result of population growth. In 1990, approximately 1.23 people out of 10,000 made a complaint. In 2004, 5.59 people out of 10,000 made a complaint. See Figure 6.1.

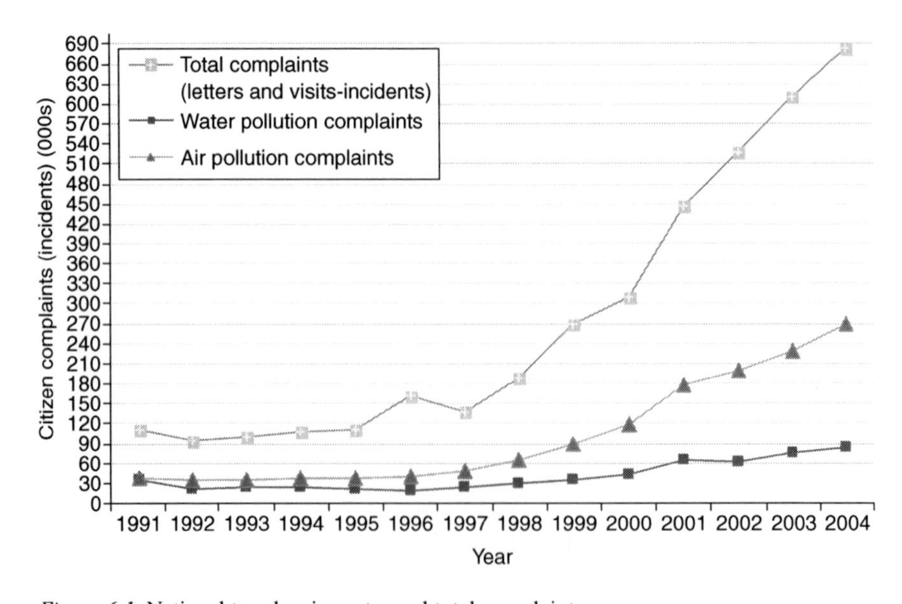

Figure 6.1 National trends: air, water and total complaints.

In recent years, aggregated at the national level, noise pollution has been the subject of the greatest number of complaints. Noise pollution complaints are often the most easily resolved, sometimes by altering hours of operation of an enterprise or by providing immediate compensation to those affected. Following noise pollution complaints in frequency are air pollution complaints, then water pollution complaints, and finally solid waste complaints. Air and water pollution complaints are sometimes not very easy to resolve and involve longer-term problems. The industries which people complain about most frequently are cement, paper, power, steel, aluminum, chemical, and fertilizer.[33]

Aggregated at the national level, the total number of water pollution complaints in China grew 256 percent between 1990 (32,654 complaints) and 2004 (83,654 complaints). Within that period, however, different trends emerged. Between 1991 and 1992 complaints soared, between 1992 and 1996 complaint numbers remained static or even decreased, and after 1997, complaints gradually increased. See Figure 6.2.

In contrast, over the 11-year period air pollution complaints steadily increased from 48,878 in 1990 to 270,249 in 2004, an increase of about 550 percent (Figure 6.3).

Aggregated at the provincial level, complaint trends are much more complex. When the total number of complaints is adjusted to account for differences in population among provinces, we find that those provinces with the absolute highest number of complaints do not always correspond to those that have the highest number of per capita complaints. Tables 6.1 and 6.2 show the provinces most often in the top ten for per capita water and air complaints and those most often in the top ten for absolute numbers of water and air complaints during the

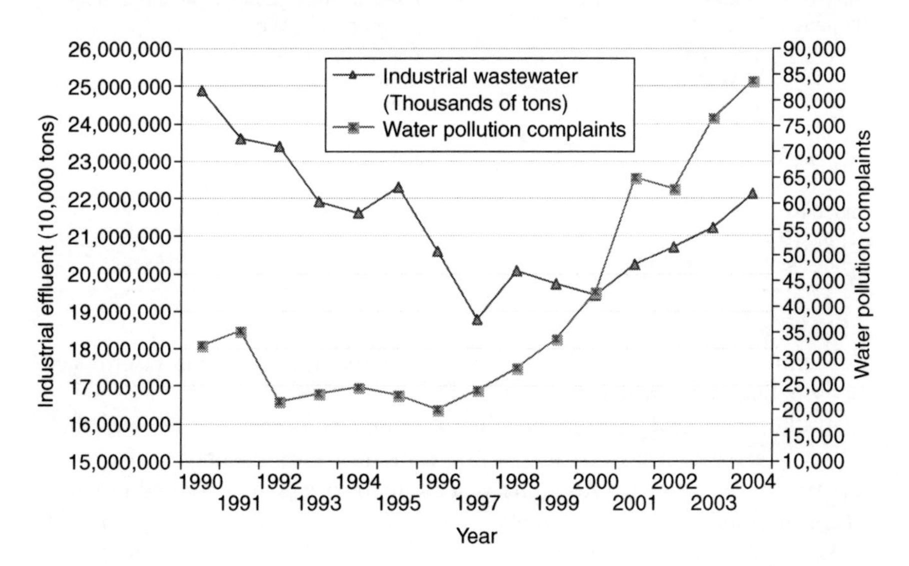

Figure 6.2 National trends: water pollution complaints and industrial wastewater effluent discharges.

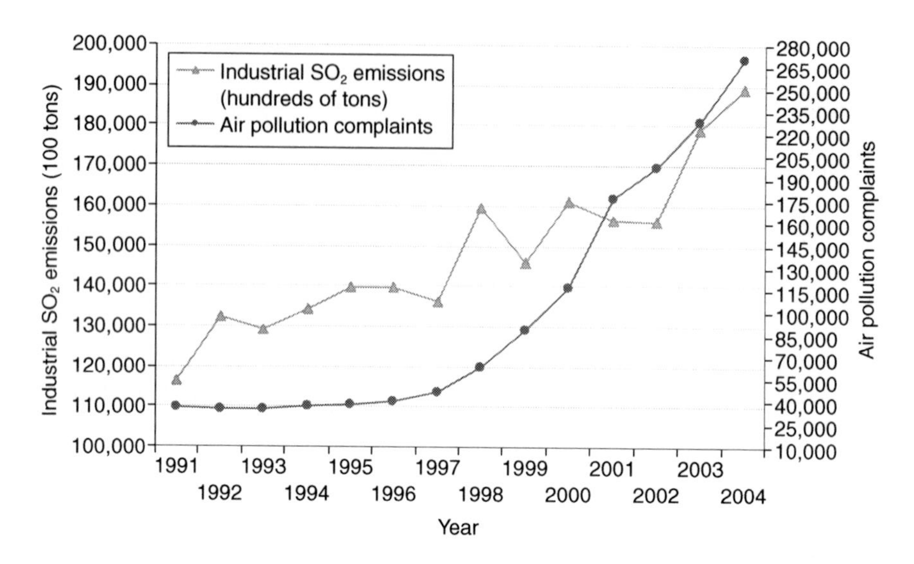

Figure 6.3 National trends: air pollution complaints and industrial SO_2 emissions.

Table 6.1 Comparing total H_2O pollution complaints and per capita H_2O pollution complaints (in order of frequency)

Per capita water complaints	*Total water complaints*
Jiangsu (13 years)	Jiangsu, Guangdong, Shandong, Sichuan, Hebei (13 years)
Zhejiang (12 years)	Zhejiang, Hunan (12 years)
Guangdong (11 years)	Guangxi (11years)
Shandong (10 years)	Henan (8 years)
Guangxi (9 years)	Liaoning (6 years)
Fujian and Sichuan (including Chongqing) (8 years)	Jiangxi, Fujian (4 years)
Shanghai (7 years)	Shanghai, Hubei (3 years)
Hunan (6 years)	Yunnan, Anhui (1 year)
Beijing (5 years)	
Hebei, Jiangxi, Hainan (4 years)	

13-year period between 1992 and 2004.[34] Figure 6.4 graphs the top 11 provinces for air pollution complaints.

At the provincial level trends in the numbers of complaints sometimes display great variance over time in the same province. Examples of extreme variation in one province from one year to the next include Liaoning province, where the total number of complaints in 1989 was 7,845, while it was only 1,852 in 1999. However, the number of complaints climbed back up in 2000 to 29,531. In another example, in Yunnan in 1998, there were 56 complaints, while in 1999 there were 4,191 (an increase of over 7,000 percent).[35]

Table 6.2 Comparing total air pollution complaints and per capita air pollution complaints (in order of frequency)

Per capita air complaints	*Total air complaints*
Beijing, Guangdong (13 years)	Guangdong, Shandong, Jiangsu, Sichuan (13 years)
Liaoning Zhejiang, Jiangsu, Jilin (12 years)	Liaoning, Zhejiang (12 years)
Tianjin (11 years)	Hebei (10 years)
Shanghai (8 years)	Guangxi (9 years)
Ningxia (7 years)	Hunan, Henan, Jilin (7 years)
Shandong (6 years	Heilongjiang (5 years)
Sichuan (5 years)	Fujian (4 years)
Fujian, Heilongjiang, Guangxi, Shanxi (3 years)	Hubei, Shanxi (2 years)
Qinghai (2 years)	
Xinjiang (1 year)	

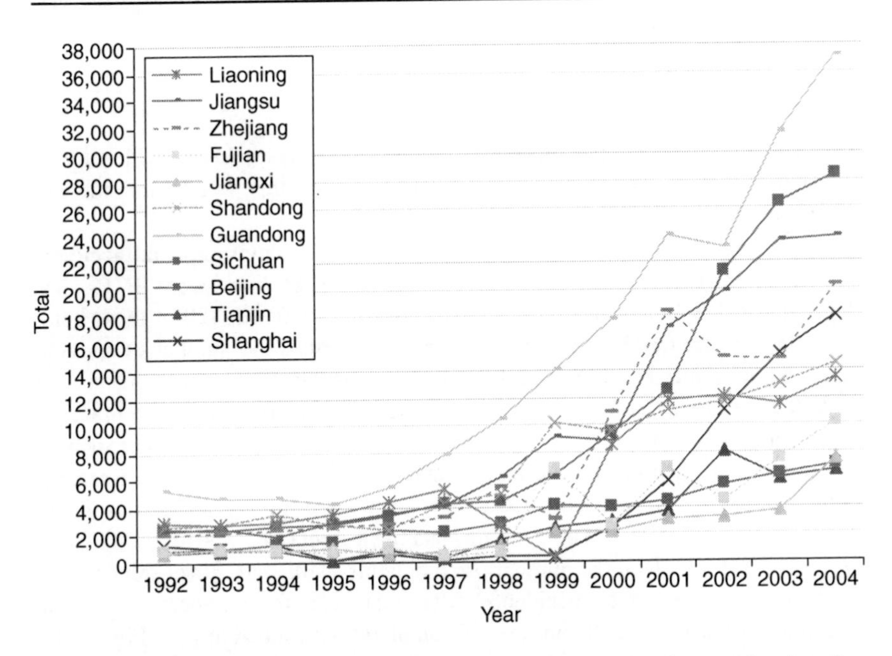

Figure 6.4 Top provinces and municipalities for combined total and per capita air pollution complaints – all years.

Data reliability

A typical reaction to these variations is skepticism about the reliability of Chinese data; however, some variation in complaint levels is common in any country and China is no different.[36] On the other hand, stories of data manipulation during the Mao period are infamous and very recent stories of "fairy-tale" river pollution cleanup campaigns remind one to be cautious of Chinese data.[37] In addition, interviews with environmental protection officials indicate that the numbers of reported complaints are undercounted.[38] Some authorities also admit that they are given guidance about the direction statistical trends should take.[39]

Ignoring Chinese data is not an option in the long run. It is only through analyzing environmental protection data along with other sources of data, and testing theoretical relationships relying on such data, that scholars and citizens can gather specific evidence with which to back up calls for data accountability.[40] Generating a catalog of general reasons why environmental protection data shows large fluctuations or is puzzling is worthwhile to improve the quality of research data over time.

One explanation for dramatic fluctuations in environmental protection data from one year to the next is unreported data. Some cities, townships, and counties sometimes do not supply relevant data to the next highest administrative level, even though doing so affects budget allocations,[41] which means the data aggregated at the provincial level could be incomplete.

Environmental accidents

The data on environmental accidents, for example, is concrete evidence that incomplete data exists and does influence aggregated totals. In addition, the data reveals other puzzles worthy of further research.

According to official statistics, the total number of environmental accidents peaked somewhere between 1985 and 1990.[42] In 1985 there were 2,716 accidents and in 1990 there were 3,462 environmental accidents, 120 of these being extremely severe (*teda*) or severe (*zhongda*).[43] Because the number of *zhongda* and *teda* accidents is one of the 13 indicators, first instituted in 1990, used to rank the environmental performance of the 32 primary cities in China,[44] it creates an incentive for a local government to undercount the number of accidents within its administrative area. Data fluctuations following performance policy changes without corresponding substantive policy or technological changes are suspect. (See Table 6.3 for data on accidents.)

Until 1994, trends in the number of complaints seemed to mirror trends in the number of environmental accidents. After that year, the number of accidents began to decrease while the number of complaints continued to rise (Figures 6.5 and 6.6). An examination of the data reveals that it is the dramatic drop in the number of accidents between 1994 and 1995 that accounts for the uncoupling. We are left with the puzzle of why environmental accidents decreased dramatically between 1994 and 1995.

There are a few reasons for the dramatic drop in the number of accidents: (1) in 1995 and subsequent years, provinces failed to submit data or officials decided not to include data in the statistical yearbook. In 1995, data on nine provinces is clearly missing.[45] In subsequent years, the yearbooks present data in such a way as to make it impossible to tell if a province failed to submit data or if there were zero accidents in a province.[46]

Table 6.3 National trends: environmental accidents

Year	*Total*	*H_2O*	*Air*	*Teda*	*Zhongda*	*People injured*	*Deaths*	*Compensation*
1990	3,462	NA	NA	NA	60	10,000	4	97,430,000
1991	3,038	NA	NA	43	106	NA	NA	41,090,000
1992	2,667	NA	NA	49	81	NA	NA	51,906,000
1993	2,761	1,431	888	64	80	1,436	12	41,240,000
1994	3,001	1,617	986	44	97	4,668	3	44,511,000
1995	1,966	1,022	732	56	84	42,765	5	38,547,000
1996	1,446	677	585	42	31	33	1	26,240,000
1997	1,992	986	752	36	77	188	1	26,940,000
1998	1,422	788	464	29	37	152	6	18,516,600
1999	1,614	888	582	29	43	261	2	21,163,200
2000	2,411	1,138	864	42	65	578	10	31,449,300
2001	1,842	1,096	576	32	52	187	2	29,487,000
2002	1,921	1,097	597	27	31	97	2	26,297,000
2003	1,843	1,042	654	20	30	416	3	19,991,000
2004	1,441	753	569	25	29	339	12	34,872,000

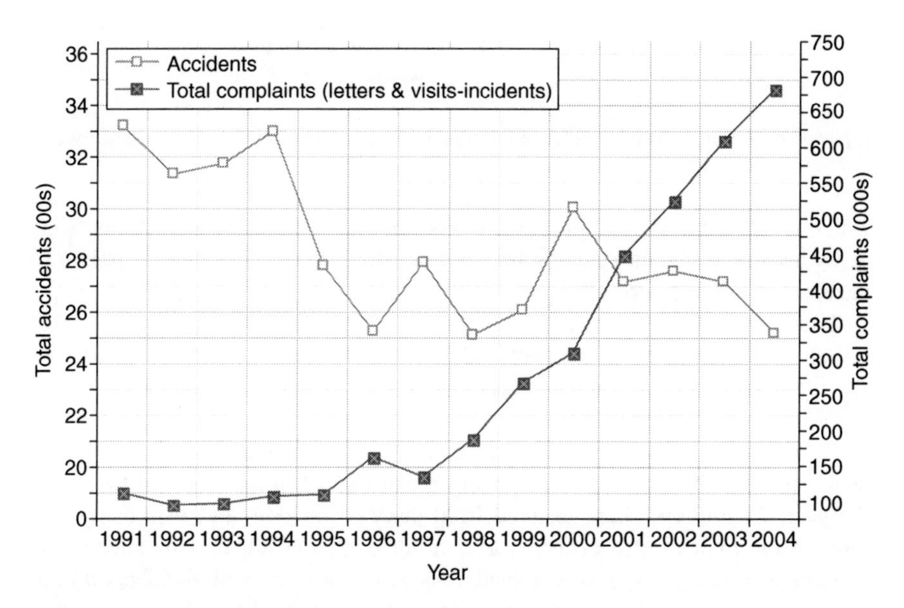

Figure 6.5 National trends: total accidents and total complaints.

Another possible reason for the decline in accidents is that government safety requirements became stricter, which improved environmental safety, especially after the serious accidents along the Huai River in 1994 drew such attention.[47]

Finally, it is possible that authorities have tinkered with the recorded numbers of environmental accidents for political reasons.[48] For example, there is a striking similarity between the overall trends in air and water pollution accidents,

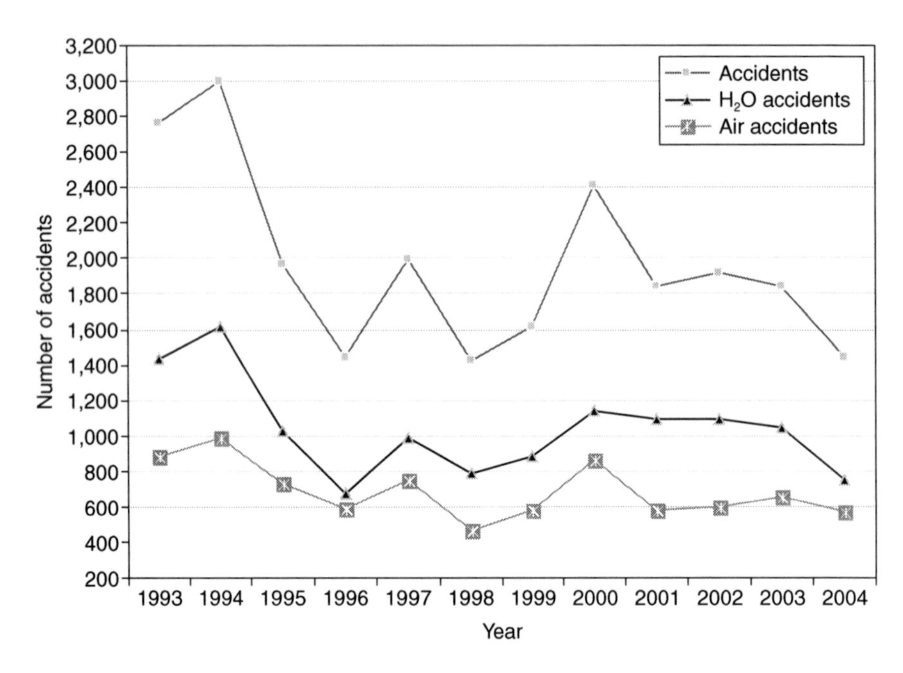

Figure 6.6 National trends: total environmental accidents, air accidents and water accidents.

which is rather puzzling. While this is not impossible (because sometimes accidents involve pollutant releases into both air and water), the finding points to puzzles worthy of future research

The data on environmental accidents aggregated at the provincial level revealed another puzzle. In 1995, 42,765 people were reportedly injured in environmental accidents, and one accident in Inner Mongolia accounts for 40,000 of those injured. In Inner Mongolia that year there were four accidents related to water pollution, 13 to air pollution, nine to noise pollution and one "other" accident. This "other" accident is interesting because most of the financial losses for the year, 25 million Yuan, resulted from this one incident and it is likely that most of the injuries also occurred as a result of this accident. This anomaly is extremely puzzling, especially considering narrative reports of accidents do not include a reference to this particular incident, which is the largest accident in China in terms of injured persons. This puzzle too is worthy of future research.[49]

Data-reporting methods such as these call into question the claims by environmental protection officials that the environment sector is an "open sector" where citizens have greater access to meaningful data and illustrate that higher-level EPBs still do not have the authority to compel lower-level EPBs to submit data relevant to their planning processes.

A complete and objectively accurate picture of variations in complaints may never be built; however, the quantitative survey in this research provides some clues about trends in complaint and other environmental protection data as they are reflected in official statistics. So, from an official Chinese point of view, the data is accurate. Testing relationships among the data and testing the data against theoretical precepts will help to analyze the quality of the data as reported and provide specific examples with which to back up requests for data accountability. Therefore it is worthwhile to conduct statistical analysis using the official data, even though it might be deficient. The statistical survey based on this official data is valid in and of itself because of the large number of cases in the provincial-level data set.

Multiple regression analysis: putting it all together

Based on official Chinese data, do levels of pollution, economic well-being, and education levels affect variations in water and air pollution complaints in China? While it is impossible to establish with complete certainty that causal relationships exist among the variables, it is possible to discover if correlations among the variables are statistically significant. An in-depth analysis of the national level data is not possible because there are too few cases with which to make valid statistical inference. However, there is sufficient data aggregated at the provincial level to obtain robust and valid results. The pooled panel data aggregated at the provincial level lends itself to linear multiple regression analysis

Bivariate analysis revealed that relationships between the dependent variables – environmental complaints (pollution-stream specific) – and all of the independent variables are significant. The correlations are all significant to the 0.001 or 0.000 level, with pollution levels having the strongest correlation. Except for the pollution levels variable, the correlations all yielded low R^2 statistics but relatively high F statistics.

The multivariate results are more complex.[50] In the air pollution complaint case, no matter how SO_2 is measured (absolute emissions, natural density, or social intensity) all three models are statistically significant to the 0.000 level and the model with the absolute emissions measure of industrial SO_2 yields the highest R^2 (0.770) and F values (87.743). This model has the best "fit." Holding all other variables constant in this model, the measure of industrial SO_2 has the highest partial correlation and t scores and the measure of general education is not significant.

In the water pollution complaint case, no matter how industrial wastewater effluent is measured (absolute effluent, natural density, and social intensity), the models are statistically significant to the 0.000 level and the model with the absolute effluent measure of industrial wastewater yields the highest R^2 (0.799) and F values (98.488). This model has the best "fit." Holding all other variables constant in this best fit model, the measure of industrial wastewater has the highest partial correlation and t scores (see Table 6.4) and, of note, again the measure of general education is not significant.

In conclusion, any of the models containing three variables is better at explaining variance in complaint numbers than any model with only one or two variables. Part of the reason for this is there are simply more explanatory

Table 6.4 Multivariate regression results – best fit models

Provincial level multivariate regression results – models with best fit			
Water pollution		*Air pollution*	
Model 1 – Log of water pollution complaints; the log of industrial wastewater effluent in tons, the log of per capita GDP, and the log of the General education variable.		**Model 2** – Log of air pollution complaints; the log of industrial SO_2 emissions, the log of per capita GDP, and the log of the General education variable.	
Determinants of Air Pollution Complaint Variation *t* values, [partial correlations], and (standard errors)		Determinants of Air Pollution Complaint Variation *t* values, [partial correlations], and (standard errors)	
Log effluent	31.482**** [0.853] (0.032)	Log SO_2 effluent	22.373**** [0.757] (0.033)
Log per cap GDP	2.294** [0.118] (0.146)	Log per cap GDP	8.909**** [0.418] (0.096)
Log Gen Ed	1.705 [0.088] (0.146)	Log Gen Ed	1.017 [0.053] (0.185)
$R^2 = 0.799$ St. Error = 0.6544217 F = 98.488**** N = 388 Degrees of Freedom = 15 ** = $p < 0.05$; *** = $p < 0.01$; **** = $p < 0.001$ Durbin-Watson = 1.883		$R^2 = 0.770$ St. Error = 0.7618310 F = 87.743 N = 374 Degrees of Freedom = 15 ** = $p < 0.05$; *** = $p < 0.01$; **** = $p < 0.001$ Durbin-Watson = 1.397	

variables, but in addition, the variables are the right ones. From these results, we can conclude it is not necessarily true that the social density measure of pollution is a better predictor of complaint variable. However, the provincial analysis is still highly aggregated and more localized analysis may yield different results. Each of the models with three variables is significant to the 0.000 level; this means levels of pollution, per capita GDP, and general education are statistically correlated to variations in complaints. However, while each of the models can explain approximately 77 to 79 percent of the variance in complaint numbers, the results show that the models cannot account for all variation.

Higher pollution, economic well-being, and education levels are not the only variables with power to explain why environmental complaints increased in China from 1990 to 2004. The next section of this chapter examines other less quantifiable variables.

Institutional, legal, and political correlates to increasing complaints

Specific variables to be examined relate to the institutional, legal, and political contexts in China, including: improvements in EPB capacity; greater institutionalization of the complaint management system; signals from top leadership that increased citizen participation should be tolerated; and encouragement by environmental protection authorities to utilize the complaint system.[51]

The environmental complaint system is a part of a broader system of complaint management in China. The practice of filing complaints with local authorities has existed in China since imperial times.[52] However, Chinese leaders did not enshrine the right to complain until the passage of the 1982 constitution.[53] Throughout the 1980s Chinese leaders "restored" the complaint bureau system and complaint levels increased. In the late 1980s and early 1990s, leaders stepped up the pace of institutionalizing the national complaint resolution system. During this time the "high tide" of complaints began to recede somewhat as authorities addressed most citizens' "historical grievances."[54] In 1992, Chinese leaders held the first "National Complaint Working Meeting" (*diyi chuanguo xinfang gongzuo huiyi*), although officials held "National Conferences on Complaint Bureaus" since 1980.[55]

In 1996, authorities promulgated the State Council *Xinfang Tiaoli* (Regulations Regarding the Management of Letters and Visits) and revised them in 2006.[56] These national level regulations and all other rules and measures related to the general complaint system serve to guide environmental complaint work, although at times the environmental complaint system has served as a trial or model case for the development of general complaint work.[57]

Environmental complaint work: regulatory structure

Central-level authorities began to develop a regulatory structure to guide environmental complaint management in the late 1980s. In 1988, Chinese leaders created the National Environmental Protection Agency (NEPA), effectively making national-level environmental protection work independently of the Ministry of Urban and Rural Construction and Environmental Protection, which paved the way for a new approach to managing environmental complaints.[58] The new agency began to directly address environmental cases and complaints raised by the National People's Congress and the Chinese People's Political Consultative Conference. In the late 1980s, in a few complaint management policy "trial cases," NEPA mandated that a handful of local EPBs at the provincial and city levels establish offices or lead working groups and standardize complaint management procedures.[59]

Authorities incorporated lessons learned from the trial cases into draft measures to regulate environmental management practices on a wide scale, and in December of 1990 the National Environmental Protection Agency (NEPA) passed the *Huanjing Baohu Xinfang Guanli Banfa* (Measures Concerning the Management of Environmental Protection Complaints), which went into effect in February of 1991.[60] Revised measures went into effect in 1997 and again in 2006. In addition, several other rules and guidelines influenced complaint management work,[61] including provincial and city EPB complaint rules (*guiding*) and "internal bureau procedures"; however, some cities have lagged behind in this work.[62]

A comparison of the revisions of the three national-level EPB Environmental Complaint Management Measures that went into effect in 1991, 1997, and 2006 respectively illustrates how Party and government perceptions of environmental complaint work evolved, how authorities institutionalized the complaint resolution system, how the rights and duties of citizens and officials developed, how complaint procedures changed, and how authorities managed complaints to higher-level offices (*shangfang*) over time.

Purpose of complaint measures

Each revision states authorities must abide by relevant laws, regulations, and policies in managing environmental complaints, so it is clear that Chinese leaders intended the environmental complaint system to be an alternative to the legal system, even after 2006.[63] The preamble of each revision outlines the basic reasons for the measure and clearly shows the priorities of officials at any given time. The purpose of the 1991 measure was to standardize and institutionalize the system of managing complaints in accordance with the 1989 Environmental Protection Law (EPL) and relevant "guiding principles." The goals of the 1997 measure were to strengthen ties between the people and authorities, protect petitioners' legal rights, maintain social stability and "order" in complaint work, and promote economic growth and development. The purposes of the 2006 measure were to standardize environmental complaint work, maintain order in the complaint system, and protect petitioners' environmental rights.

Scope and structures in environmental complaint work

The scope of allowable complaints and channels for voicing those complaints remained relatively static over time. The 1991 measures stipulate that citizens can file complaints to the EPB at their administrative level or at the next level up. Citizens can file mass complaints, but only five representatives can "visit" a complaint office at one time. Worthy of note is that incorporated into all three revisions is a clause stating that complaint management personnel need not accept a complaint regarding an issue if the petitioner has already filed a suit, an administrative review, or the issue is under mediation.[64]

The mandated governmental structures and budgetary allocation requirements for each level of government changed over time.[65] According to the 1991 measure, national, provincial (autonomous area, municipality, and city (prefectural) EPBs had to establish their own complaint management mechanisms and procedures but county (district) level EPBs merely had to follow the procedures set up at the higher level.[66] This changed in 2006; the revised measure dictated that county (district) level EPBs establish complaint management mechanisms. In addition the revised measure called on provincial (autonomous region and municipality) EPBs to establish independent complaint offices. The 2006 measure includes provisions stating publicly for the first time that complaint work is part of the official evaluation system.

Increases in complaint numbers cannot be explained by the assertion that additional administrative level statistics were included in complaint counts after 1997; however, the improvements in complaint management system capacity at the county (district) levels and above throughout the 1990s probably contributed to better record-keeping and improved capability to handle additional numbers of citizen grievances. While counties (districts) did not need to establish their own complaint management mechanisms or even assign specialized personnel until 2006, they were still charged with documenting and managing complaints within their jurisdiction (1991 measure, provision 10). This means that, since 1991, complaint data included the numbers of complaints recorded at the county (district) level.

However, the 1991, 1997, and 2006 revisions do not assign any complaint management or documentation responsibilities to township environmental protection offices (EPOs) or village-level governmental authorities, so all of the complaints to village-level officials are not counted in the annual complaint statistics. This does not mean that township EPOs and village-level authorities do not receive complaints: it simply means that township EPO officials and village-level authorities are not required by the Measures Concerning the Management of Environmental Protection Complaints to report the number of complaints they receive. This is a good indication that oversight of these authorities in relation to environmental complaints is lacking. This is also additional evidence that the total number of citizen environmental complaints reported in SEPA's annual statistics is greatly underestimated.

The 2006 revision suggested new principles guiding complaint management work as well as new processes and cooperative opportunities for managing complaints that could potentially improve treatment of petitioners.[67] For example, EPBs *in principle* should prevent complaints from occurring, should fulfill their duties according to law, and make policy according to scientific and democratic principals, as well as protect citizens' rights to access to information, public participation, public oversight, and governmental transparency (provision 4). In addition, the 2006 measure specifies that EPB officials *can* hold public hearings regarding serious, complicated, or difficult-to-resolve complaints (provision 29); however, the provision does not *require* officials to hold public hearings.

Interestingly, the 2006 measure also suggests that EPBs can utilize education, negotiation, and mediation, as well as coordinate with relevant social organizations, legal aid organizations, experts and consultants, and volunteers to resolve people's complaints.[68] There is no guarantee that EPB delegation of responsibility to other "relevant" organizations to resolve complaints will lead to better outcomes; it could lead to a more complicated and less accountable system of complaint resolution. A decrease or unexplainable variation in environmental complaints in SEPA annual statistics could accompany the use of these tangential organizations and individuals to resolve citizen complaints.

Responsibilities of officials and citizen rights

Based on each revision of the Measures Concerning the Management of Environmental Protection Complaints, officials have specific responsibilities toward citizens. Since 1997, citizens have had the right to expect EPB personnel to keep their personal information private. The 1997 revision also stipulates that officials must adhere to the guidance of the government at each administrative level and must accept citizen oversight.

Over time, in theory, officials became subject to greater scrutiny and open to stiffer punishments for not fulfilling their duties or for abusing their office.[69] In 1997, EPBs instituted procedural changes in order to reduce opportunities for corruption.[70] The 2006 measure explicitly states officials face penalties for a variety of behaviors. It clearly specifies conditions under which higher-level EPB personnel should supervise, punish, or make suggestions to lower-level EPBs, including instances where a complaint has not been dealt with in a timely manner, where personnel neglect to implement a decision regarding a complaint, or where personnel engaged in fraud or deception in managing a complaint (provision 35). The measure also more clearly stipulates that lower-level EPB officials could face administrative punishment or criminal charges if they, among other actions, infringe upon the legal rights of citizens by engaging in physical violence toward petitioners, exceeding the authority of their office, not fulfilling their required duties, misapplying the law or not following legal procedures (provisions 37–40).

Upper-level EPBs and other relevant governmental agencies supervise the complaint management work of lower-level EPBs in a variety of ways as well as pressure them to prevent disputes and be more responsive to citizen complaints.

Lower-level EPBs must file brief reports regarding complaints every three to six months and reports that are more detailed once a year. If lower-level EPBs do not submit required reports, then part of their funding could be withheld.[71]

Typically, if a higher-level EPB receives a large number of complaints from citizens from a lower-level administrative area, the EPB will call in representatives from the lower-level EPB for "training." The EPBs that are unable to resolve complaints at their own administrative level are often "criticized." However, the lack of authority of EPBs impedes the hierarchical supervisory process. Local government officials are still responsible for personnel decisions. In some cases, provincial EPBs may not even have the authority to require city EPB officials to attend training.[72]

Responsibilities of citizens

In principle, revisions to the Measures Concerning the Management of Environmental Protection Complaints relating to the scope of allowable complaints, the responsibilities of officials, and the procedures for filing a complaint, translated into improved citizen rights within the complaint system framework.

At the same time, the revised provisions expanded greatly the responsibilities of citizens and prohibitions against certain behaviors. The 1991 provision simply dictates that citizens are not allowed to threaten or engage in physical violence against EPB officials, or disrupt the order of complaint management work, otherwise they could be turned over to public security personnel. The 1997 revision elaborates upon these responsibilities and prohibitions and expands the reasons to call security officials to include instances when citizens do not respond to "criticism" and education efforts (1997 measure, provision 21).[73] The 1997 revision introduces prohibitions against doing anything to harm the interests of the state, the collective, or other individuals. In addition, citizens are to abide by complaint management procedures, respect social morals, and "take care of" property at facilities where complaint management personnel work (provision 8).

The 2006 measure expands again the list of prohibitions to include besieging a facility, blocking official vehicles, obstructing public roadways, injuring themselves, loitering or causing a commotion at complaint departments, and carrying hazardous materials. Citizens cannot incite, coerce, or employ financial inducements to entice citizens to complain, or establish ties to manipulate petitioners nor can petitioners take advantage of complaints to make money. In addition, the measure prohibits other rather innocuous or vaguely defined behaviors such as utilizing surveillance equipment (which theoretically could include a camera) passing out fliers, using slogans, wearing special clothing, engaging in extreme behavior or otherwise disturbing public order or violating public morals (provision 21).

Complaint management procedures

The procedures for filing environmental complaints and for managing them are straightforward and did not change dramatically as reflected in complaint management measures. If a citizen has a complaint, suggestion, criticism, or request, they can, among other methods, write, e-mail, call, or visit the environmental protection bureau in charge of that administrative area or the next highest level. Citizens might approach other government or Party organs with their complaint; however, these various organizations rarely directly handle environmental complaints.[74] It is more likely that personnel in these organs will instruct the citizen to take their complaint to the environmental protection authorities at the appropriate administrative level. An EPB is charged with responding to complaints regarding issues that occur within its jurisdiction, but it can be confusing as to which bureau has jurisdiction. Determining who is responsible for a complaint can be a complex issue, and can lead to certain cases "falling through the cracks."[75]

Most EPBs have similar procedures for managing complaints, although there is some variation from place to place and from one administrative level to the next.[76] The complaint processes begins when a citizen or legal entity contacts an EPB about an environmental grievance. In general, complaints to EPBs from members of the media, governmental officials, National People's Congress (NPC) and Chinese People's Political Consultative Conference (CPPCC) representatives, and other VIPs are taken more seriously and handled as "special cases."[77]

EPBs are generally equipped to take complaints 24 hours a day and many have set up "pollution hotlines."[78] Typically, during the day, staff in the Administrative Affairs Division take care of phone calls, letters, and visitors. At night, specified EPB staff answer the phones. Some EPBs instituted a "rotation system" whereby the entire staff of an EPB would rotate taking night shifts to answer phone calls. By the end of 2005, 69.4 percent of administrative divisions above the county level had established a pollution hotline.[79] The 1997 measure directs EPB staff to record personal information about the petitioner including their name, contact information, and work unit, which means if a citizen wishes to remain anonymous, he/she runs the risk that the EPB will not accept the complaint.

Officials from the Administrative Affairs Division/Section of an EPB usually record a complaint and then pass it on to the appropriate department, such as the department of pollution control or the department of development and construction, which then investigates the complaint. While the EPB director was responsible for complaint work prior to the 1991 regulation, after the regulation the director is only notified about problems that involve other administrative areas, are extremely serious, or are likely to cause social unrest.[80]

While the 1990 measure did not stipulate time limits for investigating and resolving complaints, the 1997 measure stipulates that complaints must be resolved within 30 days and the complainant must be notified of the results.

However, the 2006 measure changed this to 60 days. For complex problems the time limit is flexible, but according to the 2006 measure officials cannot take more than 30 additional days to resolve the issue. The longer time limit allowed for by the 2006 measures may give time for more in-depth investigation but it also could increase citizen frustration levels. Complaints passed down from an EPB at a higher administrative level do not have to be resolved for 90 days.[81]

If a citizen is not satisfied with the decision of the EPB based on its investigation, they can utilize a review process; however, this process can be expensive, time-consuming, and ineffective.[82] The 1997 measure allowed for a petitioner to appeal to the EPB at the next highest administrative level. The 2006 measure broadens the possible avenues of redress; a petitioner then had the right to request in writing a review of the EPB's decision from the government at the same administrative level or the next higher-level EPB within 30 days of receiving the results of an official investigation. Further, the 2006 measure stipulates that if the petitioner is not happy with the first review, they can within 30 days request in writing another review from the government at the same level or the next highest-level EPB. However, the 1997 and 2006 measure stipulates that the EPB at a higher administrative level has the choice to resolve the complaint itself or to defer the complaint back to the lower-level EPB for review (1997 measure, provision 31). In addition, officials at the higher-level EPB must consider the suggestions of the lower-level EPB (2006 measure, provision 33). Just as in the court system, a citizen has the right to appeal twice; after governmental or higher-level EPB authorities have reviewed the original investigation results a second time, the petitioner does not have the right to resubmit a petition.

Shangfang and recent developments

While environmental protection authorities have been concerned about petitioners taking their complaints to higher administrative levels (*shangfang*) since the early-to-mid 1990s (*Zhongguo Huanjing Nianjian* 1998, p. 151), in 2004, environmental protection officials established a small working group to deliberate remedies to the on-going *shangfang* problem and also began to experiment with new complaint resolution structures. The move is partially in response to central government and Party authorities' concerns about increasing numbers of petitioners traveling to Beijing.[83] In 2004, the EPB created experimental "*Huanbao Zhifa Heyiting*" ("Environmental Protection Enforcement Collegiate Benches") to improve long-term complaint management efficiency in Jiangsu, Hebei, Heilongjiang, Shandong, Qingdao, and Nanjing.[84]

Citizens still take complaints to authorities at a higher administrative level for a variety of reasons. The enterprise that is the target of the complaint could be a higher-level enterprise, requiring citizens to complain to a higher level because for example, a city EPB has no jurisdiction over a provincial-level enterprise.[85] Citizens could be dissatisfied with the action taken by a lower-level EPB or believe the lower-level EPB personnel are corrupt. They could decide to take their complaint to the highest level possible because they believe it will be more effective or they could have personal connections within higher-level governmental offices. However, higher-level EPBs rarely handle complaints in lower-level administrative areas directly. In most cases, the higher-level EPB will direct citizens to take their complaints to the EPB in their district or city. This can lead to a great deal of frustration on the part of citizens.

In summary, the three versions of the Measures Concerning the Management of Environmental Protection Complaints in 1991, 1997, and 2006 illustrate how the guiding principles of complaint work changed over time and how these principals influenced complaint levels. Authorities intended the measures to standardize and institutionalize the complaint system and at any given time, they also were meant to protect petitioners' rights, promote economic development, and maintain social stability. City (prefectural) level and above EPBs were required to establish complaint management mechanisms as early as 1991, but while county-level EPBs had to report complaint numbers, they did not need to establish complaint management mechanisms until 2006. Of note, township- and county-level EPOs and other related governmental agencies have not been required to report complaints, meaning the number of complaints recorded in the environmental yearbooks does not include complaints filed at the most basic level of Chinese society. In theory, revisions to the measures strengthened citizens' rights by outlining increasingly stricter requirements of officials and by stipulating that officials could face administrative punishment or criminal charges if, among other actions, they infringe upon the legal rights of citizens. However, at the same time, the revised provisions expanded greatly the responsibilities of citizens and prohibitions against certain behaviors including some that are ill-defined or that would expand the numbers of or enhance the solidarity among petitioners with similar grievances, such as posting flyers. Basic procedures for managing complaints remained relatively unchanged, except that the 2006 revision extended the time limit for resolving ordinary complaints from 30 to 60 days. Environmental protection officials have been concerned about *shangfang* since the early 1990s, but greater cross-sector, national-level attention to the *shangfang* phenomenon in 2003 and 2004 prompted environmental officials to explore new mechanisms to resolve environmental complaints.

Improvements in institutional capacity

As EPB institutional capacity grew, complaint management procedures became clearer, and the complaint system became more accessible, greater numbers of citizens turned to EPBs with their complaints. In the early 1990s, despite the etchings of a basic regulatory framework and central-level support, few provinces, cities, district and county EPBs had established complaint system structures or assigned specialized personnel, and citizen complaints were not a priority for most EPBs.[86] It took time, environmental education, and pressure from citizens and authorities at higher administrative levels to convince EPBs to allocate money to develop complaint system structures. By the end of the Eighth

Five-Year Plan (1995) the system for managing complaints was "becoming more perfect day by day" (*rizhen wanshan*) and most provinces, autonomous zones, and municipalities had developed mechanisms for managing complaints, standardized complaint work, and institutionalized the complaint system.[87] These institutional developments naturally led to increases in the numbers of reported environmental complaints. For example, when authorities first began to advertise telephone hotlines, the numbers of complaints increased considerably.[88] In addition, as overall EPB capacity improved and EPBs engaged in increasing numbers of environmental education programs, campaigns, and other outreach activities, citizens acted on these authoritative signals about priority environmental problems and complained in increasing numbers about related issues.[89]

Shifts in China's political and institutional contexts

In addition to greater institutional capacity and pressure from above to establish complaint management mechanisms and be more responsive to citizen complaints, higher complaint levels in the late 1990s can be partially explained by changes in the overall political context in China. In 1997 Deng Xiaoping passed away, ending a political era. Even though Deng did not occupy a formal position in the Chinese Party or governmental hierarchy, he was an elder, giving him power behind the scenes. Just as Mao's death triggered a rise in citizen complaints, so too did Deng's death. After Deng's death there seemed to be greater room for protest in Chinese society and higher official tolerance of citizen participation in society.[90] While tolerance for citizen protest under Hu Jintao seems to have subsided, officials continue to channel citizen grievances and encourage citizen participation through government-controlled outlets.

Improvements in the status of China's environmental protection apparatus in China's governmental hierarchy and encouragement by EPB officials to utilize the *xinfang* system also contributed to the rise in environmental complaints in the late 1990s and early 2000s. In 1998, the State Environmental Protection Administration was formally upgraded and the environmental protection minister, Xie Zhenhua, was given a minister's portfolio.[91] This move gave SEPA more power than before vis-à-vis other government ministries. For instance, it allowed environmental protection officials at each administrative level to call for inter-ministerial meetings and strengthened the professional relationship between EPBs at higher and lower administrative levels. The official change also signaled to lower-level authorities and to citizens that environmental protection was becoming more of a Party and governmental priority, meaning there would be more political space to complain about environmental problems. Indeed, many environmental protection officials welcomed complaints because complaints helped officials to locate unregistered polluting enterprises and gave them more leverage when trying to convince local government officials to shut down or relocate polluting enterprises. Environmental protection officials encouraged complaints because it helped them do their job better.[92] Some areas experimented with a reward system that provided citizens with financial incentives to report pollution problems. The environmental complaint system is also typically the first stop for citizens with environmental grievances, and acts as a pressure valve for such grievances. Without the complaint system, environmental protection and public security authorities would have more contentious disputes to manage and the court system would likely face a larger case load. As argued by Carl Minzner, the complaint system is an alternative to the legal system.[93]

Linkages: grassroots environmental grievances and environmental movement emergence

Not only has the complaint system become an alternative to the legal system, it has become an outlet for environmental grievances, which has robbed momentum from a potential anti-pollution movement in China. The partial success of the complaint system to provide a channel of public participation in environmental policy implementation is one such reason for the loss of momentum toward an environmental movement.

Some evidence indicates the general complaint system is too flawed to be successful.[94] First-hand evidence gathered by human rights organizations indicates that China's national complaint system fails to resolve citizen grievances.[95] In addition, one 2004 Chinese Academy of Social Sciences survey found that only 0.2 percent of all types of citizen complaints are resolved.[96]

However, other evidence suggests that at least some environmental protection officials became more responsive to complaints over time. In the early 1990s, personnel who managed complaints often had difficulty in garnering the support of office or bureau leaders for victims of pollution.[97] In contrast, in 1999 one district-level official declared "higher management cares about complaints now more than they used to, so they are easier to resolve."[98]

Evidence from interviews suggests that authorities have taken citizen opinion into consideration in infrastructure planning and when they determine which enterprises to reorganize, relocate, or shut down.[99] In addition, according to an empirical study conducted by researchers at the World Bank that sought to discover the determinants of the enforcement of pollution charges in China and the relative bargaining power of firms with environmental authorities, some firms pay the entire pollution charge levied while other firms only pay a portion of the pollution charge levied. They found that a firm that was the target of more citizen environmental complaints had less bargaining power.[100]

Another reason why the rise in environmental grievances has not yet led to an anti-pollution movement is the particularistic nature of environmental complaints. Citizens that complain tend to focus on their immediate situation and do not try to frame their grievances as universal. In other words, petitioners are concerned with "not in my backyard" problems and not China's overall industrial pollution problems, which leads to isolated instances of contention and not a widespread movement.

Second, petitioners have not yet begun to co-operate en masse with each other or with environmental organizations, which inhibits the growth of a grassroots anti-pollution movement. The mobilizing structures that sometimes help individual citizens to band together to collectively contend with enterprises or officials in China are often the affected work unit, residents' committees or other local Party organizations, and local EPBs. These organizations are not typically thought of as social movement organizations. China's environmental social organizations for the most part were concerned with environmental degradation issues until the early 2000s and did not often join in solidarity with local groups of pollution victims.[101] In other words, grassroots petition and protest movements remain largely separate from the more organized, elite environmental groups.

Third, the efforts of authorities to prevent petitioners from organizing inhibit the growth of an anti-pollution movement. Petitioners that attempt to organize, even peacefully, face the threat of repression.[102] Environmental organizations that reach out to assist local grassroots movements are likely to be targeted by Chinese authorities and possibly accused of attempting to incite petitioners. These political threats increase the costs of collective action making an organized anti-pollution movement less likely. In addition, the 2006 Measures Concerning the Management of Environmental Protection Complaint restrict citizens from utilizing slogans, passing out fliers, wearing special clothing, and other behaviors, which could build group solidarity and focus of purpose, and further the development of an environmental movement.

While the environmental complaint system may have had some measure of success in satisfying people's grievances and likely had some localized impact on environmental quality, the suppression of a more organized anti-pollution movement weakens and diminishes public demand for a cleaner environment, which will ultimately slow down environmental protection achievements.[103] In addition, the weaknesses in the complaint system could trigger the exact behaviors that officials are trying to restrict, i.e. environmental protests.

Weaknesses in the complaint system

While officials have attempted to institutionalize and, at least until 2004, popularize the complaint system and some environmental protection authorities have been more responsive to citizens, the complaint system is far from perfect.[104] Remaining problems are: exclusion of township EPOs in the complaint system; official brutalism and disregard for and non-resolution of complaints often due to local-level economic protectionism and corruption; lack of accountability in the system; and procedures that lead to arbitrary decisions by officials.

Township EPOs are not required to submit records of environmental complaints nor are they required to establish complaint management mechanisms or procedures. The three versions of the complaint management procedures simply ignore government complaint mechanisms at the township and village levels, greatly reducing accountability at the most basic level. Township EPO officials are pulled in two different directions because they are assigned industrial development as well as environmental protection duties and their job advancement typically depends upon their achieving industrial development – making the choice between environmental protection and industrial development easy.[105]

Central-level officials acknowledge that complaints are ignored or are never resolved. Authorities have considered non-resolution of complaints at the local level a problem since at least the mid-1990s because it has led to increasing numbers of petitioners taking their complaint to higher levels of government.[106] In one case in Taiyuan, citizens complained for years about the air and water pollution from a nearby power plant. The EPB representative commented "*meiyou banfa*" (nothing can be done), particularly since the power plant is not a "*zhongdian*" (priority) with the provincial government.

Sometimes officials say nothing can be done because they seek to protect local economic interests and governmental income from enterprises. In the Taiyuan case, for example, if the local coal-fired electrical power plant were shut down, it would lead to power shortages and economic hardships; the plant provides a necessary service to the community. Economic hardship is often used as an excuse for failing to implement environmental protection measures or for refusing to provide compensation to victims of pollution.

One official admitted the common attitude of officials is that if a citizen does not sue, it is not necessary to take the complaint seriously ("*bugao, buli*").[107] Perhaps this attitude springs from the belief that they have no authority to resolve a complaint, or it could be that they simply do not care. Either way, adopting an attitude of "*bugao, buli*" is a sign that there is not enough accountability built into the complaint system at the local level.

Official brutality and the stifling of environmental petitioners is a serious problem that turns petitioners into protesters, creates tensions between officials and citizens, and in the long run harms social harmony and environmental quality. One only needs to read the cases of Sun Xiaodi, Fu Xiancai, the CITIC Dameng Mining Industries Ltd. manganese pollution, the Pubugou Dam, or Huaxi, among many others to see how brutality and the repression of petitioners negatively impacts the integrity of the petitioning system.[108]

At this point, there is insufficient oversight and accountability built into the complaint system. If a citizen takes a complaint to a higher administrative level, the EPB officials at the higher level can either decide to investigate the complaint or send the complaint back to the lower level for review. This flexibility allows for arbitrary decisions and contributes to inconsistent oversight from higher authorities.

At the national level, there is little or no oversight of complaints directly received at the national level. These complaints were not included in the total number of complaints reported in the China Environment Yearbook until the 1999 edition, making them impervious to public oversight.[109] Until the late 1990s, there was only one official working full-time in the national level complaint office. One official who worked part-time in the national level complaint office reported that no one kept track of the complaints coming in and there were "complaint letters all over the floor." The SEPA complaint office in Beijing receives complaint letters from all over the country. These complaints are typically those that have not been resolved at the local level, and are therefore extremely serious. Not all complaints fell through the cracks at the national level. Many were resolved, but usually only after a long period of time.[110]

Conclusion

Using complaints as a means of interest articulation and expressing grievances is not new in China. However, the way in which authorities frame and deal with complaints has changed significantly. When Deng Xiaoping took power and rectified some of the wrongs perpetuated during the Cultural Revolution, Chinese leaders re-framed citizen complaints as being less political and so they became more permissible.[111] In 1982, the revised constitution granted citizens the right to criticize and make suggestions to authorities, theoretically without reprisal. However, the national complaint system did not become institutionalized and standardized until the early 1990s. It can be inferred that authorities institutionalized the complaint system in part as a response to citizen complaints and demands for reform that culminated in the Tiananmen massacre in 1989.[112]

In line with and sometimes leading national trends, officials in the environmental sector strengthened, institutionalized, and expanded the administrative regulatory and legal framework for managing environmental complaints. Environmental protection authorities officially began collecting data on environmental complaints in 1989 but select locations have data from 1983. The National Environmental Protection Agency passed the *Huanjing Baohu Xinfang Guanli Banfa* (Regulation Concerning the Management of Environmental Protection Complaints), which went into effect in February of 1991, effectively institutionalizing complaint management. Environmental protection authorities revised this measure in 1997 and again in 2006.

As institutionalization of the complaint system proceeded, the number of environmental complaints increased. The total number of environmental complaints aggregated at the national level increased 513 percent between 1990 and 2004. The provinces that have the highest overall and per capita complaints regarding air and water pollution are: Guangdong, Shandong, Jiangsu, Zhejiang, Sichuan, and Fujian.

Provincial-level quantitative data supports the assertion that variables reflecting levels of pollution, economic development, and general education taken together in one statistical model are significantly correlated with this rise in citizen environmental complaints. The models containing all three variables have more explanatory power than any of the individual variables alone. In both the water and air pollution cases, the absolute measure of pollution rather than the measures of natural density and social intensity of pollution has more explanatory power in accounting for variations in complaint levels. In addition, levels of pollution are better predictors of variations in complaints than either levels of education or economic development. The numbers ultimately tell us that a model with all three independent variables should be used when seeking to predict variations in complaint levels and that reducing pollution in those areas with the highest absolute level of pollution will have a greater impact on complaint numbers.

Not only are levels of pollution, education, and economic well-being correlated with the rise in complaints in China, but the institutional, legal, and political contexts within which the complaint system developed influenced variations in complaint numbers. Specifically, institutionalization and standardization of the complaint management system, improvements in EPB organizational capacity, pressure from central-level officials to be more responsive to citizen demands at the local level (for the sake of protecting the legitimacy of the Party and the government), and encouragement to utilize complaint channels by environmental protection authorities contributed to increasing numbers of environmental complaints.

Institutionalization and standardization of the complaint management system and improvements in EPB organizational capacity contributed to the measurable rise in environmental complaints in several ways. In the early 1990s, few provinces, cities, and county EPBs had established complaint system structures or assigned specialized personnel to manage complaints because they simply were not a priority for officials. In addition, as environmental protection offices at the city level and below became independent of the Ministry of Construction, personnel could focus on environmental protection exclusively. By the end of the Ninth Five-Year Plan (2000), most provinces and cities had established specialized complaint management mechanisms and assigned specialized staff, leading to better record-keeping and increased capacity to manage complaints. By the end of 2005, 69.4 percent of administrative divisions above the county level had established and advertised a pollution hotline leading to greater citizen recognition of the option to complain to EPBs. The more dramatic increase in complaints in the late 1990s is not due to a change in the levels of administration required to report complaints, i.e. EPBs at the county (district) level and above have had to report complaints since 1990 and village and township EPOs have never had to report the number of complaints they receive. The change is more likely due to rising pollution as well as the institutionalization and popularization of the complaint system, which made it more convenient for citizens to file complaints.

Political pressure from top Chinese leaders to become more responsive to environmental complaints signaled to citizens that there was more political space to complain and created the perception that local officials would be more likely to respond to their complaints. Throughout the 1990s, references to improving governmental responsiveness to environmental complaints are included in speeches by Chinese leaders and in the Chinese Agenda 21. High-level authorities have supported the complaint system because it reinforces the continuation of "top-down" supervision of governmental officials, i.e. higher-level authorities can use environmental complaint data to learn which local-level officials require training in legal enforcement or citizen–official relations. In addition, when lower-level officials are more responsive to citizens, it boosts the legitimacy of the government and the Party.

Officials have encouraged use of the complaint system to resolve environmental grievances. Encouraging use of the complaint system allows Party and government leaders to channel grievances by people who might otherwise choose other less desirable paths to resolve their pollution problem, such as initiating a dispute, protesting, going on strike, or other such disruptive behavior. In other words, the complaint system is used by the state to prevent environmental disputes and uncontrolled citizen protests.

Environmental officials especially have encouraged citizens to resolve their environmental grievances through administrative channels and some EPB personnel have become more responsive contributing to the rise of complaints. There is greater understanding on the part of environmental officials about the hazards and unintended consequences of industrial pollution so they are therefore more likely to understand pollution victim complaints. In addition, environmental officials realize that encouraging complaints benefits their *danwei* and environmental protection efforts. When more citizens complain, it strengthens the authority of environmental protection officials and helps them to implement environmental laws and policies. Citizens act as the eyes and the ears of environmental officials and are the ground troops in the battle to compel enterprises to comply with environmental laws and policies. While citizen complaints may not help environmental protection authorities ferret out all pollution problems, complaints are still a valuable tool to protect the environment.

While the complaint system has afforded ordinary citizens at the county level and above an outlet for their environmental grievances, the system is plagued by serious problems and many doubt its effectiveness. Remaining problems include: exclusion of township EPOs in the complaint system, official brutalism toward petitioners and disregard for and non-resolution of complaints, lack of accountability in the system, and procedures that lead to arbitrary decisions by officials. Despite legal protection, retribution against petitioners is particularly damaging and turns petitioners into protesters. These problems lead to low confidence in the government's ability to enforce laws, respect citizens' rights, and to protect the environment, which has had a negative effect on government legitimacy.

Some evidence indicates many complaints are left unresolved and it is still unclear whether citizens are satisfied with the way officials are protecting the environment in general. In a survey conducted by SEPA and researchers from Beijing University, 86 percent of citizens polled in 2000 believed that the government was not sufficiently enforcing environmental laws.[113] Likewise in a survey of Beijing University students, 91.2 percent felt the government was not doing enough to protect the environment.[114]

The environmental complaint system has channeled momentum away from the development of an anti-pollution movement in China. Individual complainants or groups filing petitions to resolve "not-in-my-backyard" type problems have not acted collectively in a sustained manner to try to mobilize broad support for wide-reaching policy changes. Part of the reason for this is that complaint regulations allow for group complaints, but seek to control and even criminalize collective action. Another reason why environmental grievances have not led to a more widespread movement is the paucity of environmental social movement organizations and the lack of the involvement of such groups with local anti-pollution efforts. So far, except in a few cases, environmental social organizations have not actively utilized the complaint system on behalf of pollution victims. One such effort by the leader of Defenders of Lake Tai, Wu Lihong, likely led to his arrest in 2007. Nor have China's environmental social organizations united specific groups of pollution victims to mobilize an anti-pollution movement for several reasons: the anti-development overtones of such efforts; the personal preferences of organized environmental social organization members; the lack of communication between pollution victim and the more elite and organized environmental groups;[115] and the threat of repression from authorities. However, in the future, the complaint system could be one avenue for environmental social organizations to mobilize support for broader policy change to protect the environment.

Notes

1 For more information about environmental degradation and environmental grievances in Chinese history see: M. Elvin and T. Liu, *Sediments of Time: Environment and Society in Chinese History*, Cambridge: Cambridge University Press, 1998; M. Elvin, "The Environmental Legacy of Imperial China" in R.L. Edmonds (ed.) *Managing the Chinese Environment*, New York: Oxford University Press, 1998; P. Perdue, *Exhausting the Earth: State and Peasant in Hunan, 1500–1850*, Cambridge, MA: Harvard University Press, 1987; B.V. Rooij, "Implementation of Chinese Environmental Law: Regular Enforcement and Political Campaigns," *Development and Change*, Vol. 37, No. 1 (2006), pp. 57–74; J. Shapiro, *Mao's War Against Nature*, New York: Cambridge University Press; 2001; R. Marks, *The Origins of the Modern World: A Global and Ecological Narrative*, Lanham, MD: Lexington-Rowman and Littlefield, 2002; R. Marks, *Tigers, Rice, Silk and Silt: Environment and Economy in Late Imperial South China*, New York: Cambridge University Press, 1998; V. Smil, *The Bad Earth*, Armonk: M.E. Sharpe, 1984, and *China's Environmental Crisis*, Armonk: M.E. Sharpe, 1993; A. Brettell, "China's Pollution Challenge: The Impact of Economic Growth and Environmental Complaints on Environmental and Social Outcomes" in S. Guo (ed.) *Challenges Facing Chinese Political Development*, Lanham, MD: Rowman and Littlefield, forthcoming 2007.

2 The term *xinfang* is extremely broad. The formal Chinese term *xinfang* includes both the less risky appeal and the more risky adversarial activities. *Xinfang* translated here as "complaint" includes: *shenqing* (file a petition – a formal request for something from authorities; *xunwen* (inquiry); *fanying* (response, reaction); *jiefa* (expose, unmask, bring to light); *piping* (criticize); *biaoyang* (praise, commend); *jianyi* (suggestion); and comments on professional performance (*hangye zuofeng lei*). The formal categories in the Environmental Yearbooks are *xinfeng* (letters), *laifang* (visits), *faming jianyi lei* (inventions and suggestions), and *hangye zuofenglei* (professional performance). For more discussion about the term *xinfang* see C.F. Minzner, "Xinfang: An Alternative to Formal Chinese Legal Institutions," *Stanford Journal of International Law*, Vol. 42, No. 2 (Winter 2006), pp. 115–16. Some

scholars such as Tianjin Shi have categorized the process as "particularized contacting"; T. Shi, *Political Participation in Beijing*, Cambridge, MA: Harvard College; 1997. Citizens primarily use letters and in-person visits to communicate complaints, but there also are other methods including telegrams, pictures, phone calls, tapes, and videos.

3 These figures include only complaints and suggestions people made to environmental protection offices, bureaus, and SEPA. It does not include environmental complaints sent to other ministries, or government, legislative, and Party offices.

4 A short list of other resources on this subject include E.J. Perry and M. Selden, *Chinese Society: Change, Conflict and Resistance*, 2nd edition, New York: Routledge, 2003; K.J. O'Brien and L. Li, "The Politics of Lodging Complaints in Rural China," *The China Quarterly*, No. 143 (1995), pp. 757–83; J. Wasserstrom and E. Perry, *Popular Protest and Political Culture in Modern China*, Boulder, CO: Westview Press, 1994; D. Salinger, *Peasant, Migrants, and Protest Contesting Citizenship in Urban China: Peasant, Migrants, the State, and the Logic of the Market*, Stanford, CA: Studies of the East Asian Institute, 1999; T. Shi, *Political Participation*; P. Thornton, "Framing Dissent in Contemporary China: Irony, Ambiguity and Metonymy," *The China Quarterly*, No. 171 (September 2002), pp. 661–81; J. Unger, "Power, Patronage, and Protest in Rural China" in W. Tyrene (ed.) *China Briefing 2000: The Continuing Transformation*, Armonk, New York: M.E. Sharpe, 2000, pp. 71–94; T. Wright, "State Repression and Student Protest in Contemporary China," *The China Quarterly*, No. 157 (March 1999), pp. 142–72; and T. Bernstein and X. Li, *Taxation Without Representation in Contemporary Rural China*, Cambridge: Cambridge University Press, 2003.

5 See J. Dawson, *Eco-Nationalism: Anti-Nuclear Activism and National Identity in Russia, Lithuania, and Ukraine*, Durham, NC: Duke University Press, 1996; D. Fisher, "The Emergence of the Environmental Movement in Eastern Europe and Its Role in the Revolutions of 1989" in B.J. Webster (ed.) *Environmental Action in Eastern Europe: Responses to Crisis*, Armonk NY: M.E. Sharpe, 1993; M.I. Goldman, "Environmentalism and Nationalism: An Unlikely Twist in an Unlikely Direction" in J. Massey Stewart (ed.) *The Soviet Environment: Problems, Policies and Politics*, Cambridge: Cambridge University Press, 1990; and C. Ziegler, "Environmental Politics and Policy Under Perestroika" in J.B. Sedaitis and J. Butterfield (eds) *Perestroika from Below: Social Movements in the Soviet Union*, Boulder, CO: Westview Press, 1991.

6 For information on Japan's environmental movements see M.A. McKean, *Environmental Protest and Citizen Politics in Japan*, Berkeley: University of California Press, 1981; M. Schreurs, *Environmental Politics in Japan, Germany, and the United States*, Cambridge: Cambridge University Press, 2002; H. Weidner, "An Administrative Compensation System for Pollution-Related Health Damages" in S. Tsuru and H. Weidner (eds) *Environmental Policy in Japan*, Berlin, Germany: Edition Sigma Rainer Bohn, 1989; D.E. Apter and N. Sawa, *Against the State: Politics and Social Protest in Japan*, Cambridge: Harvard University Press, 1984; and F. Upham, *Law and Social Change in Postwar Japan*, Cambridge, MA: Harvard University Press, 1987.

7 R. Weller, *Alternate Civilities: Democracy and Culture in China and Taiwan*, Boulder, CO: Westview Press; 1999; J.F. Williams and C. Chang, "Paying the Price of Economic Development in Taiwan: Environmental Degradation" in M.A. Rubenstein (ed.) *The Other Taiwan*, Armonk, NY: M.E. Sharpe, 1994; S. Tang and C. Tang, "Democratization and the Environment: Entrepreneurial Politics and Interest Representation in Taiwan," *The China Quarterly*, no. 158 (1999), pp. 350–66.

8 S. Lee, "Environmental Movements in South Korea" in F. Shiu and A.L. So (eds) *Asia's Environmental Movements: Comparative Perspectives*, Armonk, NY: M.E. Sharpe, 1999; Norman Eder, *Poisoned Prosperity: Development, Modernization, and the Environment in South Korea*, Armonk, NY: M.E. Sharpe; 1996.

9 *Huanjing Jiufen Fangfan Yu Chuli Shiwu Quanshu* [Environmental Disputes: A Practical Guide for Prevention and Management], ed. M. Wang, Beijing, China: Zhongguo Yanshi Chubanshe, 1999.

10 Responding to citizen environmental grievances could include systematic changes such as fulfilling responsibility system contractual promises, strengthening the enforcement of China's already comprehensive set of environmental laws and regulations, increasing pollution fees and fines to make them meaningful, rationalizing environmental damage compensation systems, and satisfactorily responding to a large enough percentage of complaints. The Party and state do not improve their legitimacy by taking the suggestions of citizens and improving enforcement or cracking down on local officials but then detaining, brutalizing, harassing, or putting in prison the citizens that originated the suggestions or complaints, as occurred in the Chen Guangcheng, Pubugou Dam and SARS cases, with Dr Gao Yaojie in the Henan HIV/AIDs issue, and other cases.

11 I focused on air and water pollution complaints because air and water pollution problems are typically, although not always, more egregious than noise and solid waste issues. Noise pollution complaints, although more pervasive than other types of complaints, are often more easily resolved. Some measurements of water pollution have begun to decline while most air pollutants are still on the rise, which increases the variability of the pollution measurements. In addition, environmental protection authorities place more emphasis on water and air pollution problems.

12 The qualitative analysis is based on interview and other data gathered at the national level and in three case cities, Beijing, Chengdu, and Taiyuan.

13 *Zhongguo Huanjing Nianjian, 1998* [China Environment Yearbook, 1998], Beijing: Zhongguo Huanjing Nianjian Chubanshe, 1998, p. 336.

14 See T. Shi, *Political Participation in Beijing* and W. Alford, R. Weller, L. Hall, K. Polenske, Y. Shen and D. Zweig, "The Human Dimensions of Pollution Policy Implementation: Air Quality in Rural China," *Journal of Contemporary China*, Vol. 11, No. 32 (2002), pp. 495–513.

15 In previous research, variables measuring education more generally were found highly correlated with variations in complaints. See S. Dasgupta and D. Wheeler "Citizen Complaints as Environmental Indicators: Evidence From China," Washington, DC: World Bank, Policy Research Department, 1997, pp. 11–12. Tianjin Shi found that higher education levels brought about by economic development were highly significant in his analysis of the factors correlated with citizen participation in Beijing between 1988 and 1996. See Shi, *Political Participation in Beijing*, p. 168.

16 Among the works discussing the relationship between environmental awareness and action see especially Dunlap (1989) and Ostman and Parker (1987: 3–9).

17 Brechin and Kempton (1994) "Global Environmentalism: A Challenge to the Post-materialism Thesis?" and Kaimowitz (1996) "Social Pressure for Environmental Reform". Research has shown that the density of pollution, the amount of pollution divided by the relevant land area, was highly correlated to Japanese society's response to its air pollution problems. Jeffery Broadbent measured Japanese society's response to its air pollution problems as the rapidity of the response (indicated by the percentage of reduction in sulfur dioxide), and the thoroughness of the response (indicated by its percentage reduction at the 20-year mark). The "social density" of pollution is the amount of pollution times the population density. These concepts are similar to those used by Broadbent. See J. Broadbent, *Environmental Politics in Japan*, Cambridge: Cambridge University Press, 1999.

18 See S. Dasgupta and D. Wheeler, "Citizen Complaints as Environmental Indicators: Evidence from China," Washington, DC: World Bank, Policy Research Department, 1997, pp. 11–12. The generalizability of that research is questionable, however, because it used a measure of the total number of complaints while including only

SO_2, particulate emissions and chemical oxygen demand, (COD) measurements for pollution in the regression analysis.

19 W.J. Baumol and W.E. Oats, *Economics, Environmental Policy, and the Quality of Life*, Englewood Cliffs, NJ: Prentice-Hall, 1979, Chapters 12 and 13; P.A.G van Bergeijk, "International Trade and the Environmental Challenge," *Journal of World Trade*, Vol. 25, No. 6 (1991), pp. 105–15; and G.M. Grossman and A.B. Krueger, "Environmental Impacts of a North American Free Trade Agreement" in P.M. Garbier (ed.) *The Mexico–U.S. Free Trade Agreement*, Cambridge, MA: MIT Press, 1993; G.M. Grossman and A.B. Krueger, "Economic Growth and the Environment," *Quarterly Journal of Economics*, Vol. 112 (1995), pp. 153–77.

20 R.E. Dunlap, George H. Gallup, Jr, and A.M. Gallup, "Of Global Concern: Results of the Health of the Planet Survey," *Environment*, Vol. 35, No. 9 (November 1993), pp. 7–39.

21 For an excellent discussion of the relationship between the demand for environmental quality and trade see P. Thompson and L.A. Strohm, "Trade and Environmental Quality: A Review of the Evidence," *Journal of Environment & Development*, Vol. 5, No. 4 (1996), pp. 363–83.

22 W. Ruttan, "Technology and the Environment," *American Journal of Agricultural Economics*, Vol. 53 (1971), pp. 707–17; I. Walter, "International Economic Repercussions of Environmental Policy: An Economist's Perspective" in S. Rubin (ed.) *Environment and Trade*, Totowa, NJ: Allanheld, Osmum, 1982. L.W. Milbrath and M.L. Goel provide an excellent typology of factors that affect participation in general. The categories are (1) variables in the larger environment, such as the social system, the political system, political culture, etc.; (2) variables in the immediate environment, i.e. stimuli in the microenvironment that can be specific to the type of participation and the type of policy in question; (3) "life position factors" including education, age, race, sex, etc.; (4) variables internal to an individual, including attitudes, beliefs, personality traits, etc. See L.W. Milbrath and M.L. Goel (1977) *Political Participation: How and Why Do People Get Involved in Politics?* Chicago: Rand McNally College Publishing, 1977, pp. 24–34.

23 Evidence from the western literature on participation suggests that economic levels are influential factors in determining participation levels; however, in some studies "personalized contacting" does not follow this pattern; Verba *et al.* (1978) *Participation and Political Equality*, p. 63. Also, other research points out "socioeconomic status and political activity show different levels of correlation from nation to nation"; Milbrath and Goel, *Political Participation*, p. 92.

24 S.P. Huntington and Joan Nelson, *No Easy Choice: Political Participation in Developing Countries*, Cambridge, MA: Harvard University Press, 1967, p. 132.

25 W. Alford *et al.*, "The Human Dimensions of Pollution Policy Implementation."

26 K. O'Brian, "Rightful Resistance," *World Politics*, Vol. 49, No. 1 (1996), pp. 31–55; E. Perry and M. Selden, *Chinese Society: Change, Conflict and Resistance*, 2nd edition, New York: Routledge, 2003.

27 Perry and Selden, *Chinese Society*, Introduction.

28 This measurement does not take drop-out rates into consideration. The author chose this measurement because there are too many gaps in data on illiteracy rates.

29 The nature of available data dictated that levels of industrial pollution needed to be used because the Chinese stopped reporting total levels of sulfur dioxide emissions in 1996 and only supplied industrial emissions.

30 The answers should help policy-makers to determine which pollution problems are the higher priority, those with the absolute highest pollution levels or those in more densely populated areas.

31 Complaints are counted in two different ways: by the number of people that complain (total complaints) and by the number of incidents people complain about (incidents). The first method would count the number of people who complained about

an event (*renci*), while the second method would count incidents (*pici*). However, the dual method of counting is only used for visits; it is not used for letters to the EPB, so the total number of incidents is unknown. This method of counting illustrates that officials only value counting the number of people complaining per incident when complaints involve visits to EPBs, and is related to officials' concern for social stability.

32 These figures include only complaints and suggestions people made to environmental protection offices, bureaus, and SEPA. It does not include environmental complaints sent to other ministries, or government, legislative, and Party offices.

33 Interviews 31, 45, and 49, Fall 1999 and Winter 2000.

34 Out of the 29 provinces on the combined top ten lists for H_2O pollution complaints and per capita H_2O complaints, 11 provinces (38 percent) are on both lists. These are: Jiangsu, Guangdong, Shandong, Zhejiang, Sichuan, Hebei, Hunan, Guangxi, Jiangxi, Fujian, and Shanghai. Out of the 32 provinces on the combined top ten lists for air pollution complaints and per capita air pollution complaints, ten provinces (31.25 percent) are on both lists: Guangdong, Shandong, Jiangsu, Sichuan, Liaoning, Zhejiang, Jilin, Heilongjiang, Fujian, and Shanxi. The provinces that have the highest overall and per capita numbers for both air and water complaints are: Guangdong, Shandong, Jiangsu, Zhejiang, Sichuan, and Fujian.

35 Other provinces with large shifts in complaints include Hunan, Yunnan, Ningxia, Shaanxi, Qinghai, and Hunan.

36 For example, in Japan, environmental protests were nearly non-existent until 1965 but peaked in 1973 with over 125,000 protests before dramatically dropping to less than 30,000 by 1974. Jeffrey Broadbent, *Environmental Politics in Japan*, p. 102. The changes in the numbers of complaints at the national level is more gradual, but complaints at the local level reveal an erratic pattern; in one area they jumped from around 220 in 1972 to 280 in 1973, and 250 in 1987 to 320 in 1988. Broadbent, *Environmental Politics in Japan*, p. 327.

37 For dramatic stories about data inflation during the Mao period see Shapiro, *Mao's War Against Nature* and *Mao's China and After: A History of the People's Republic*, 3rd edition, New York: Simon and Schuster Free Press, 1999. In November of 2006 a letter came to public light that told the story of how hundreds of people participated in a county-government-sponsored "10,000 strong Diao River pollution management campaign," which in reality was a big show for the media of people moving rocks and sand from one side of the river to the other so authorities could boast about how they had cleaned up the Diao River. See J. Martinsen, "A Fairy Tale in Guangxi," Danwei Website, at www.danwei.org/state_media/a_fairy_tale_in_guangxi.php.

38 Interview 45, Spring 2000.

39 Interview with former health and human services statistician, Winter 2007.

40 It is not worthwhile to disregard Chinese data based on the assumption that it is unreliable. It is only through examining and testing it that scholars can know for sure how reliable the data really is. It is only after specific contradictions or irregularities are isolated that these data issues can be brought up with Chinese authorities and scholars, so that over time data-reporting can improve. For example, the data on accidents is incomplete because lower-level EPBs do not report statistics. Knowing this, citizens then have a basis for demanding data accountability.

41 If an EPB does not report complete statistics to the next highest level its budget allocation can be reduced. Interview 21, Fall 1999.

42 It is unclear exactly when the number of accidents reached a peak because the data for 1986 through 1989 for each province is not publicly available.

43 There are several categories of accidents: common, large, severe, and extremely severe. Common environmental accidents include those where there are economic losses of over 1,000 but less than 10,000 Yuan. Large environmental accidents are those that include at least one of the following: (1) there is economic damage of over

10,000 but lest than 50,000 Yuan; (2) people are poisoned; (3) the accident causes a dispute between the enterprise and the people; (4) the surrounding environment is harmed. Severe environmental accidents are those that include any of the following: (1) there are economic losses of over 50,000 but less than 100,000 Yuan; (2) people find evidence of poisoning, radiation injuries, or anything else that could lead to deformities or handicaps; (3) people are poisoned; (4) environmental pollution threatens the peace and stability of society; and (5) there is significant environmental damage. Extremely severe environmental accidents are those that include any of the following: (1) there are economic losses of over 100,000 Yuan; (2) there is obvious evidence of poisoning or radiation exposure; (3) someone dies as a result of the accident; (4) the economic or social life of the locale is severely affected; and (5) the environment is seriously harmed (*Baogao Huanjing Wuran Yu Pohuai Shigu de Zhanxing Banfa* (Provisional Regulation regarding Notification of Environmental Pollution and Accidents), National Environmental Protection Administration, 10 September 1987.)

44 *Zhongguo Huanjing Nianjian, 1990* [China Environment Yearbook, 1990], Beijing: China Environment Yearbook Press, 1991, p. 136. Of note is that the total amount paid out to citizens in the form of compensation for damages sustained in environmental accidents annually decreased dramatically. In 1990, the state recorded payments of 97,430,000 Yuan in compensation to victims of pollution, while in 2001 only 29,487,000 Yuan was paid out. This is partially as a result of the fewer severe (*zhongda*) and extremely severe (*teda*) accidents, as well as fewer less damaging accidents in later years. However, it is likely that a lower number of accidents is not the only explanation; this is one important issue for future research.

45 These provinces are: Shanxi, Liaoning, Zhejiang, Fujian, Guangxi, Hainan, Yunnan, Gansu, and Ningxia.

46 Prior to 1996, the yearbook listed a zero if there were no accidents in a province and a dash if there was no information. However, after 1996 (the 1997 yearbook) both notations are scrapped and instead there is simply a blank line.

47 See E. Economy, *The River Runs Black*, Ithaca, NY: Cornell University Press, 2004, Chapter 1, for detailed information about the numerous disasters along the Huai River. The accidents along the Hui River politicized accidents, which could be another reason why data reporting changed.

48 As mentioned earlier in this chapter, authorities are sometimes given guidance about which direction data trends should go. Reasons for this tinkering could include the need to satisfy higher-level authorities' expectations, to illustrate that certain policies are working or not working, to acquire a higher budget allocation, or to further an individual's career.

49 The author has searched for more information on the accidents in Inner Mongolia, but so far little concrete information is available. It is unlikely that the reference is a typographical error because of the enormous economic damages associated with the "other" accident in Inner Mongolia that year. The injured could refer to people affected by a major dust storm that year. In the narrative section on environmental accidents in the yearbook, some information is given on just such a storm. However, the yearbook lists economic losses of 56 million Yuan associated with that storm. This total is inconsistent with the amount of losses due to accidents recorded in the statistical section of the yearbook, which lists only 25 million Yuan in damages. In addition, the narrative section does not indicate that the storm inflicted injuries to people. If the statistics do refer to dust storms then it would seem odd that the data on the huge dust storm of 1998 did not lead to an equally large number of injured. The relevant yearbook does list enormous economic losses, however, due to six accidents in the "other" category for 1998 in Inner Mongolia, which could be associated with large dust storms. The world disaster databases record no significant accidents and the Inner Mongolian newspapers do not reveal any large-scale accident that year.

50 Before the author conducted regressions, she graphed the dependent variables (and air and water pollution complaints) and the independent variables (levels of industrial effluent and industrial SO_2, per capital GDP, and the measure for general education) to determine the curve fit. After examining the variables, the author transformed the variables by taking their logs. This made the variables more normally distributed, clarified the shape of their relationships, and decreased the variance in the error terms. Scatter plots of each independent variable with the dependent variables displayed linear relationships, so the author used an OLS linear regression model. In addition, the author added dummy variables for each year (minus one) to help with the inherent problem of autocorrelation in the time-series data. The author tested for multicoliniarity among the independent variables and found that multicoliniarity was not a problem. In all cases the tolerance and variance inflation factors (VIF) for each regression of the independent variables with each other between 1.0 and 3.8. In addition, none of the auxiliary R^2 values (each independent variable regressed with each dependent variable) exceeded the overall R^2 of the multivariate regression. The author also calculated the Durbin–Watson statistic (*d*) to test for autocorrelation in the multivariate equations and found that in both the air and water pollution cases the *d* statistic indicated no significant autocorrelation existed.

51 After 2004, anecdotal evidence suggests authorities became less tolerant of some forms of participation, but as of early 2007 complaints continued to rise.

52 For a discussion of the general complaint system in China see L.M. Luehrmann, "Facing Citizen Complaints in China, 1951–1996," *Asian Survey*, Vol. XLIII, No. 5 (September/October 2003), pp. 845–66; Y. Cai, "Managed Political Participation in China," *Political Science Quarterly*, Vol. 119, No. 3, (Fall 2004), pp. 425–51; C.F. Minzner, "Xinfang: An Alternative to Formal Chinese Legal Institutions," pp. 103–47; Human Rights Watch, "The Petitioning System: A Report from Human Rights Watch," December 2005; online, available at http://hrw.org/reports/2005/china1205/4.htm; E. Michelson, "Justice from Above or Justice from Below? Popular Strategies for Resolving Grievances in Rural China," *The China Quarterly*, Vol. 192 (Winter 2007); E. Michelson, "Climbing the Dispute Pagoda: Grievances and Appeals to the Official Justice System in Rural China," forthcoming in - *American Sociological Review* 2007 (pre-copy edited version available at www.indiana.edu/~emsoc/Publications/Michelson_DisputePagoda.pdf;); Gongmeng, "Report on Appeals in China" [*Zhongguo Xinfang Baogao*], online, available at www.gongmeng.cn (accessed 29 October 2006); T. Shi, *Political Participation in Beijing*; K. O'Brien and L. Li, "The Politics of Lodging Complaints in Rural China," *The China Quarterly*, Vol. 143 (1995), pp. 756–83; L. Li and K.J. O'Brien, "Villagers and Popular Resistance in Contemporary China," *Modern China*, Vol. 22, No. 1 (1996), pp. 28–61; K. O'Brien, "Collective Action in the Chinese Countryside," *The China Journal*, Vol. 48 (2002), pp. 139–54; K. O'Brien, *Rightful Resistance: Contentious Politics in Rural China*, Cambridge: Cambridge University Press, 2006; I. Johnson, *Wild Grass: Three Stories of Change in Modern China*, New York: Pantheon, 2004.

53 According to the Chinese constitution of 1982, "Chinese citizens have the right to criticize and make suggestions to any state organ or functionary. Citizens have the right to make to relevant state organ complaints or charges against, or exposures of, any state organ or functionary for violation of the law or dereliction of duty; but fabrication or distortion of facts for purposes of libel or false incrimination is prohibited. Citizens who have suffered losses as a result of infringement of their civic rights by any state organ or functionary have the right to compensation in accordance with the law"; Article 41 Constitution of the People's Republic of China.

54 For an in-depth discussion of the development of the national complaint system between 1951 and 1996, see Luehrmann, "Facing Citizen Complaints," pp. 845–66.

An administrative management manual from the early 1990s clearly categorizes managing complaints as "professional work," designates which ministries are responsible for which type of complaints, and states that all Party, People's Congress, and governmental bodies at the provincial, municipal, autonomous region, city, prefecture, and county levels also were supposed to establish offices (or assign personnel) to manage complaints; S. Weiben, *People's Republic of China Administrative Management Encyclopedia,* Beijing: Renmin Ribao Chubanshe, 1992, pp. 256–66. The State Council Administrative Office (Guowuyuan Bangongting) is the body within the State Council that manages citizen complaints. Within this Office, there is a special department devoted to managing complaints called the Zhong Ban Guo Ban Xinfangju. It answers to the Chinese Communist Party Central Committee General (Administrative) Affairs Office (Zhongguo Gongchandang Zhongyang Weiyuanhui Bangongting or Zhong Gong Zhong Yang Bangongting) and the State Council General (Administrative) Affairs Office. It passes on important information about citizen contacts to the Party Central Committee and the leaders of the State Council. It takes care of any complaint affairs requested by the leaders of the Central Committee or leaders of the State Council. It offers guidance (*zhidao*) regarding the complaint work of ministries and locations. It is responsible for managing difficult inter-ministerial or cross-boundary complaints. This organization is then the "last stop" for complaints within the formal government and Party structures; *Zhongyang Jilu Jiancha Weiyuanhui*; and the Supreme People's Procuratorate (*Zuigao Renmin Jianchayuan). Guowuyuan Jigou Gaige Gailan* [General Reader on the Restructuring of the State Council], Beijing: China News Publishing, 1998, pp. 26–8. In the early 1990s, authorities, likely People's Congresses, conducted surveys in various places to better understand the conditions at the local level regarding citizen complaints and governmental complaint management processes. One such survey took place in Taiyuan in the early 1990s; *Taiyuan Nianjian, 1994* [Taiyuan Yearbook 1994], Taiyuan: Shanxi People's Press, Shanxi Renmin Chubanshe, 1995, p. 80. Early guidance to authorities for managing complaints could be found in an administrative management encyclopedia in the form of "guiding principles." These principles were dubbed the "three services" (*sange fuwu*; (1) *wei dang de zhongxin gongzou fuwu*, (2) *wei chunzhong fuwu*, (3) *wei lingdao gongzuo fuwu*); S. Weiben, *People's Republic of China Administrative Management Encyclopedia.*

55 Luehrmann, "Facing Citizen Complaints," p. 854.

56 The State Council regulation outlined the basic rights and responsibilities of citizens as well as governmental authorities responsible for managing complaints. See Minzner, "Xinfang: An Alternative to Formal Chinese Legal Institutions," pp. 133–7.

57 For example, in 1996, the NEPA complaint office was recognized by the central government for its "outstanding" complaint work and heralded as a model for other sectors *Zhongguo Huanjing Nianjian, 1996* [China Environment Yearbook, 1996], p. 252.

58 For information on development of China's environmental protection apparatus see A.R. Jahiel, "The Organizations of Environmental Protection in China," *The China Quarterly*, No. 156 (December 1998), pp. 81–103. Prior to these structural changes several locations kept records of the numbers of environmental complaints, starting in 1983, and some local EPBs had already drawn up their own guidelines for managing complaints; Interview 21, Fall 1999. According to available records, from 1983 to 1989, there were 6,108 complaints in Beijing, 22,913 in Shanghai, 13,504 in Tianjin, 456,339 in Sichuan province, 5,893 in Guangzhou, 4,453 in Chongqing, and 3,515 in Changchun; *Zhongguo Huanjing Nianjian, 1990* [China Environment Yearbook, 1990], Beijing: Zhongguo Huanjing Kexue Chubanshe, China Environmental Sciences Press, 1990, p. 216. On the other hand, many Environment Protection Offices did not keep environmental complaint data until they became independent from the Ministry of Construction and were renamed EPBs.

59 These include Henan and Heilongjiang provinces, and the cities of Shanghai, Chongqing, Qingdao, Zhangtaikuo, and Mapishan.

60 The administrative regulation outlines the bare framework of the system and stipulates the duties of EPBs at each administrative level. It stipulates that EPBs at the county level and above should set up a system for managing complaints or assign personnel to be responsible for managing complaints. EPBs are responsible for handling environmental complaints relevant to the scope of the EPB, or personnel should direct citizens to the appropriate authority. For example, some environmental complaints related to railways would be directed to the Ministry of Railways. In addition, EPBs must investigate cases passed on to them from higher-level EPBs. EPBs are required to cooperate with other governmental ministries to manage complaints that involve more than one agency.

61 In 1992, NEPA implemented the *Huanjing Xinfang Gongzuo Youguan Guiding* [Rules Regarding Environment Complaint Work] to strengthen and clarify the administrative regulation passed in 1990. In 1994, Chinese leaders used the Chinese Agenda 21 to declare their commitment "to improve the system for making and investigating (environmental) complaints, and expand the public's role in the enforcement of sustainable development laws so as to ensure that individuals, groups and organizations with legal standing have reliable channels for participating in the enforcement process to protect their legal rights and public interests." China's Agenda 21: White Paper on China's Population, Environment, and Development in the 21st Century (adopted at the 16th Executive Meeting of the State Council of the People's Republic of China), Beijing, China: Zhongguo Huanjing Kexue Chubanshe, 1994, p. 20. Finally, the Ninth Five-year Plan called for the development of a system of public participation in the environment sector by the year 2000. It reads "in accordance with the whole process of developing the democratic and legal system, [we will] gradually establish a system for public participation in environmental protection. Implement a trial system holding public hearings in environmental impact assessment of construction projects. Give due attention to letters and visits of the public, provide timely adjudication of pollution incidents and disputes to safeguard citizens' environmental rights and interests"; *The National Ninth Five-Year Plan and the Long-Term Targets for the Year 2010 for Environmental Protection*, Beijing, China: China Environmental Science Press, 1997, p. 70.

62 For example, the Chengdu EPB had not submitted its local *xinfang guiding* (complaint regulations) to the Sichuan EPB for review by 2000. Interview 36, Spring 2000. Interview, 45, Spring 2000.

63 For a discussion about how the complaint system replaces the regular legal system see C.F. Minzner, "Xinfang: An Alternative to Formal Chinese Legal Institutions."

64 The 1997 revision goes further and instructs environmental protection officials to tell the petitioner to abide by relevant laws and regulations related to the court or organ that is managing their grievance.

65 Of note, the 1997 and the 2006 measures require EPBs to supply the equipment and budget for EPB officials to fulfill their complaint management responsibilities (1997 measure, provision 12).

66 However, county (district) EPBs were still responsible for investigating and managing or assisting higher-level EPB personnel to investigate and mange complaints within the county's jurisdiction.

67 The 2006 measure also introduces other new practices. The 2006 measure is the first time that EPBs are required to publicize their address and contact information as well as establish a "complaint day" system. It is also the first time they are required to post laws, regulations, and policies relevant to complaint work on their websites or make them available to visitors (2006 measure, provision 11).

68 Given the new Environmental Protection Federation has been involved in dispute

resolution, it will be interesting to note how this organization cooperates with EPBs in the future regarding complaint work.

69 According to the 1991 and 1997 measures, complaint personnel had to investigate complaints in earnest and "discover" the truth. They had to be hospitable and do their best to resolve complaints, especially mass petitions.

70 For example, beginning in 1997, two or more investigators were required to check out a complaint, and if the enterprise was in violation of an environmental regulation, then the investigators must contact the Policy and Law Division at the EPB to exact fines against the enterprise. In addition, enterprises no longer pay fines directly to EPB officials. They must send such payments directly to the EPB's bank account. Interview 47, Spring 2000.

71 Interview 21, Fall 1999.

72 In Sichuan province, the EPB is struggling to get legislation passed that would enable them to require "lax" lower-level EPBs to participate in special training about managing environmental complaints. Interview 31, Spring 2000.

73 Citizens cannot pester, insult, hit, or threaten the environmental protection personnel that are trying to serve them. Nor can they fabricate stories, lie, falsely accuse, or frame someone.

74 The other authorities to which people often direct complaints include local Party organizations (residents' committees, unit representatives, or the relevant level of the Party secretary's office); community leaders; local government; local People's Congress members; local People's Consultative Congress members; unit leaders; the polluting enterprise or the enterprise's supervisory agency; and newspapers, television, or other media organizations.

75 Interview 21, Fall 1999.

76 Procedures for managing complaints are similar across all sectors of society and are clearly laid out in an administrative management manual from the early 1990s. Complaint work procedures follow a particular order as follows: (1) *junbeijieduan* (preparation stage); (2) *yuexin* (reading period); and (3) *banli guocheng* (management process). Similar procedures are outlined for managing in person complaints.

77 Interview, Beijing, 1999.

78 Hotlines are advertised in newspapers and on television and radio. While such hotlines do not guarantee a citizen will see his/her complaint resolved, it does improve access and political opportunities. There is still some inequality of access for rural residents as shown by less enthusiastic promotion of pollution hotlines in outer districts and counties.

79 "Chinese Government Receives 1.148 Million Complaints on Environmental Pollution since 2003," *Xinhua*, 4 June 2006; online, available at http://english.sina.com/1/2006/0604/79614.html.

80 Interview 45, Spring 2000.

81 *Huanjing Baohu Xinfang Guanli Banfa* (Regulation Regarding the Management of Environmental Protection Complaints, *Huanjing Baohu Xinfang Guanli Banfa* (Regulation Regarding the Management of Environmental Protection Complaints),

82 Despite the difficulties associated with taking one's complaint to higher administrative levels, petitioners often continue their quest for justice. "Only 5.8 percent responded that they would give up petitioning. Over 91 percent responded that they would never give up even if they saw no results. The survey also shows that experienced petitioners often become organizers: 85.5 percent responded that they would begin 'publicizing [*xuanchuan*] government policies and laws to move/inspire [*fadong*] the masses to protect their own rights,' 68.2 percent responded that they would set up an organization to legally protect the rights of farmers [*chengli zuzhi, yifa weihu nongmin de hefa quanyi*], and 70.2 percent responded that they would 'organize the masses to open a dialogue and speak directly to the government' [*zhijie zhao zhengfu duihua, tanpan*]. The survey also pointed out that many

petitioners are willing to go beyond organizing: 53.6 percent responded that they would 'do something to scare the cadres a little' and 87.3 percent responded that they were in a life or death struggle with corrupt cadres"; Human Rights Watch Report "We Could Disappear: Retaliation and Abuses Against Chinese Petitioners," at http://hrw.org/reports/2005/china1205/4.htm.

83 In August of 2003, the Central Office of the CCP convened in conjunction with the State Council Office a national meeting to study and resolve the problem of mass petitions and the influx of petitioners to Beijing. Unfortunately, the efforts to stifle the flow of petitioners coming to Beijing could be related to the 2008 Olympics and the fear that mass numbers of petitioners will descend upon the city, thereby smashing the illusion Beijing creates for visitors.

84 Environmental protection officials began experiments with the "*Huanbao Zhifa Heyiting*" ("Environmental Protection Enforcement Collegiate Bench") for the purpose "to manage prominent problems and to utilize the pollution permit regulations to help with complaint problems." The Nanjing, Taizhun District People's Court, established the first collegiate bench as an experimental channel to resolve environmental complaints. Collegiate benches are typically formal bodies established by the court to accept cases; however, this particular arrangement seems more like an adjudication committee or informal dispute resolution panel in the CCP mediation committee tradition. Officials might not count complaints managed by these new collegiate benches in their annual totals. *Zhongguo Huanjing Nianjian, 2005* [China Environment Yearbook, 2005], Beijing: Zhongguo Huanjing Kexue Chubanshe, 2005, p. 323, and pictures at the beginning of the book.

85 Interview 49, Spring 2000.

86 *Zhongguo Huanjing Nianjian, 1991* [China Environment Yearbook, 1991], Beijing: Zhongguo Huanjing Keshue Chubanshe, 1991, p. 216.

87 *Zhongguo Huanjing Nianjian, 1998*, p. 149.

88 One example of this was in Shanghai. See Mara K. Warwick, "Environmental Information Collection and Enforcement at Small-Scale Enterprises in Shanghai: The Role of the Bureaucracy, Legislatures and Citizens," PhD dissertation, Stanford University, 2003, p. 247.

89 Other research has found that as environmental protection officials undertook campaigns to resolve specific pollution problems, citizens became more aware of those types of problems and were more apt to complain when they saw a similar problem. Warwick, "Environmental Information Collection," p. 233.

90 Frank N. Perry and Fons Lamboo, *Inventory of the Collection of the Chinese People's Movement, Spring 1989. Volume 1: Documents at the International Institute of Social History*, Amsterdam: Stichting Beheer IISG, 1990, Introduction.

91 While SEPA was still half a step below other ministries, it still had greater authority than before.

92 Interviews, 30 and 45, Spring 2000.

93 See Minzner "Xinfang: An Alternative to Formal Chinese Legal Institutions," p. 1.

94 For example, out of 111 people who were concerned about pollution problems in Anqing, Anhui, 37 believed taking action to resolve the problem would be useless; Alford *et al.* "The Human Dimensions of Pollution Policy Implementation: Air Quality in Rural China," p. 507. "Of the forty-nine petitioners who were interviewed in person or whose cases were documented by Human Rights Watch, eight had received a letter from a national office directing local officials to take care of their cases. Only one of those had been able to resolve his case as a result of the letter, and then only after he and a group of his neighbors physically threatened the local official with violence if he did not resolve the case. One petitioner described the response he got from a local official in Shanxi: I gave them the letter, and the official laughed at me and threw it right into the garbage. I said, 'How can you throw that in the garbage?' He said, 'What does this have to do with us? We don't care'"; Human

Rights Watch Report, "We Could Disappear: Retaliation and Abuses Against Chinese Petitioners" at http://hrw.org/reports/2005/china1205/4.htm (accessed October 2006).

95 See Human Rights Watch Report, "We Could Disappear".

96 Minzner, "Xinfang: An Alternative to Formal Chinese Legal Institutions," p. 105.

97 Interview 21, Fall 1999.

98 Interview 21, Fall 1999. If officials have became more responsive to complaints, then it is logical to infer that the numbers of complaints will increase, because if citizens have more hope that they will find redress, they are more likely to utilize the complaint system. Assessing citizen satisfaction with the environmental complaint system on a widespread scale is an area ripe for future research.

99 Interviews 45, Fall 1999; 15, Spring 2000; and 30, Spring 2000.

100 H. Wang, N. Mamingi, B. Laplante, and S. Dasgupta, "Incomplete Enforcement of Pollution Regulation: Bargaining Power of Chinese Factories," Policy Research Working Paper, Washington, DC; World Bank, Development Research Group, Infrastructure and Environment, 2002. Other evidence that citizens have had some impact on environmental outcomes can be found in the following resources. In Guangzhou as early as 1991, residents' complaints regarding street food stalls were addressed by the local government through stricter regulations. Citizen complaints regarding noise and air pollution were also addressed through strengthened enforcement efforts. See C. Lo, and S.L. Wing. "Environmental Agency and Public Opinion in Guangzhou: The Limits of a Popular Approach to Environmental Governance," *The China Quarterly*, No. 163 (September 2000), p. 699. Public support appears to promote enforcement efforts in Guangzhou and Chengdu. See C.W.H. Lo and G. E. Fryxell, "Enforcement Styles Among Environmental Protection Officials in China, *Journal of Public Policy*, Vol. 23, No. 1 (2003), pp. 81–115. However, other evidence suggests that even if the public is pushing for a cleaner environment, attitudes of supervising governmental officials color environmental authorities' approach to enforcement. C.W.H. Lo and G.E. Fryxell, "Effective Regulations with Little Effect? The Antecedents of the Perceptions of Environmental Officials on Enforcement Effectiveness in China," *Environmental Management*, Vol. 38, No. 3 (2006) pp. 388–410; and C.W.H. Lo and G.E. Fryxell, "Governmental and Societal Support for Environmental Enforcement in China: An Empirical Study in Guangzhou," *Journal of Development Studies*, Vol. 41, No. 4 (2005), pp. 558–88, p. 582. Also see M. Warwick, "Environmental Information Collection".

101 While NGOs such as the Center for Legal Assistance to Victims of Pollution provide legal assistance to victims of pollution that want to file court cases, the center does not organize or collaborate with victims' groups to build an environmental movement.

102 Authorities sentenced five people – Huang Jin, Mo Zhensheng, Mo Zhenning, T'an Heshan, and Xu Yugao – to prison terms in early December 2006 for organizing a peaceful sit-in outside the construction zone of a manganese plant to be built by the partially state-owned CITIC Dameng Mining Industries in Daxin County, Guangxi province. The sit-in occurred because company, Lei Township, and Daxin County officials were unresponsive to citizen complaints about the proposed construction of yet another manganese plant in the area, which had already become heavily polluted since Daxin Manganese Co. dumped waste directly into the Heishui River, the only water source for local residents; Human Rights in China Press Release, 4 December 2006, online, available at www.hrichina.org/public/index.

103 For a more in-depth discussion of this assertion see Brettell "China's Pollution Challenge: The Impact of Economic Growth and Environmental Complaints on Environmental and Social Outcomes" (forthcoming 2007).

104 Authorities recognize some of the problems leading to on-going complaints or to petitioners taking their complaint to higher levels including issues with the "working

style" of environmental protection officials that do not respond to citizen complaints in a timely manner, that do not respond at all, that protect local enterprises because the official has a stake in the enterprise, that accept bribes, that do not provide petitioners with monitoring statistics or provide inaccurate statistics about pollution from an enterprise, and that approve an environmental impact assessment for a project that is not in compliance with regulations; *Zhongguo Huanjing Nianjian, 2005* [China Environment Yearbook, 2005], Beijing: Zhongguo Huanjing Nianjianshe, 2005, p. 321.

105 M. Warwick, "Environmental Information," p. 118.

106 *Zhongguo Huanjing Nianjian, 1998* [China Environmental Yearbook, 1998], Beijing: Zhongguo Huanjing Nianjian Chubanshe, 1998, pp. 148–152.

107 Interview 47, Spring 2000.

108 For more information on these cases and a discussion about the effects that brutality and the suppression of petitioners has on social stability and environmental quality, see Brettell, "China's Pollution Challenge".

109 In 1998, SEPA received over 1,200 complaint letters and 110 visits directly. *Zhongguo Huanjing Nianjian 1999* [China Environment Yearbook 1999], Beijing: China Environment Yearbook Press, 1999, p. 91.

110 Interview 7, Fall 1999. This official related one story about a family living in a farming village in another province, downstream from a ceramics factory. The family had a fishpond that became polluted by the wastewater from the factory. The daughter loved to eat fish, so she ate it quite often. She was the first person in the family to get sick. The son later got sick. The little girl lost all the strength in her limbs. The mother thought it might have something to do with the wastewater getting into the pond. The doctor at the local hospital even made some kind of certification to that fact but the mother's complaints to the local EPB had no impact, supposedly because the ceramics factory owner had "gotten to them." The factory owner also "took revenge" on the family for complaining and beat up the mother. Eventually the daughter died. Finally, the woman came to Beijing to complain to SEPA. Nothing was done. The mother lived on the streets of Beijing for some time. Later, she got sick and returned home. The woman and her husband remained sick and the son eventually died.

111 With Hu Jintao in power, in the run-up to the 2008 Olympics, and in the aftermath of the "color revolutions" in Eastern and Central European countries, petitioning in China has regained a political hue.

112 While researchers have examined why the democracy movement failed after Tiananmen, little research has been undertaken to examine the long-term *policy* changes that took place as a result of Tiananmen.

113 *Zhongguo Huanjing Nianjian 2000* [China Environment Yearbook 2000], Beijing: China Environment Yearbook Press, 2000, p. 347.

114 Koon-Kwai Wong, "The Environmental Awareness of University Students in Beijing, China," *Journal of Contemporary China*, Vol. 12, No. 36 (August 2003), p. 532.

115 See A. Brettell, "The Politics of Public Participation and the Emergence of Environmental Proto-Movements in China," PhD dissertation, University of Maryland, 2003, pp. 415–18.

7 Not against the state, just protecting residents' interests

An urban movement in a Shanghai neighborhood[1]

Jiangang Zhu and Peter Ho

Introduction

In April 1999, a construction team moved into Shanghai's Green Garden New Village neighborhood park and prepared for the construction of a new building, a community center for the elderly, right in the park. Many residents objected, and after a long period of struggle and negotiation, the Street Council, the lowest level of local government, and an alliance between 12 home-owners' committees (*yezhu weiyuanhui*) finally reached an agreement that the government could build the center in the park under the condition that the size would be less than 650 sq m. The residents stood by and silently watched the work of the laborers. It seemed that a lingering conflict between the government and the residents would finally be resolved. However, just two days later, the residents found that the groundwork for the building was substantially larger than 650 sq m. People quickly spread the news that this center was not for the elderly in the community but for retired government cadres. The anger of the people boiled over. A certain Shen Xin, the leader of the community action, decided to take direct action after the Street Council had turned away his queries, and had sped up the construction in order to make it irreversible. At 20:00 on 25 May, Shen led about 100 residents to congregate outside the construction site to wait for the government representative for negotiations. However, by 21:00, no one had come. Shen swept his hand towards the construction wall barrier and said: "If you want to pull it down, just do it. If an accident happens, I will be responsible." So people moved into the construction site and pulled down the wall. Shen and his comrades had carefully calculated this action. He had discussed it with a lawyer and some of his allies in government circles. They suggested pulling down the wall, but refraining from destroying any machinery. The activists followed this advice closely, with the result that even some alarmed policemen who had entered the scene had nothing to say when they witnessed that the action was done in an orderly fashion. Shen hoped to make local government officials afraid of the residents' potential "violence" so that they would halt the construction activities – with success. The next day, the Street Council stopped the construction once more.

The account above is just one small episode from the last nine years of a residents' movement in a neighborhood in Shanghai. The contestation over community space in China is closely intertwined with issues of urban renewal, housing commercialization, and subaltern resistance. Since the 1980s, the implementation of the post-Mao open policy and reforms has initiated a rapid process of urban renewal. As a result, the relations between urban residents, the state, and real-estate developers have increasingly been in conflict. Faced with a wide range of growing problems such as pollution, forced resettlement, and land confiscation, residents not only complain to the local government but also get organized and take collective action to defend their residential rights.[2] Many protests and actions have taken place during the process of urban renewal, but few come to the attention of outsiders. Some are suppressed, while others succeed. The media and official discourse often represent these movements and actions as "social problems" or "conflicts among the masses" (*shequn maodun*). Since they appear non-political, they are distinguished from political protests and movements and do not get attention from the mainstream media and the upper levels of government.[3] Yet for the local government and for the residents themselves, this friction often develops into serious conflict, sometimes even into violent struggle. In the official discourse, the activists are called "troublesome people" (*diao min*) and these resistance groups and organizations are labeled "illegal organizations."

How is it that these residential movements take place nonetheless? Why do these movements and actions persist even when they lack resources and support from civil society? And, how are these collective actions possible in a semi-authoritarian setting (see also the introduction of this volume), given that they directly confront the state's local agents? These questions lead us to examine more closely the dynamics and strategies of collective action in urban grassroots China.

These movements resemble the new grassroots urban movement that sprang up in the West in opposition to top-down urban planning. As Castells argued, they were the impetus for the lower class to alter the unequal urban structure (Castells 1977, 1983).[4] In these movements, residents and non-governmental organizations (NGOs) are opposed to capitalist developers and to technocratic government intervention. Their conflict and contestation produce an "urban question" by which the resources related to urban space can be accessed through competition[5] (cf. Castells 1977, 1983; Mollenkopf 1983; Pickvance 1975; Lowe 1986; Pahl 1969; Rex 1968). The capacity to change policy depends not only on social movements, but also on the strength of local authorities, and the effectiveness of non-governmental organizations (Pickvance 1975).[6] The neighborhood movements in urban China are obviously related to the urban question, including land confiscation, forced resettlements, and environmental degradation. However, the Chinese context is markedly different from the European and North America cases described by urban sociologists. Most movement participants do not adopt a disruptive strategy against developers and local government because they cannot obtain a strong support from NGOs and civil society. On the contrary, in this study we argue that residential activists employ non-disruptive strategies and mobilize resources from "traditional" social ties, as well as from the state, which makes such collective action assume a different form from those in a Western context. This critical difference in movement dynamics is what in this volume is termed "embedded environmentalism." In this sense, environmental and residential activism in China is "not a social activity with a fair degree of autonomy and self-regulation, but is actually enmeshed in a web of interpersonal relations, formal rules, and shared ideas" (see the introduction to this volume).

The formation of social movements has been debated by new social movement theory[7] and Resource Mobilization theory scholars for a long time. Recently, the construction of collective identity as conceived by new social movement theorists and political opportunity and mobilizing strategy as emphasized in resource mobilization theory have been integrated. From this comprehensive perspective, social movements are understood as constructed socially in an ongoing process of cultural discourse and interaction[8]. Opportunity structure, action strategy, and cultural framing are new conceptual tools for research on social movements[9]. This conceptual framework is also useful for the study on resistance and conflict in contemporary China[10] (Perry and Selden 2000).

In this contribution, we will follow this model by highlighting the opportunities, strategies, and the discourse produced by the different levels of government, the media, and the movement participants themselves. First, we will provide an ethnography of a residential movement in a Shanghai neighborhood and analyze why this movement occurred, how it was able to sustain itself, and how it progressed and evolved. The ethnography will be done from the perspective of Shen Xin, the key-person in the movement. Second, and more importantly, we will demonstrate that the transition of state hegemony and the construction of a civil rights awareness brought about by legal reforms, privatization of housing, and a growing ecological discourse, led to the formation of an embedded, grassroots social movement in urban China.[11] Finally, we will show how this kind of movement is transforming ideas of citizenship in urban China.[12] This article consists of four sections. In the first section, we will provide a brief account of Green Garden New Village's (Lüyuan Xincun)[13] spatial and social organization. Then, we will describe the origins and development of collective action in Green Garden, as well as the perceptions of this collective action by the various stakeholders. In the third section the identity construction and negotiations in the neighborhood, and its relation to state transformation, will be briefly examined. Finally, it is argued that this type of movement can be an agent of social change in urban China because it reconstructs the urbanites' identity in ways that will bear significantly on the relationship between the neighborhood community and the state.[14]

Urban life in Shanghai: present and past

Green Garden New Village is a new neighborhood situated in the Pudong New District in Shanghai. During urban renewal at the beginning of the 1990s, the

Shanghai municipal government built 12 new residential communities to resettle citizens from the old town. Green Garden was one of them. At the beginning of the 1990s, in response to an increasing Chinese ecological discourse (see Calhoun and Yang in this volume) and a rapidly rising pollution, Shanghai redesigned its urban development strategy with a shifting focus from solely promoting economic growth to one that emphasized a "harmonious development of human and nature." Therefore, natural greenery was given a prominent place in the spatial planning for the Green Garden neighborhood. According to the standard of the "new village" at that time, greenery had to account for 10.16 percent of the land area, which means 1.28 sq m of greenery for every resident. When walking through the gate of the walled neighborhood, the amount of greenery is directly noticeable. Among the trees and lawns, there are 12 apartment blocks, each one over 20 stories high. Around 10,000 people live in these high-rising apartment buildings. The buildings surround a huge park, covering 8,000 sq m, which is considered to be the most scenic place in the neighborhood. The park was an important reason why most residents were willing to buy an apartment when the state since 1995 encouraged tenants to buy public housing at under-market prices.[15]

Before the resettlement, most residents in Green Garden lived in traditional "*lilong*" neighborhoods (a characteristic, traditional residential community in Shanghai comparable to the Beijing *hutong*)[16] where most houses are old, decrepit and sometimes dangerous. Therefore, for them, moving from a *lilong* to a "new village" meant a significant improvement of their life. On the negative side, however, resettled residents also need to rebuild their neighborhood life in a new, and unknown suburban area.[17] For a considerable time, resettled residents lacked a neighborhood identity. One person stated that he didn't know others in the same building and only kept in contact with old neighbors who have also moved to the "new village." But in the north of the "new village," a majority of the residents in the 23-story buildings are "collectively resettled" from a *lilong* at the Nanjing East Road. That is, they originated from the same *lilong* and currently live near their former neighbors. The face-to-face interactions and familiar relationships are still maintained in the new location. In fact, such "traditional" ties have been instrumental in mobilizing networks.

Following the Singapore model, Shanghai enacted a law in 1997 prescribing that owners should have a home-owners' committee to represent the home-owners in negotiations with real-estate management companies and developers. The committee is a legally recognized organization. Each of the 12 buildings in Green Garden has had a committee since 1997. At the beginning, people were not active in participating in the committee since they thought it was just a tool for real-estate developers and could not represent the interests of the apartment owners. However, as an increasing number of people bought their apartments and conflicts between residents and developers arose, the committees gained in importance.

Before the founding of the home-owners' committees in the late 1990s, there were only the Residents' Committees (*jumin weiyuanhui*) to administer the neighborhood (*shequ*). The Residents' Committee is an institutional heritage of China's socialist past, during which urban life was strictly regulated and controlled by the state.[18] Although the Residents' Committee is officially defined as an autonomous organization of the local residents, it is accountable to the Street Council or Street Bureau (*jiedao weiyuanhui* or *jiedao banshichu*), the lowest level of the district government. Comparable to the embedded structure of Green Students' Associations (see the first contribution in this volume), the Residents' Committee overlaps with the local branches of the Chinese Communist Party: in other words, the members of the Residents' Committee are concurrent member of local party cells.[19] In Maoist China, the Residents' Committee used to be the sole representative of citizens in a neighborhood, but with the establishment of the home-owners' committees an alternative representative institution emerged. As we shall see below, in the movement to protect the park the home-owners' committee gained considerable autonomy and leverage, and became a powerful, alternative representative of the neighborhood. In fact, the movement's main institutional basis of support proved to be an alliance of 12 home-owners' committees, pitted against the Street Council.

As Figure 7.1 shows, the neighborhood is constituted of four kinds of organizations: (1) the Residents' Committee; (2) the Home Owners' Committee; (3) the real-estate management company; and (4) informal organizations organized by the residents. The Residents' Committee, under the Street Council, is the most powerful organization on paper, as it is the official representative of residents. With the commercialization of housing, the office of the real-estate management company has emerged to run businesses related to the real estate such as cleaning, security and greenery protection, most of which were run by the governmental agencies before the reforms. Home-owners' committees are even

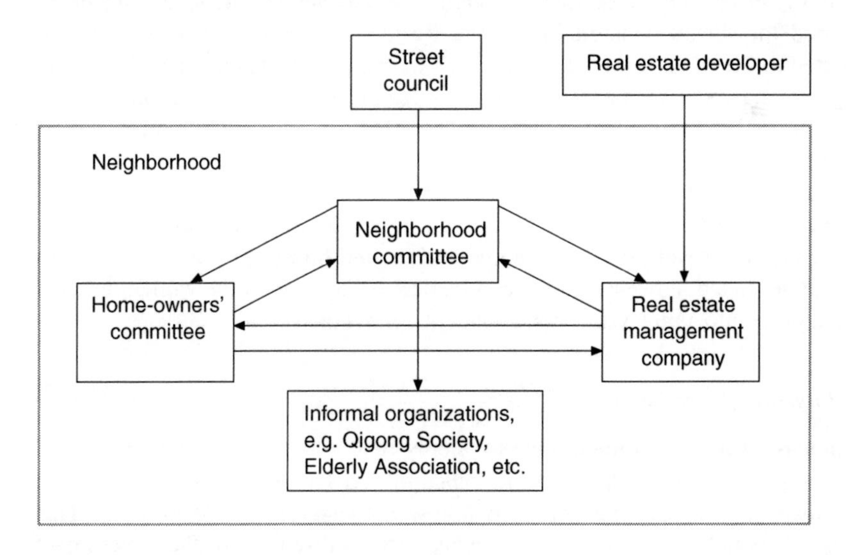

Figure 7.1 Organization structure of the neighborhood.

newer, as we will see. These three organizations do not have a hierarchical relationship with each other but interact horizontally with each other (Read 2000).[20] They are described as the "troika in a community," which controls formal power in the neighborhood. Many residents keep their distance from these organizations, and people find that some employees in these organizations are completely indifferent to community affairs. In response, residents in Green Garden organized their own informal associations, such as the *Qigong* Practice Society and the Elderly Association, to enjoy a community life. Most of these informal organizations are officially headed by the Residents' Committee, but its control is rather slack and these informal associations often act with a fair deal of autonomy.[21]

History of the movement: fighting with the developer

Finding a charismatic leader

Before the conflict with the Street Council erupted in 1999, the residents had already resisted several attempts to confiscate the park. The most serious conflict was with a real-estate developer who planned to build a new building in the middle of the park. The developer obtained permission by purposely using an outdated planning map. One day in 1993, shortly after the residents moved in, the developer began construction. Some older women, who were retired but still active in informal organizations in the neighborhood, discovered the construction work when they met for their daily physical exercises in the park. Fearing that the new building would obstruct sunlight, negatively affect the ventilation of their buildings, and destroy the scenery, the women mobilized others. A group of 39 residents mobilized through informal neighborhood organizations reported the illegal construction to the Residents' Committee, which in turn reported to the Street Council, after which the real-estate development company was notified. After repeated complaint letters did not yield any response from the company, the group of women sought to find someone who could lead their resistance against the developer. From their old neighborhood life, they knew of a resident, Shen Xin, who had a good reputation in dealing with neighborhood conflicts informally, a type of person known in Shanghai as an "uncle" (*niangjiu*). They went to Shen's home and invited him to join them. Shen Xin initially hesitated, however, in the end he decided to support the women in what would grow into a nine years' long residents' movement.

Legitimizing the resistance

Shen was aware it was impossible to simply rush into the construction site and halt the project. He needed some rationale or "reason" (*shuofa*) as he put it to legitimize his actions and convince fellow activists and his opponents. This "reason" should not go completely against official discourse or the government would disallow it and suppress their actions. At the same time, this reason should defend their rights and interests unambiguously. He made the interesting move to draw upon environmentalism as the discourse of the action, and called on the residents to protect their environmental rights. This turned out to be a clever strategy, as environmental protection received high political priority in China's development strategy. In the 1980s, environmental protection was legitimated and passed into law, and it was discussed in the press. The problem was that the laws were ignored at the local level. For example, according to the Shanghai Green Administrative Regulations (*Shanghai Shi Lühua Tiaoli*), "Building in public green space is forbidden," but this regulation was often ignored in real life. Yet, Shen dug it out and highlighted it to the real-estate developers and the government. Shen bought books on law, checked legal clauses, and drew upon them in his resistance against the confiscation of the park. During the negotiations, Shen often carried these law books and construction maps with him. In front of the government officials and developers, he recited these laws and emphasized the significance of environmental protection. According to those who attended these meetings, his inquiries and debates often brought the officials into awkward situations because they themselves were not fully aware of these clauses. Some of the core participants called this an "authenticating strategy" (*jiao zheng*), forcing government officials to comply with their own laws and regulations, even when it is widely known that the laws are enforced badly or not at all. Shen's authenticating strategy is similar to O'Brien's "rightful resistance" which emphasizes an inventive use of laws and ideology to resist "disloyal" elites in the local government (1996).[22] At a later stage in the movement, this strategy proved very successful, as the government found it difficult to deny the residents' claims and complaints, while the activists even won sympathy from several senior officials.

By experience, Shen and his fellow activists were well aware of the various conflicting interests and discrepancies between different government departments, and attempted to use these to their advantage. Within the local government, the Urban Construction Committee and its real-estate development company[23] focused on the district's economic growth. Yet, the Bureau of Park Management and the Bureau of Urban Planning were primarily concerned with environmental protection. Seeing that his appeals to the Street Council were to no avail, Shen decided to give up on it, and directly report the illegal construction to the former two departments. The Bureau of Park Management and the Bureau of Urban Planning were sympathetic to his cause, but shrank from a confrontation with other government departments. However, this informal support in itself was sufficient to embolden Shen to continue his fight, and he decided to organize the residents in filing a collective complaint (*jiti shangfang*) to the Shanghai municipal government.[24] Filing such complaints was not without risks, as it was not uncommon for them to be turned down, after which government retaliation is often a likely outcome.

Reaching out to the media

The activists also hoped to get support from the media. In recent years, the Chinese media have obtained greater leverage and room to disclose problems at the grassroots in cases when these are not regarded as politically sensitive in official eyes (see discussion in Calhoun and Yang in this volume). Shen contacted a female reporter named Li from the *Wenhui Bao*, one of Shanghai's most influential newspapers. Li recounted that at the time she was writing a report on Green Garden's community development, Shen had come in person to her office and had invited her to Green Garden to see what had really happened. When she arrived at the residential estate, the residents rushed to her and agitatedly told her about the confiscation of the park. Li herself is active as an environmentalist, and was therefore sympathetic to the activists' cause, for which she wrote several reports. With Li's assistance, the news was delivered via internal channels to a vice-mayor of the municipal government. He instructed to handle the issue "according to the law" (*yifa banshi*). Although this instruction was open to various interpretations, it actually provided sufficient leeway for the Bureau of Park Management to call an official halt to the construction project.

Taking direct action

Even with this explicit government order by the Bureau of Park Management, the real-estate developer still hoped to continue building clandestinely. But the volunteers, acting as a neighborhood watch, quickly found out about this intention and informed Shen. At that time, Shen believed that his "reason" (*shuofa*) had already won sufficient legitimacy in the eyes of the government as he had learnt of the vice-mayor's instruction through friends in the government bureaucracy. However, he also knew that if the activists did not stop the construction themselves, the bureaucrats – hampered by various procedures – would not be able to stop it in time. On the evening of 15 June 1994, when the residents came back home from work, he organized younger residents to break into the construction site and literally take down the scaffolding. When asked why they dared to do this, Shen replied: "The government had already forbidden the construction but they still went on, so what they were doing was illegal. We rushed into the site just to stop the illegal work. That's why we dared to do it."

Another core activist, Mr Wang, a university professor, told us that another reason for their daring was the fact that the opponents belonged to a company rather than the government, as a result of which the risk was relatively low. At most, it would be regarded as a "conflict between an enterprise and the masses," and not as a political conflict between the people and the state. For these reasons, almost 1,000 residents pulled down the walls around the construction site. Meanwhile, the residents also called the Bureau of Urban Planning and reported the illegal construction there as well. A month later, several major newspapers in Shanghai reported the heated struggle in Green Garden. The media reports generated a positive response from the municipal government.

Finally, the Bureau of Urban Planning formally invalidated the issued construction permit. Confronted with joint pressure from the media, government, and residents, the company had to abandon the project and withdrew from the park. This success tremendously inspired the residents. For them, it meant that the general public recognized their struggle and confirmed that the park belonged to them. As a result, this experience has become part of the collective memory of the community and contributed to the building of an alternative identity of the neighborhood. Through this action, Shen became a well-known figure. The case was a historical event that rallied the residents and later on allowed them to oppose the Street Council.

Due to this action, Shen Xin had won the respect of the residents of the building where he lived and was elected as the first director of the home-owners' committee when it was established in 1997. From this action, Shen and other residents also learnt about the importance of greenery protection for the community. Shen organized a team to protect the greenery, and through his insistence and work the park was finally completed in 1995 and opened for the well-visited Chrysanthemum Exhibition of the Pudong district. The exhibition was so successful that it instantly turned the park into the central symbol of Green Garden New Village. Inspired by this, the residents – ironically, together with the real-estate management company – erected a stele with the edifying inscription "It is better to eat no meat than to live without green."

Renewed struggle: confronting the local government

Negotiation with the government

The account above does not merely end here, as the contestation for the park actually continued. In fact during the next dispute, residents confronted the local government directly, although it had been their ally during the previous protest against the real-estate developer.[25] According to the original spatial planning, the Street Council was allowed to build a 135 sq m community center for the elderly in the park. However in 1997, the Street Council and the Organization Department (*zuzhi bu*) of the district government decided to jointly build a 2,600 sq m center for "old cadres." This term (in Chinese: *lao ganbu*) actually refers to a special group of retired, but influential, Party and government officials who were in office prior to, or shortly after the founding of the People's Republic in 1949.[26]

Dissatisfied that the function and size of the planned building had completely changed, the residents rose in opposition once more. They argued that the park was a public facility that belonged to the community. Since the residents owned the buildings around the park, this park should also belong to the residents, they maintained. Repeating the former environmentalist discourse, which had won them so much success on a previous occasion, the residents added that greenery was protected by the state and should not be destroyed randomly. The local government, on the other hand, regarded the park as state property, and deemed itself as the representative of the state (which in fact is illegal according to the Land Administration Law).[27] When we interviewed the head of the Street Council, he insisted that the government has the power to use the park without compensation. After several rounds of talks between the Street Council and the homeowners' committee, the negotiation reached an impasse. After discussions with the other core activists, Shen decided to continue to use environmentalist discourse and ask for help from the media. The resulting news reports led to instructions from a municipal official that the Street Council had to temporarily stop building the center for old cadres, though it was permitted to build a larger center for the elderly in the community compared with the original design. The Street Council even encouraged some old men in the neighborhood to go to Shen's apartment and accuse him of obstructing the building of the center. Shen was facing rising pressure from these neighbors. Although he knew it was instigated by the government, he and other members of the homeowners' committee compromised and agreed to allow the building of a bigger center for the elderly in the community. On 13 April 1998, the Street Council and 12 homeowners' committees reached the agreement that the size of the center could increase to no more than 650 sq m and that the center was only for the elderly in the community, and not for "old cadres."

For the Street Council, this agreement was just a clever tactic, as its aim remained to build a center for "old cadres." In an interview, Shen told me: "I had thought that even if the Street Council would obtain our agreement they could not build the center anyway, because the law stipulates that no one has the right to build on planned public greenery. Unexpectedly, they did obtain permission." After a half-year's efforts and protracted negotiations with concerned government departments and influential senior officials, the Street Council managed to obtain all the necessary permits to build the center for old cadres, and even more, at a size of 1,960 sq m. This fact, of course, was carefully kept secret from the residents. In April 1999, when the builders went into the park, the events described at the beginning of this contribution took place.

Petition on 4 June

Concerning the land ownership of the park, both sides – the residents and the Street Council – lacked legal grounds, since the ownership rights of public land are intentionally left undefined in national law.[28] However, due to their earlier victory, a group of young residents wanted to abandon negotiations, and take radical actions to mobilize people instead. The youngsters hung up a huge banner with the text of the stele that "It is better to eat no meat than to have no green" on the wall of a nearby building. In addition, also 30 people were organized to send a petition to the municipal government just before 4 June, the most politically sensitive date in China due to its connotation with the June 4th Movement in 1989. Although he supported the youngsters' relatively "radical activism," Shen did not want to enrage the municipal government. The particular tactics he chose, are a clear illustration of China's embedded context. Shen told the residents: "We should make them notice us but should not let them lose face." As a result, the activists did not visit the government offices during the day when passing crowds would see them, but in the evening. Considering his personal security, Shen did not go. Instead, another activist leader, a college professor living in the same building, led the people to knock at the doors of the municipal government. This face-saving tactic helped Shen and the attendees not to be seen as committing a political crime.

Yet, the radical action itself still startled the Street Council. Soon police came and surrounded the building. Shen and some chief petitioners were warned that organizing the petition was illegal in this sensitive period around 4 June. Meanwhile, Mr Hu, the Street Council official in charge of the construction project, organized the Residents' Committee and the retired Communist Party members in the community with good connections in the Residents' Committee to hold a neighborhood meeting and criticize the activists. The Street Council and Residents' Committee argued in public that Shen and his fellow activists were selfish residents trying to preserve their narrow interest at the expense of the wider community interest. In a more vicious vein, the participants were even portrayed as rioters dissatisfied with the rule of the Communist Party and the government. The Residents' Committee even exposed personal matters pertaining to Shen's family – that he allegedly had a strained relationship with his mother and his younger brother, which in Confucianist China would make him appear unfilial in the eyes of the residents.

In spite of the activists' caution to depoliticize the matter, the Street Council consciously treated the conflict as a political question. Fearing being clamped down on, the residents' movement became divided, and their resolve started to waver. A director of one of the homeowners' committees, who formerly had been firmly supporting Shen, changed his attitude under the mounting pressure of the Residents' Committee. Li, the female director of the Residents' Committee, told me: "I used my personal ties [*guanxi*] to persuade the directors of the homeowners' committees other than Shen's. They are Communist Party members and trust our Party Branch. So I first corrected their views."[29] Through her doing, some activists became afraid of the political risks and withdrew their complaints. Ultimately, the alliance of the 12 homeowners' committees, which had been the strongest base of support for Shen, was broken up. After the police surrounded the building and warned the participants, Shen wanted the residents to sign a complaint letter protesting against their treatment by the police, but several homeowners' committees refused to sign. Contrarily, others were outraged by the actions of the police and the local government. It is important to note that those who remained committed to the cause were living in the same building as Shen and came from the same *lilong*. Former social ties thus proved critical for the movement to stand their ground against the Street Council.

They felt seriously insulted that the police visited their home only because they filed a legal complaint and asked the Street Council to obey the former agreement. As emotions rose, some activists advocated even more radical action by sabotaging the entire construction site. Shen felt that this disruptive strategy could never win widespread support from the other residents and officials, who would be making the ultimate decision. Shen responded to the activists' call for radicalism by saying: "It is easy to destroy the construction site, but will it work? And what about our life afterwards? Jail?!"

De-politicizing what the state politicizes

The movement had reached deadlock. After several days' of reflection, Shen decided to continue the struggle, but strongly rejected the idea of challenging the state. "No matter whether they threaten or try to bribe me," Shen said, "I won't stop until we win completely." Yet, despite his determination, Shen did not feel fully reassured about a good outcome of the dispute. He knew that some officials of the Street Council detested him and were desperately trying to collect incriminating evidence against him. One single wrong step and he would be arrested. At the same time, the Street Council alternately tried to bribe and blackmail him through the construction company. The company promised to give him money and allow him to move away, but if he did not agree they would send hooligans to beat him up. Shen and the movement had landed at a critical juncture: the movement's leadership was facing a potential split among its supporters and heavy pressure from the government, as well as threats and bribes from the developers. Yet, Shen decided to press on and attempted to find a new strategy.

Shen undertook several actions in the struggle against the Street Council. First, he bombarded the Bureau of Park Management and the Bureau of Urban Planning with reports, emphasizing that the decision of the Street Council to build the center for old cadres was illegal and infringed upon the residents' environmental and property rights. He suspected that the Street Council and district government had put pressure on the other two bureaus to stay out of the conflict. Shen's counter-strategy was to keep on exposing abuses virtually every week; in his words: "We ordinary people [*laobaixing*] have little we can do against officials but we also lose little if we go against them. We can spend lots of time and energy to pester [*chan*] and annoy [*fan*] them." At times when few residents went to protest at the Park Management Bureau, Shen even mobilized his wife to go. The continuous actions forced the two bureaus to deal with the matter in a serious way, instead of ignoring it. Shen even sent complaint letters to the central government and the prime minister. The letter obviously never reached the hands of the prime minister, but it certainly did result in an enquiry from the central government and added more pressure on the local government. Moreover, Shen also sought media attention again. Some influential newspapers reported the dispute by highlighting the environmental aspect of the conflict, rather than portraying it as a conflict between the government and residents. In the media's interpretation, the activists' action mainly exposed a socio-economic problem caused by the unlawful behavior of the government. The residents were identified as the victims of an illegal confiscation. The influence of the media was not only important to win the sympathy from the public, but also to gain the attention of high officials in the municipal government. As a final resort, Shen prepared a lawsuit with the help of several lawyers, even though he feared its outcome.

An interesting feature of the residents' movement is the manner in which they chose to frame their conflict towards the government and media. In their meetings, Shen and his fellow activists came to the conclusion that it was insufficient to draw upon the ecological discourse to legitimize their movement. They had to find a new way to legitimize and frame their resistance. At the time, the state promoted legal reforms and attempted to establish a so-called "nomocracy," i.e. to establish the rule of law instead of the rule by administrative government. In reaction, the activists decided to use the rule of law as their authenticating strategy. In countering the local government, Shen frequently used articles and clauses from laws, which he combined with official rhetoric, such as "Environmental protection is the basic strategy of the state" and "The Party should govern the people by law." By making use of this rhetoric, he attempted to build a counter-discourse which he subsequently disseminated through his contacts with the media, government and residents. He consulted with a lawyer on the specific wordings in relevant laws, and used these to his own advantage. When he mobilized the 12 homeowners' committees, he claimed that

> The land belongs to the state. But the representative of state is not the government, but the law. The law stipulates that the use right to green land belongs to the home owners and not to the Street Council.... We defend the law. We are the people, so we stand on the side of the state. We know we are not against the state, but are just protecting our interests.

In effect, Shen made a clever combination of the proper ecological and socialist discourse in order to defend the residents' legal rights and economic interests.

It is noteworthy to examine the contradictory interpretations of Shen's action by the different sectors in the government. In the eyes of the media and officials sympathetic to the Green Garden activists, Shen was labeled a hero. The Bureau of Park Management even endowed Shen with the honorary title of "Shanghai Guardian of Greenery" in 2000 and 2001 and thus legitimized his effort. Later, Shen and the officials in the Street Council stated that this honorary title prevented him from being arrested. Of more interest, is the fact that the Street Council without exception defined the residents' activism as "criminal behavior," and even went as far as to suggest that they were a "third force"[30] challenging the rule of the Chinese Communist Party. When the head of the Street Council was interviewed, he described the residents as far more dangerous than the Falun Gong Sect. With a sense of drama he stated: "The Falun Gong is easy to repress but these bad citizens [*diao min*] are much more troublesome. Some media also played a misguiding role and promoted their resistance." The Street Council's actions demonstrates that the (local) state – in discrediting or clamping down on a movement – consciously attempts to *politicize* state–society conflicts, even when the nature of the conflict is overtly economic: a dispute over green space. Interestingly, the Green Garden activists' choice for a strategy to adhere to what Ho has termed a "de-politicization of politics" (see this volume) has thus become the bone of contention itself.

Shen's perseverance finally paid off when in 2000 the Bureau of Park Management and the Bureau of Urban Planning ordered the Street Council to stop the construction. Initially, the Street Council tried to delay implementing the government order, but this action only further eroded its legitimacy, tilting the balance to the protesters' advantage. In the end, the Street Council backed down. Today the park is still a deserted construction site, though the residents have good hope that they can keep their park.

Identities: shifts and continuities

During the socialist era, the identity of residents in urban China critically depended on a sense of belonging to the formal neighborhood organizations (*jumin weiyuanhui*) or, in the case of Shanghai, the *lilong*. At the time, a resident's identity was a function of his recognition as a member of the neighborhood organization and more informal relations with the members of a traditional *lilong* community.[31] As we have seen above, before the housing reforms, the Residents' Committee was the dominant neighborhood organization and served as an intermediary institution between the Street Council and the residents. Due to the strictly organized nature of the neighborhood, its residents merely regarded it as an administrative organ of the state. In fact, to many residents the neighborhood was tantamount to the Residents' Committee. As law-abiding citizens, residents had to submit to the Residents' Committee's administration. However, most residents lacked a strong neighborhood identity as their loyalty and ties primarily lay with the *danwei* (work unit), which provided them with medical insurance, retirement pension, a social network, and housing, of course. It is thus no wonder that residents during the socialist period felt few ties with the Residents' Committee. Yet, with the housing reforms and the subsequent transformation of the neighborhood organizations since the 1990s – notably, the emergence of the so-called "homeowners' committees" – residents changed their identities and ideas of citizenship.

An important development is that people gradually became more concerned about the neighborhood as it was closely related to the commercial value of their newly acquired apartments. In addition, as community management also became increasingly commercialized, residents started to demand better services (waste management, security, and green space planning) since they paid for it. Privatization and commercialization thus created residents that were conscious of their residential and environmental rights. What we witness from the nine years' struggle of the Green Garden residents is a substantial strengthening of a collective identity versus "outside invaders." As the movement gained momentum, the neighborhood ceased to be merely a unit of state control, yet evolved into a space and place which residents deemed their own(ership). Although illegal, as urban land by law belongs to the state, some core activists even argued that since they had bought their apartments, the property right of the greenery should also change from state-owned to collectively owned, similar to the reform of state-owned to private enterprises.

The establishment of the homeowners' committees is a second element that induced a change in neighborhood identity. The homeowner committee is a legal entity stipulated in the newly proclaimed Administrative Regulation on Real Estate Management. Very few residents are aware of this regulation and those that were aware had little faith in it, as they thought it was just another administrative regulation not to be taken seriously. In fact, the Street Council has no legal power to administer the homeowners' committees, as a result of which a considerable autonomy was their share. During the movement, Shen and his lawyers discovered the potential of the homeowners' committees, and subsequently formed an alliance of 12 such committees to challenge the Street Council's authority. This alliance proved a viable new representation of the neighborhood, and quickly undermined the position of the Residents' Committee as the community's official representative. The collective action itself was the most important force in changing the former neighborhood identity.

Whereas the first action against the developers and its success forged a common identity of the neighborhood, the following action against the Street Council led to its breaking up. Some residents who chose to support the Residents' Committee desired to maintain a socialist image of the neighborhood as a stable component of the socialist state. Others who remained loyal to the original aims of the movement actively rejected the official interpretation of neighborhood identity, and attempted to construct an alternative identity based on residential rights and local autonomy. It is critical to remark that those activists that stayed committed to the cause shared not only the same building with the movement's leader, but also the same "traditional" social ties dating back to their days of the *lilong* community. In this sense, both new identities (through the homeowner committee and a newly perceived community), as well as past identities (the *lilong*) mingled, and gave the movement its specific dynamics and features.

The contested definition of neighborhood identity is intertwined with the changing view of the state among residents. During the socialist era, the state was equal to the government and the Party, which controlled all resources and power. Since the 1980s, although not many dared to challenge the state's legitimacy openly, people did start to question the legitimacy of the government, in particular the local government, as the representative of the state. In the Green Garden dispute, the officials of the local government claimed that they represented the state and were entitled to appropriate the land in the name of the state. This, of course, was contested by the movement's participants. In countering the local state's argument, Shen used the official socialist discourse to claim that "the people" (*min*) form the root of the state, and *not* the government. In his words:

> The will of the people is expressed by the law made by the Party. The Central Party demands to govern by law. However, the Street Council has violated the Greenery Regulations, Real Estate Management Regulations, and Urban Planning Regulations. Thus, it has betrayed the Party, and betrayed the people. It is not entitled to represent the state.

By detaching the local government (the Street Council) from the state, Shen and his supporters could successfully separate neighborhood governance from the control of the local government in favor of the homeowners' committee. It is important to note that in all our interviews the movement participants, including Shen, claimed that they absolutely did not wish to go against the state. In their view, they attempted to assist the state in "correcting the improper behavior of the local government." Moreover, the activists took care to limit their actions and demands strictly within the neighborhood as they wanted to prove that their movement was far removed from anything political.

Concluding observations

In this contribution, we have shown how the privatization of housing, a growing ecological discourse, and the emergence of new residents' institutions (the homeowner associations) have provided the breeding ground for a small but ultimately successful urban movement in a Shanghai community. The Green Garden dispute involved a group of residents, a real-estate company, and a local state organization (the Street Council) fighting over the rights to a park near the apartment complex. The dispute evolved in two stages: during the first stage, a real-estate developer wished to expropriate a portion of the park to construct a new building. By making full use of the media, the environmental discourse due to the "greening of the state", and the new legal space open to citizens (e.g. filing complaints), this dispute ended in success for the residents of Green Garden.[35]

A critical factor in this success was the movement's conscious choice to depoliticize its actions. The movement stated openly that it did not want to challenge the authority of the Party-state. For instance, the participants avoided using the word "movement" (*yundong*) to define their actions, as it might be misinterpreted as popular protests against the state.[32] The residents also strictly limited the scope of their activism to their own local neighborhood. Due to this strict localization the movement could avoid being seen as challenging Party rule, and thus gained support from the state as well as a wider group of residents. Lastly, by insisting that environmental protection was at stake rather than political freedom, the Shanghai authorities were willing and able to side with the residents rather than with the real-estate company. This is what one might call "embedded activism" in full practice.

Based on case-studies from Taiwan and mainland China, Weller casts doubt on the universality of civil society and maintains that civil associations and their activities in mainland China and Taiwan are more strongly associated with "traditional" and informal ties, such as through kinship, religion and local community (Weller 1999).[33] In our case, we can actually see that the Green Garden movement is related to "traditional" *lilong* neighborhood ties, as Weller argued. As the core activists in Green Garden were close neighbors living in old Shanghai *lilong* communities, they shared a common history, trust and identity.[34] However, through our study it has become clear that the movement also led to the formation of a newly perceived identity, a process in which the newly established institutions of the homeowner committees played a vital role.

The second stage of the dispute posed a substantially greater challenge to the activists, because it involved not a commercial but a local state actor: the Street Council. During this conflict, the Street Council planned to build a center for retired Party cadres in the park. The movement's leadership soon found out that old strategies which won them victory on the first occasion were no longer sufficient. The former strategies were intensified: seeking greater media attention, writing an "avalanche" of complaint letters, and taking legal action (in this case through a lawsuit). Yet, of greater interest is that the movement was prompted to change its framing tactics due to a change in its opponent. The Street Council, as a state actor, knew the rules of an "embedded game," and attempted to discredit the activists by consciously politicizing their actions. Using political discourse that branded the movement as dangerous to Party rule, and comparing it to the Falun Gong Sect, the Street Council's attacks posed a veritable challenge for the activists. The original depoliticized, environmental discourse was no longer useful. In response, the movement's leader built a counter-discourse on the basis of official rhetoric which allowed him to claim that the residents represented the nation's people; that the people are the central state's allies in defending the law; and that as such, they were justified in correcting the *local* government which had transgressed the law. In this sense, it is interesting to see that activists' conscious choice for an embedded strategy of "de-politicized politics" has itself become a stake in a state–society conflict.

Notes

1 Warm thanks to John Logan, Lin Meng, Benjamin R. Read, and particularly Joseph Bosco for their comments on an earlier draft of this contribution. We gratefully acknowledge the Small Grant Program, funded by Urban China Research Network that supported this research. Zhu Jiangang is an associate professor in the Department of Anthropology at the Sun Yat-Sen University in Guangzhou, China. He is also the director of the Research Institute for Civil Society at the Sun Yat-Sen University. His mailing address is Department of Anthropology, The Sun Yat-Sen University, 135 Xin Gang West Road, Guangzhou 510275, Guangdong province, China. Peter Ho is professor in International Development Studies and director of the Centre for Development Studies at the University of Groningen, The Netherlands.

2 See also L. Tomba, "Residential Space and Collective Interest Formation in Beijing's Housing Disputes," *The China Quarterly*, No. 184 (December 2005), pp. 934–51.

3 Only recently, in 2003, has the conflict over urban resettlement been given high attention by the media and central government due to several self-immolation incidents in Nanjing and Beijing. See Mao Wenxian and Ju Tao, "*Tuitujixia de Beiju*"

[The tragedy in front of the bulldozer] at http://news.21cn.com/domestic/guoshi/2003/09/03/1254412.shtml, 2003; and Yan Xiaojuan, "*Anhui qinyangxian nongming Zhu Zhengliang Tiananmen zifen zhengxiang*" [The fact of the self-immolation in Tiananmen of Zhu Zhengliang, a peasant from Qingyang County in Anhui province], at http://news.sina.com.cn/c/2003–09–18/1337775762s.shtml, 2003.

4 See M. Castells, *The Urban Question*, London: Edward Arnold, 1977; and M. Castells, *The City and the Grassroots*, Berkeley: University of California Press, 1983.

5 Ibid. Also see J. Mollenkopf, *The Contested City*, Princeton, NJ: Princeton University Press, 1983; C.G. Pickvance, "On the Study of Urban Social Movements," *Sociological Review*, Vol. 23, No. 1 (1975), pp. 29–49; S. Lowe, *Urban Social Movement: After Castells*, London: Macmillan Education, 1986; R.E. Pahl, "Urban Social Theory and Research," *Environment and Planning*, Vol. 1, No. 2 (1969), pp. 143–53; J.A. Rex, "The Sociology of a Zone of Transition" in R.E. Pahl (ed.) *Readings in Urban Sociology*, Oxford: Pergamon Press, 1968, pp. 211–31.

6 C.G. Pickvance, "On the Study of Urban Social Movements," *Sociological Review*, vol. 23, No. 1 (1975), pp. 129–49.

7 Since the 1970s, similar social movements have erupted in Western and South American countries. They contrast with old types of labor or peasant movements by the character of identity politics instead of classical struggles (A. Touraine, *The Voice and The Eye: An Analysis of Social Movement*, Cambridge: Cambridge University Press, 1981; and A. Melucci, *Nomads of the Present: Social Movements and Individual Needs in Contemporary Society*, London: Hutchinson Radius, 1989). These movements are loosely organized, vaguely class-based and associated with mobilizations for peace, for the autonomy of women and gay persons, and against environmental pollution. Therefore, the site of struggle changed from the workplace to communities and to cultural representation. Nonmaterial needs were highlighted. The movements engaged in small-scale, antihierarchical actions and experimented with direct democracy within shifting and loosely networked core constituencies (B. Klandermans, "New Social Movements and Resource Mobilization: The European and the American Approach Revisited" in D. Rucht (ed.) *Research on Social Movements*, Boulder, CO: Westview Press, 1991; B. Beccalli, "The Modern Women's Movement in Italy," *New Left Review*, No. 204 (1994), pp. 86–112). On the negative side, these New Social Movements find it difficult to form strong organizations and lack a stable social base (Beccalli 1994). In NSM theory, the issue of identity gained prominence (M. Zald, "Looking Backward to Look Forward: Reflections on the Past and Future of the Resource Mobilization Research Program" in A. Morris and C. Mueller (eds) *Frontiers in Social Movement Theory*, New Haven, CT: Yale University Press, 1992, pp. 326–48). According to these theorists, what participants of these new social movements seek are not economic interests, but new forms of culture, which require new representations of subjectivity (e.g. the gay movement's redefinition of queer).

8 W. Gamson, "The Social Psychology of Collective Action" in A. Morris and C. Mueller (eds) *Frontiers in Social Movement Theory*, New Haven, CT: Yale University Press, 1992.

9 See D. McAdam, J.D. McCarthy and M.N. Zald (eds) *Comparative Perspectives on Social Movements: Political Opportunities, Mobilizing Structures, and Cultural Framings*, Cambridge: Cambridge University Press, 1996; and S. Tarrow, *Power in Movement: Social Movements and Contentious Politics*, Cambridge: Cambridge University Press, 1998.

10 E.J. Perry and M. Selden (eds) *Chinese Society: Change, Conflict and Resistance*, London: Routledge, 2000.

11 In the field research in Guangzhou, Beijing and Changsha, instances have been found of similar collective actions that have taken place in neighborhoods. They share a similar strategy, discourse and ways of acting.

12 See also M. Goldman and E.J. Perry (eds) *Changing Meanings of Citizenship in Modern China*, Harvard: Harvard University Press, 2002.
13 The name of the neighborhood and all residents are pseudonyms.
14 Protest movements in China are relatively sensitive and studying them can therefore be difficult. Fortunately, however, environmental issues at the local level in China are not perceived as political problems but as "social" problems. Furthermore, the tactic of activists is not to engage in street demonstrations but to stay in the neighborhood. Such "self-imposed censorship" brings difficulties for academic research as well as for surveillance from the government. Anthropologists' fieldwork can, however, overcome such invisibility. By living long-term in the community and establishing trust with partners, there was the opportunity to encounter such collective actions "by accident." Since local government and movement leaders all see these accidental collective actions as a social or administrative problem (that is, not a political problem), the local government acquiesced to the research. When the leaders of the movements learnt that outside scholars, were interested in them, they were glad to provide information and to be interviewed because they hoped more people would know about and support their struggle. Therefore, the research itself was not considered politically sensitive. Even so, a neutral stand and low profile was kept by the researchers in order to avoid any political risk.
15 At the time, residents paid between 40,000 and 100,000 *yuan* for each apartment. Sixty percent of the residents living in these 12 buildings have bought their apartment. For Shanghai people, at that time, buying an apartment meant the family would spend almost all its savings on the deposit. Many saw it as their lifelong achievement.
16 The *lilong* is the smallest unit for residential control in Shanghai. The *lilong* is often said to be a relic of traditional society. That is, the sense of community developed in the *lilong* is said to resemble village patterns, the personal relation among neighborhood is seen as face-to-face and inter-assistance among neighborhood. The character *li* means neighborhood and *long* means alley; together they are read *lilong*.
17 Long after these buildings have been built, there are still not many shops nearby or buses to downtown. Residents even complain about the lack of street lamps and that the hospital and government agencies are far away from the community.
18 The urban administration in China is divided into various administrative levels: municipality, district, sub-district, and neighborhood. At the sub-district the state is represented by the Street Council, and at the neighborhood this is the Residents' Committee. A discussion and full overview of the urban administrative structure is given in J.Q. Cheng, J. Turkstra, M.J. Peng and P. Ho, "Urban land administration and planning in China," *Land Use Policy*, Vol. 23, No. 4 (2006), pp. 604–16. For a historical review of urban administration, see also M.K. Whyte, "Urban Life in the People's Republic," in R. MacFarquhar and J.K. Fairbank (eds) *Cambridge History of China, The People's Republic, Part 2: Revolutions within the Chinese Revolution 1966–1982*, Cambridge: Cambridge University Press, 1991, p. 697. In Bestor's *Neighborhood Tokyo* (Stanford: Stanford University Press, 1989) the employee also does not participate in community life but the core of the neighborhood is the shopkeepers' families rather than those active in the neighborhood organization.
19 The Secretary of the Party Branch is the highest leader normally. The Street Work Committee of the Communist Party (the Party's organization in the street subdistricts) often nominates him. Normally he is also a resident in the neighborhood or nearby. Local identity and state agency are intertwoven in these organizations.
20 B. Read, "Revitalizing the State's Urban 'Nerve Tips'," *The China Quarterly* (2000), 806–20.
21 When the protest movement started, the participants in these informal associations also played an important role in the movement's communication and provided a social space for people to act. For instance, it was a group of women of the Elderly Association which first alerted the residents about the illegal construction in the early 1990s.

22 Kevin O'Brien, "Rightful Resistance," *World Politics*, Vol. 49, No. 1 (1996), pp. 31–55.

23 The municipal government owns this company and the Urban Construction Committee directly administers it.

24 It is worth pointing out that according to government policy, collective complaints do not imply that the construction project is stopped with immediate effect. The handling of the complaint and the construction can take place at the same time. In fact, in present-day China, even if the complaint is found valid by the government the project might still continue as the project manager can claim that he did not receive the instruction in time, thus making the construction irreversible.

25 Although the local government at the time did not do much more than taking a sympathetic attitude, for the residents it was sufficient for them to regard it as their ally.

26 Superficially, old cadres have no formal power any more. However, because of their close relations with those in authority, old cadres still possess influence in politics. For some officials, making friends with old cadres is a short cut to reaching a higher position.

27 See also P. Ho, *Institutions in Transition: Land Ownership, Property Rights and Social Conflict in China*, Oxford: Oxford University Press, 2005.

28 P. Ho, "Who Owns China's Land? Policies, Property Rights and Deliberate Institutional Ambiguity," *The China Quarterly*, Vol. 166 (June 2001), pp. 387–414.

29 An interesting case-study on the role of *guanxi* in creating the sense of a community is described by J. Farrer, "'Idle Talk': Neighborhood Gossip as a Medium of Social Communication in Reform Era Shanghai," in T. Gold, D. Guthrie and D. Wank (eds) *Social Connections in China: Institutions, Culture and the Changing Nature of Guanxi*, Cambridge: Cambridge University Press, 2002.

30 "Third force" is borrowed here, referring originally to social groups who sought to differentiate themselves from the Communist Party and Guomingdang or Nationalist Party during the Civil War in the 1940s.

31 J.G. Zhu, "*Guo yu jia zhijian: Shanghai linli de shimin tuanti yu shequ yundong de minzuzhi*" [Between the family and the state: an ethnography of the civil associations and community movements in a Shanghai lilong neighborhood], PhD Thesis, Department of Anthropology, Chinese University of Hong Kong, 2002, pp. 85–91.

32 In China the word *yundong* is related to mass campaigns used by the state to control and mobilize people at the urban grassroots. However, besides campaigns, it might also refer to protest movements against authorities, such as the April 5th Movement or the students' movement of 1989. The boundary between campaign and movement is therefore ambiguous (see also, for example, Pieke 1996).

33 R. Weller, *Alternate Civilities: Democracy and Culture in China and Taiwan*, Boulder, CO: Westview Press, 1999.

34 It supports Snow and Morris' argument that, in a movement, collective identity is based on face-to-face relations and informal organization. See D. Snow and R. Benford, "Ideology, Frame Resonance and Participant Mobilization" in B. Klandermans, H. Kriesi and S. Tarrow (eds) *From Structure to Action: Comparing Social Movements Across Cultures*, Greenwich, CT: JAI Press, 1988, pp. 74–118; and A. Morris, "Black Southern Student Sit-in Movement: An Analysis of Internal Organization," *American Sociological Review*, Vol. 46 (1981), pp. 744–67; A. Morris, *The Origins of the Civil Rights Movement*, New York: Free Press, 1984.

35 For information on greening in China, see Pete Ho and Eduard B. Vermeen (eds), *China's Limits to Growth: Prospects for Greening State and Society*, Oxford: Blackwell Publishers, 2006.

8 Caged by boundaries?

NGO cooperation at the Sino-Russian border[1]

Yanfei Sun and Maria Tysiachniouk

Introduction

The international nature of many environmental problems calls for both regional and global cooperation. As many nation-states are, by themselves, inadequately prepared to deal with transboundary environmental problems,[2] the potential solutions lie not only with multinational governmental treaties, but also in the emergence of a transboundary civil society, particularly through non-governmental organizations (NGOs).[3] Scholarly work has highlighted the role of elite international environmental NGOs, such as Greenpeace, WWF, and Friends of the Earth, who have played a huge role in global environmental politics and networks.[4] Yet often such research has paid insufficient attention to the agency of local NGOs in this global activist structure. Studies on transnational social movement organizations have noted the growing role of transnational local people's movements.[5] However, most of these studies selected cases in which local NGOs were actively seeking and/or successfully integrated into the transnational movement network, possibly leading to the underestimation of the difficulties entailed in forming and maintaining cross-boundary cooperative ties. The cases studied in this paper demonstrate that despite the transnational nature of many environmental problems, Chinese and Russian NGOs are still largely confined within national boundaries.

The majority of the literature that discusses the involvement of NGOs from developing countries in the global network is about international environmental activist groups joining forces with local environmental groups to protest the commercial activities of multi-national corporations or the gigantic national development projects in developing countries that will hazardously affect the ecology and the indigenous people.[6] Nonetheless, increasingly many cross-border environmental problems are taking place between developing countries themselves. The aggressive economic development is accompanied by serious environmental problems, which impact neighboring countries as well. Are environmental NGOs in the developing countries able to cooperate in face of this? What factors contribute to the success or failure of cross-boundary network formation? What roles do international NGOs play? Through a case study of the environmental NGOs in the Sino-Russian border regions, we hope to address the above questions.

After describing the kinds of environmental problems existing on this border, we look at the environmental NGOs in the border regions in both countries. Discussions of the structures and operations of these NGOs are followed by their perceptions of and actions towards the cross-border problems. We then proceed to offer explanations for the lack of success over transboundary network formation so far. In the end, we discuss the hopes and potentials for Chinese and Russian NGOs to cooperate in order to combat cross-border environmental problems.[7]

Transboundary environmental problems on the Sino-Russian border

The border between China and Russia measures approximately 4,300 km.[8] The region extends from the southwestern Primorsky region (Maritime Province) in the Russian Far East (RFE) to Chita borders on China's Jilin and Heilongjiang provinces, and the Inner Mongolia Autonomous Region. This paper focuses its attention on the section where the RFE borders on China's Northeast, formerly known as Manchuria. See Figure 8.1.

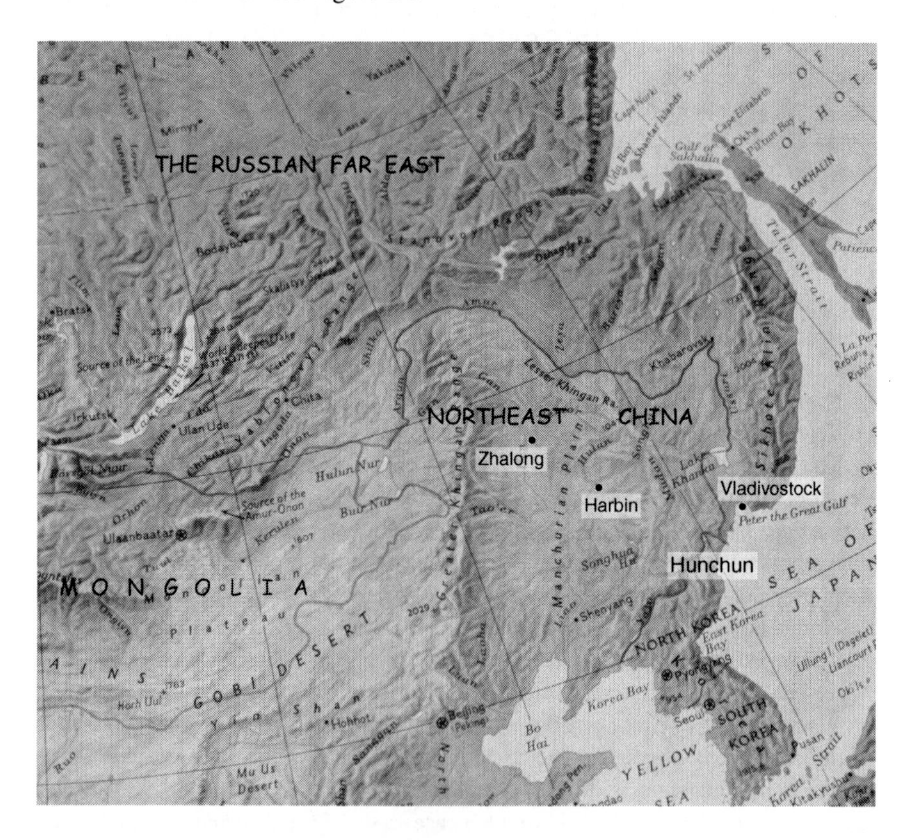

Figure 8.1 The Sino-Russian border regions in the Eastern Section.

The RFE border region can be rightfully claimed as a world-class ecological treasure. Besides being the largest relatively intact pristine region and primary forest in the world, the RFE forest possesses immeasurable biodiversity and is home to a number of endangered species including the Siberian tiger, Far East leopard, and musk deer. The Sino-Russian border rivers, especially the Amur River (Heilongjiang) and the Ussuri River (Wusulijiang), provide sources of drinking water, irrigation and fishery to people of both countries. The border river basins are the habitats of an extraordinary array of temperate flora and fauna, including many red-book species. The Amur wetlands, Lake Khanka (*Xingkaihu*) and the Tumen River Delta in particular are crucial points for migrating birds along East-Asia/Australasian flyways.

As eco-systems cross the political border, many of the environmental questions are transboundary in nature. Furthermore, since the normalization of Sino-Soviet diplomatic relations in 1988, border trade and contacts between China and the then Soviet Union (and now Russia) have grown tremendously. The influx of Chinese labor, commodities and capital to the RFE border regions, while boosting the local economy, also bring negative impact to the environment. Three transboundary environmental problems have loomed up as the most conspicuous ones: (1) illegal logging and timber trade; (2) illegal harvesting of other plants, poaching, and trafficking; and (3) river pollution.[9]

Illegal logging and timber trade

As a result of momentous economic growth, housing reforms, the booming construction and furniture industry, and improved consumption capacity, China's demand for wood products has increased drastically since the late 1980s. The limited domestic timber supply pushed China to turn to other countries to meet its expanding timber demand. China's appetite for timber imports was further expanded due to the state's 1998 logging ban on domestic natural forests and tariff reduction on forest products. In recent years, over 40 percent of total commercial timber consumed domestically has been imported and China has become second only to the United States in total import of forest products.[10]

Russia is one of China's major timber suppliers, ranking among the first three in the last two decades and ranking first in recent years. In 2002, China's log imports from Russia reached 14.8 million cu m, up by 1,460 percent since 1997.[11] The majority of the timber trade is in raw logs. About a half of this timber came from provinces of the Amur basin.[12]

The RFE timber industry, due to the shrinking Russian domestic market and the higher transportation cost to the European part of Russia, is growingly dependent on exports to Asian markets. The timber industry has become the sole revenue source of many enterprises in the RFE. Many accessible forests, particularly those around railroads and near population centers, have been overlogged. Ineffective law enforcement and rampant corruption have made illegal logging practices widespread.[13] Illegal logging and trade are found not only among organized crime networks, but also among villagers who resort to them to make a living. In sum, illegal logging and the unsustainable timber trade are devastating the forests of the Far East, especially the valuable coniferous forests. Their adverse economic and social effects are manifold: they extract natural resources, reduce regional revenue, jeopardize the local economic structure and exacerbate the living environment of the local community.

Illegal harvesting wild plants, poaching and trafficking

In a similar fashion the growing Chinese demand for certain wild plants and animals which are traditionally believed by the Chinese to have healing powers or to be culinary delicacies poses great threats to the wildlife and ecosystems in the RFE. Illegal harvesting of wild plants and poaching occur in all parts of the RFE along the Chinese border. In some cases, Chinese men cross over borders to illegally harvest and poach themselves; in other cases, Chinese traders purchase the wildlife products from Russian locals. Wild ginseng (*panax ginseng*), extinct in Asia except in the Primosky Kray and the south of Khabarovsk territory of the RFE, is among the illegally harvested Red Data Book[14] plants. According to expert estimates, in 1996–8 the volume of wild ginseng roots illegally procured and smuggled into China has reached an annual level of 500–600 kg (in crude weight) while the Russian government export quota was set at 50 kg in 1997.[15]

Wild animals being poached and trafficked from the RFE to China range from endangered rare species, such as tiger, leopard, and musk deer, to more common species, such as fish, frog, sea cucumber, and sea urchin. In recent years, the Chinese demand for Siberian frogs (*Rana Chensinensis*), which are used in cosmetics, traditional medicine, and gourmet dishes in China, has generated poaching and trafficking of this kind of amphibian widely distributed in the RFE. In several incidents, Chinese poachers crossed borders to catch frogs by spreading herbicides in the rivers and thus threatened the ecosystem and the public health of local residents.[16] Interception by Russian customs of illegal wild animal products bound for China occurred from time to time. However, the confiscated volume, already very astonishing, was only the tip of the iceberg.[17]

River pollution

The Amur River's major tributaries include the Sungari River (Songhuajiang) and the Ussuri River. *Tichookeanskaya Zvezda* (Pacific Ocean Star), a regional newspaper of the RFE, highlights the condition of the Amur River, which is polluted by phenol compounds. In Khabarovsk, which is located near where the Ussuri River joins the Amur River, the pollution is so heavy that swimming in the river and drinking tap water is forbidden. Fish caught in the lower part of the Amur River have a specific smell.[18] In the lower reaches of the Amur River where indigenous peoples live, many suffer from multiple diseases caused by phenol compounds.[19] The Sungari River, which runs through an industrial zone of Northeast China and several major cities, is extremely polluted. The Tumen

River acts as a border between China and North Korea in its upper reaches, and between North Korea and Russia at its mouth where it flows into Peter the Great Bay. Water quality in the middle to lower reaches of the Tumen River is not fit for industrial and municipal use. Fishing in Peter the Great Bay is also affected. The pollution of the border rivers has already become a diplomatic issue between Russia and China.

Except for downstream overfishing and dam-building on tributaries, Russia does not cause major natural resource-related problems for China. Admittedly, it is China causing Russia environmental problems more than the other way round in the eastern section of the Sino-Russian border. This is attributable to the asymmetrical socio-economic conditions between the two sides. There are far more people and more intense human activities in China's border area. China's side also is more developed agriculturally and industrially. Furthermore, China's economy remains soaring since it took off in the 1980s. China's importation of raw materials has increased greatly in recent years. It has increasingly become the source of acid rain and maritime pollution in neighboring countries. Its dam-building activities on the upper streams of international rivers have raised international concern. On the other hand, Russia's economic development has been sluggish. The RFE in particular suffered a severe economic crisis when the Soviet Union collapsed. The Putin government's relaxation of regulation over preservation and protection of natural resources, together with the lawlessness of the post-Soviet RFE, has resulted in rampant illegal logging, poaching, trade and trafficking.

Indeed, without China's cooperation, curtailment of environmental problems in the RFE by Russia alone is not feasible. The Chinese government and Russian government have signed several treaties on cross-border environmental cooperation.[20] While such treaties provide a framework within which collaboration over specific cross-border environmental issues can be discussed, both sides have been slow to put them into application. Furthermore, with transborder environmental issues being of low priority for Beijing and Moscow, it is apparent that the solution of transboundary environmental problems cannot fully rely upon inter-governmental cooperation. It needs broad participation of civil societies beyond the border, especially that of the environmental NGOs.

Limited cooperation across borders

In face of the severe environmental problems troubling the border areas, cross-border NGO cooperation has been rare. A handful of visible attempts were made by international NGOs without significant long-lasting positive impacts on the environment. International NGOs, such as Pacific Environment (PERC), Wildlife Conservation Society (WCS), the International Crane Foundation (ICF), WWF, Greenpeace, and Global Greengrants Fund (GGF), are operating in both China and Russia. The networks in which these organizations are nested provide them with natural advantages in working on cross-border environmental issues. For example, the offices of TRAFFIC[21]-Russia and TRAFFIC-China exchange information and frequently prepare joint publications on the illegal cross-border trade of musk deer glands. Similarly, the ICF facilitated a joint environmental program for school children in both Russia and China. It also implemented the Global Environmental Facility project on crane preservation, in which both Russia and China were participants.

WWF-RFE and WWF-China have tried to work together on several issues: for example, the creation of a transborder nature reserve around Lake Khasan in 1996. However, the cooperation was not well coordinated and did not last. Some links between WWF-Russia and WWF-China have occurred through IKEA, a multinational furniture corporation with subsidiaries in both Russia and China. IKEA partnered with WWF in globally promoting sustainable forestry. It has a policy of buying only legal and transparently harvested wood, and therefore is interested in promoting a sustainable, legal wood trade between Russia and China. The WWF–IKEA project in China involves activities along the border. However, it has not produced any results at the time of writing.

In addition to limited collaboration between branches or offices of the same organization, international NGOs also try to bring the local NGOs of the two sides together and help them to construct collaborative networks. Regrettably, no significant outcome has come out of such endeavors – no effective ties between local NGOs from the two sides have been forged. Perhaps the initiative with the greatest potential to leave a lasting impact is the Amur/Heilong Ambassador Campaign, jointly sponsored and organized by WWF-RFE and WWF-China in the summer of 2005. It selected and funded student environmental groups from the border regions of both countries to conduct summer camps in different sites in the Amur River valley with research tasks. It encouraged and directed student groups to pay attention to cross-border environmental issues and rewarded groups with good collaborative plans. The international student workshop at Lake Khanka in October 2005 was specifically planned to promote future cooperation between student groups. Six concrete joint plans on cooperation on cross-border environmental issues were drafted when the workshop concluded. WWF-RFE's consultant, who was studying Mandarin in the Northeast Forestry University and trying to befriend local NGOs and Chinese student groups in Heilongjiang, played a key role in coordinating this campaign. However, it is too early to say how long this project will last and what it will yield.

To analyze the constraints and potentials for the two sides to cooperate, we need to learn their situations first. Below, we will survey environmentalists' activities in the border regions.[22]

Student environmentalism in Northeast China[23]

In terms of environmental activism, Northeast China, and in particular Heilongjiang province, portrays a different picture than the rest of the country. From the other contributions in this volume, we have learnt that green NGOs are active in a wide variety of places, such as Beijing (see Ru and Ortolano in this volume), Shanghai (see Zhu and Ho), Qinghai (Morton) and Yunnan provinces (Calhoun and Yang). However, in Heilongjiang environmental NGOs are remarkably absent. One can find government-organized NGOs (GONGOs), but these enjoy much less autonomy from the government, as elsewhere in China, being structurally and administratively inseparable from their host state institution. In fact, the only major NGOs in Heilongjiang are student environmental organizations.

There are more than a dozen university student environmental groups in the province. Student environmental groups concentrate in the provincial capital, Harbin, where most higher education institutions are located. Almost every university or college has established at least one student environmental group, such as Green Union (*lüse xiehui*) at Harbin Institute of Technology (HIT), Green Angel Union (*lüse shizhe xiehui*) at Northeast Forestry University (NFU), and Green Civilization Association (*lüse wenming xiehui*) at the Harbin University of Commerce (HUC). The first two student organizations are the most influential and dominant of all student environmental organizations in Heilongjiang. The capacity and influence of the latter, however, is comparable to that of the average green student groups in the province.

While Green Angel Union of NFU and Green Union of HIT were respectively founded in 1996 and 1997, the majority of the rest of the student environmental groups in Heilongjiang came into existence only after 2000. With a longer history and some advantages, the two aforementioned groups have acquired capacity, resources, influence, and social recognition unrivalled by the others.[24] Both Green Angel and Green Union are well connected to the networks of environmental NGOs, especially student green organizations, in the country. They participated in national student environmental forums and national student summer camps. Indeed, quite often, news and information from environmental organizations outside of Heilongjiang were disseminated to other Heilongjiang student green groups through these two organizations. Financially, they were also better off than the rest of the student green organizations in Heilongjiang.[25] Both organizations have received wide media coverage and have reaped a chain of honors.[26]

A more profound dissimilarity among these student organizations, however, lies in the degree of independence that they have from the school authority. As illustrated above, nominally all student organizations were, in the terms of Ho (this volume), "embedded" in a web of multilevel control, especially under the direct leadership of the Communist Youth League (CYL). In practice, however, some organizations have managed to carve out a space of independent growth. The degree of autonomy from school authority greatly shaped the nature of the student organizations. While Green Union of HIT so far has successfully asserted its relative autonomy, Green Angel has been very much under the management of the CYL of NFU. There are various reasons why the school authority of NFU has an interest in retaining control over Green Angel. The strong linkage between environmental protection and many of the academic disciplines of the university won legitimacy and favor for Green Angel in the eyes of the school authorities. The latter feel not only obliged to endorse it, but also motivated to take the initiative to build and promote Green Angel as a "brand name," at a time when environmental protection is gaining prominence in the state agenda, media coverage and popular consciousness. The differential degree of control of the school authority over the two student organizations resulted in two distinct organizational cultures. In their interactions over time, this divergence became a source of discord. Green Union tends to think of Green Angel as not much different from a student union, fretting over the latter's focus on angling for fame instead of doing practical things. Green Union views Green Angel with a fair deal of suspicion – similar to the view Chinese NGOs take towards GONGOs. Green Angel felt the slight and took it as the sign of the arrogance and snobbery of students of HIT, a first-tier university in Heilongjiang. Frictions also arose from competition over resources and recognition.

The disunity between the two dominant student green groups in Harbin was an important obstacle standing in the way of the formation of any close networks of local student environmental organizations. The Green Longjiang Environment Alliance (hereafter referred to as Green Longjiang) is a case in point. In 2003, it was proposed that Green Longjiang (*lüse longjiang*) be created as a platform of inter-group communication and cooperation, a resource- and information-sharing center, and a discussion forum for the 12 existing student environmental groups in Harbin. This was in accordance with the nationwide trend of student environmental groups forming regional unions or networks.[27] However, the above-mentioned dislike between the two "big brothers," as well as other friction, brought about year-long dissension and conflicts over multiple issues. Green Union even threatened to withdraw from Green Longjiang altogether in April 2004. They did not agree to join Green Longjiang until early 2005.[28]

A major factor standing in the way of transboundary cooperation is the simple fact that the focus of green students' groups tends to be local. This is partly due to difficulties in linking with their Russian counterparts (as we will see below), and partly the result of perception. Although the RFE is much closer geographically to the students than the Tibetan Plateau, it is psychologically much further away. Moreover, cross-border environmental problems have received relatively little attention in the Chinese media, while students' news sources are usually limited to the Chinese NGO circle and media. In other cases in which China is creating environmental problems for neighboring countries, Chinese environmentalists did talk about borderless environmentalism. However, it is a reality that their environmentalism is, most of the time, caged within national borders.

NGOs in the Russian Far East

Environmental NGOs are much greater in numbers and more diverse in the RFE than in the China border region. In addition to green student groups, which we also found in Heilongjiang, the RFE features many international NGO offices and grant-giving organizations, NGO resource centers, and well-established local NGOs. Furthermore, we also find hundreds of smaller environmental groups in the RFE. Funds to support such groups are flowing into this region specifically because the international NGOs are promoting the development of civil society in this biodiversity hot spot. However, many of these groups operate only during the times when they are given a grant for a specific project. These groups are thus very unstable. In some places in the border region, where the majority of the population is involved in illegal trade with China, such groups also tend to be unpopular and isolated as they try to impede the informal economy of the villagers.

International NGOs and the established NGOs are the major players in environmental protection in the RFE. We can find the highest concentration of environmental NGOs in Primorsky Kray, as Vladivostock is the biggest city in the region. Most organizations have found it convenient to set up offices here. The high concentration of NGOs in Primorsky Kray is also a response to the diverse environmental problems here. The majority of timber exported to China comes through Primorsky Kray and there is a larger population of endangered species, such as the leopard and the Amur tiger. There are multiple national NGOs in Khabarovsky Kray, only a few in Amur Oblast, and fewest in the Jewish Autonomous Oblast. As we have seen above, there is a trend towards excessive dependence on Western funds. The competition for Western grants is fierce, and partly contributes to the lack of unity among NGOs in the RFE.

Survivors of socialism: Nature Protection Corps and GONGOs

The so-called "Nature Protection Corps" were a significant force for environmental activism in the USSR in the 1960s, but today they have lost much of their former mobilization capacity and energy. As Nature Protection Corps might emerge as a renewed movement in the future – currently larger NGOs attempt to revitalize them – we will provide a short historical review of them here. Nature Protection Corps appeared in universities all over the USSR as an important force of Soviet environmentalism[29] under Khrushchev's relatively liberal regime in the 1960s.[30] In the USSR, there was a general trend to reproduce the kind of institutions of the center in the periphery.[31] This mechanism allowed Nature Protection Corps to proliferate up to 150 groups. For many years, Nature Protection Corps were a powerful conservation force with support from the academic community.

Across the USSR, Nature Protection Corps were involved in similar kinds of activities, such as assisting nature reserves (*zapovedniks*[32]) to fight poaching and illegal harvesting, carrying out small research projects and environmental education. The organizational structure of Nature Protection Corps was similar to that of the Chinese student green groups today: Nature Protection Corps operated under the CYL and had faculty advisors, and most of them were part of state-led organizations, such as the All-Russia Society for the Protection of Nature (VOOP),[33] or the Society of Naturalists. Similar to China today, some Nature

Protection Corps in the 1960s and 1970s gained relative independence from both the CYL and the university administration while others did not. During this period, CYL leaders just as in China were trying to promote their career advancement through promoting student engagement in nature protection initiatives. Usually conflicts between Nature Protection Corps and CYL occurred when CYL imposed their own ideas on how Nature Protection Corps should operate.[34]

The policy of *glasnost*[35] and *Perestroika*[36] introduced by Gorbachev in the late 1980s ushered in the freedom of speech and democratization. Nature Protection Corps soon declared their independence from both CYL and VOOP.[37] Despite that, the movement declined rather quickly. As in China today, Nature Protection Corps in the 1960s and 1970s represented a safe outlet for public engagement against the backdrop of the juggernaut of the party-state apparatus – it was truly "a little corner of freedom" at the time. With *glasnost* unleashing broader social forces that were formerly quiescent, mass rallies and all kinds of social organizations erupted, and Nature Protection Corps lost their special status. Moreover, traditional Nature Protection Corps activities with a focus on nature conservation became insignificant in the face of the more pressing environmental issues such as radioactive pollution, especially after the Chernobyl accident. Yet most of the Nature Protection Corps failed to adapt to the new public concern and melted away. Many former members joined other causes, sometimes overtly political ones.[38]

During the Soviet era, there existed another source of environmental activism: critical, scientist-led professional societies for nature conservation and protection. We have already mentioned the Society of Naturalists and VOOP. These organizations played an important role in the Soviet environmental movement. The state itself initiated these organizations and controlled their operations. VOOP was the most prominent and in terms of its organizational structure and function the most comparable with current GONGOs in China. VOOP was founded in 1925. The People's Commissariat for Internal Affairs approved VOOP's charter along with financial support and mandated it to be a mostly educational society. VOOP was authorized to be a membership organization and to collect membership dues. The purpose of the state was to use membership dues to support nature reserves and other environmental protection activities. VOOP was the largest society of the USSR, as almost the entire population of the Soviet Union were mandatory members. VOOP worked closely with the Young Pioneers and CYLs, hosted many Nature Protection Corps brigades and was always at the core of government-organized environmental initiatives. It had a highly hierarchical structure with the head office in Moscow and branches all over the country. VOOP capacity and significance declined dramatically during Gorbachev's *Perestroika* years, after which they almost totally lost their state funding and membership. VOOP was not prepared to do fundraising when funding ended. Therefore, many of the VOOP branches that survived became independent NGOs, which maintain close ties with governmental agencies. The RFE branch of VOOP survived the reform period and is currently focusing on local educational activities with high school students. It does not play any significant role in the RFE environmental scene and at present does not have much potential to get engaged in cross-border cooperation with the Chinese NGOs.

Currently, the *zapovedniks* or nature reserves are weakly supported by the government. Although they continue to be part of federal state structures, they often act as NGOs. Many of them are supported by grants from international NGOs, such as WCS, WWF, or directly funded by foreign foundations. Throughout the history of *zapovedniks*, there was a constant fight between conservationists and other state authorities over their existence. During Putin's presidency this fight continues and *zapovedniks* face the threat of being converted into national parks. Currently the mechanisms for public participation are in place. Therefore, *zapovedniks* have turned to NGOs and civil society for support. *Zapovedniks* have public outreach programs. They work with universities and schools, and administer Nature Protection Corps. For example, Ussuriiski *Zapovednik* is very actively involved in environmental education of high school and university students. Its public outreach program works with many schools and universities. It even organizes commercial tours. It promotes ecological tourism, has its own museum and hotel, and manages an ecological trail in the *zapovednik* buffer zone.[39] Ussuriiski *Zapovednik*, like many others, maintains a Soviet tradition of hosting and educating Nature Protection Corps activists who help rangers during the summer. Today, support for such activities comes not from the state, but from international NGOs. In the future, *Zapovedniks* and other reserves can be potentially good partners for Russia–China collaborative initiatives, especially for joint educational activities.

Russian and international NGOs in the RFE

In the RFE, the two most powerful activist forces are the international NGOs, as well as the established Russian NGOs. During *Perestroika*, numerous NGOs emerged in the RFE, thanks to the opened public sphere and the huge influx of Western funds. Compared to China, NGO registration is relatively easy, although much paperwork has to be done, and the NGO is required to have a bank account in order to get registered. Yet, since the Russian Duma (the national parliament) passed a new law on NGOs on 21 December 2005, control over NGOs has been stepped up. Under a revised version, international NGOs no longer have to register as a Russian organization as originally stipulated. The approximately 450,000 NGOs in Russia, however, will have to reregister with the Federal Registration Agency, which can close down organizations if their missions or activities damage "Russian sovereignty, independence, territorial integrity, national unity, cultural heritage, and national interest."[40] It is uncertain what the effect of the new NGO law on NGOs will be.

During the early years of NGO development, many were evanescent, although a number of them grew, matured over time, and went through an increasing process of professionalization. The Bureau of Regional Public

Campaigns (BROC) is one of such Russian NGOs in the RFE region. It actively works with the media and has considerable experience in organizing mass awareness campaigns in the press. The group was created by a journalist and almost all of its staff members are young university graduates or students. BROC is known as a very radical environmental organization, which can be seen in their statements, for example, "We are at war, at war for our environment."[41] It has never flinched from openly berating the government, industry, and even WWF-RFE. Believing that WWF-RFE failed to honestly address current forestry problems, BROC often criticized WWF's "compromising" partnership building, and not-so-radical stance on environmental issues. BROC solicits funding directly from Western foundations and does not rely on international NGOs operating in the RFE. It has its own network with multiple Western sources of funding, as well as local small student groups and journalists. For this reason, it is able to harshly criticize other NGOs without fear of losing funding. Most often they criticize NGOs for ineffectiveness in using Western money. As one activist stated in an interview:

> Thanks to Western Foundations, the money per day spent on every tiger in the Far East is around $250. Wouldn't you like to have such a salary? I guess, yes! However, tigers still do not flourish and multiply. This is because NGOs need money, and if the tiger population grows, the problem is gone, and money from the West stops coming. What will happen with the NGOs then?

BROC activists often operate as detectives, secretly following illegal traders heading for China or corrupt Russian state employees. When they acquire secret information, they disclose it to the public. BROC consistently releases "hot" (i.e. interesting and often scandalous) information related to various environmental violations. BROC is one of the few Russian NGOs that has actually made attempts to cope with China–Russia border environmental problems.

In 2000, BROC, in conjunction with the Russian NGO Dauria and several American NGOs (Forest Trends, Coastal Rainforest Coalition, and PERC), organized a conference on the Russia–China timber trade. They also included the participation of Chinese researchers and Chinese NGO representatives. Experts from BROC participated in a study on the Russia–China timber trade, initiated by Forest Trends in 2004. The conference assessed the current situation and discussed possibilities of organizing a campaign to promote sustainable forestry and responsible trade between the countries. A few other examples of NGOs active in the RFE, are Ecodal and the Phoenix Foundation. The former consists of scientists and lawyers in the Khabarovsky region. They work on amendments to the Forest Code of Khabarovsky Kray, conduct scientific research, initiate lawsuits in order to protect citizen's environmental rights and act on behalf of pollution victims. The Phoenix Foundation has, uncharacteristically, grown to be a grant-giver itself. This NGO solicits funds from Western foundations and redistributes them in the form of small grants to small grassroots groups. Most of its support is related to preservation of rare and engendered species, especially the Amur tiger. They organize annual tiger festivals, support tiger education programs, and work with anti-poaching brigades that were originally created by WWF.

Apart from the Russian NGOs, the RFE is also the working area of international NGOs. In fact, virtually all international foundations consider the RFE a hot spot for biodiversity conservation. In particular, they focus on the preservation of old growth cedar forests and endangered species, such as leopards and tigers, and their habitats. There is a high concentration of international NGOs in Vladivostok. Many foundations invest in conservation within this region, as well as the empowerment of the Russian NGO community to promote preservation. WWF-RFE, the Initiative for Social Action and Renewal (ISAR) of the Far East, Reproduction of Lessons Learned (ROLL),[42] WCS, PERC, GGF and ICF are the major funding agencies of environmental initiatives in the region. Each of these funding agencies is a node around which a whole variety of smaller NGO groups operate due to the small grant programs.

WWF-RFE is one of the most prominent NGOs in the region. WWF's work in the RFE began in 1994. They have four major programmatic themes: forest, marine, preservation of rare and endangered species, and protecting freshwater ecosystems of the Amur River. In the mid-1990s, WWF created a state inspection team called Tiger to prevent illegal poaching and trade with China. Later on, its focus shifted to disrupting criminal networks and mafia control of the region's forest sector. For this purpose, another brigade, Kedr, was created.[43] Initially, Tiger and Kedr were affiliated with and supervised by government agencies, but all funds, equipment, training, legal counseling, and salaries came from WWF-RFE. WWF made considerable efforts to promote sustainable forestry practices in the RFE by promoting model forest territories, fostering forest certification and assisting and monitoring companies that decided to take the certification path. WWF also actively works with media to notify the public about its activities.

Similar to the Chinese NGO community, competition for funding among environmental activists fosters conflict and has diminished the mobilization capacity of NGOs in coping with severe cross-border problems. The following statements from our interviews demonstrate this isolationist tendency among NGOs: "*We* are working, but *they* [other NGOs] are only washing money";[44] "*we* are working, but *they* are only making noise,"[45] while another interviewee stated: "*we* are working, but *they* ignore our work and think that they are the only ones working" (italics added).[46] The isolationist character of the NGO community in the RFE can be considered a negative externality to the foundation's hot-spot strategy, as is also shown by other Russian regions which host much less conflict within the NGO community.

International NGO branches of the RFE, primarily WWF, TRAFFIC, ISAR, ROLL, ICF, mainly network with their Moscow offices, Western offices, Western partners, and with NGOs in the Far East to whom they provide support. Those international NGOs that do not have offices in Moscow, such as PERC and WCS, have created a network of grant recipients in the target region. Cooperation within the international NGO community in the RFE exists, but is limited. NGOs conduct joint actions, meet at events, and have personal contacts with each other's staff. For example, Phoenix and ISAR send their press releases through WWF, thus using WWF infrastructure when working with the media. However, generally NGO networks mainly stretch to Russia's capital and to the West.

Russian perceptions of China: constraints to cooperation

The NGOs that are expected to take the lead to forge ties with China are understandably the two most powerful forces in the RFE, namely the established Russian NGOs and the international NGOs. Most NGO representatives acknowledge the cross-boundary nature of environmental problems in the border region, but simultaneously frame it as a Chinese problem. In almost every interview with NGO representatives in the RFE, statements such as "our major environmental problem is the Chinese," "closeness to China," or even "the Chinese invasion" were blurted out.

Some attribute the environmental problems in the RFE to the asymmetrical population density between the Chinese and Russian border zones. As one interviewee put it: "Environmental problems are created by people, and on our side of the border we have ten times fewer people than in China. That is why the Chinese consider our region practically a *zapovednik.*" Others pointed to imbalanced human activities, as an activist stated: "If you come to the bank of the Ussuri River, you can see that on our side there is forest, while on their side there are only agricultural fields, roads, settlements, and towns."[47] Russian environmentalists also note that China is increasingly becoming a source of pollution for Russia as a result of the rapid Chinese urbanization and industrialization. One of them lamented:

> The city of Fuyuan has developed very quickly in the last decade. The Sungari River from China flows into the Amur River. Along the Sungari River, many paper and metallurgical plants have been during this time. They discharge everything without treatment plants ... Everything comes to us in Khabarovsk.[48]

And indeed, the words of this interviewee became a grim reality in November 2005, when due to an explosion in a state factory in Jilin tons of toxic benzene spilled into the Sungari, and killed all aquatic life.

Many of our respondents from environmental organizations blamed much of the environmental problems on the influx of Chinese businessmen into the RFE. One interviewee depicted the Chinese businessmen as people who carried a suitcase full of money, in search of cheap timber. One respondent from WWF-RFE said: "It is a kind of a slithering expansion of China into the RFE. They are coming here, legalizing their visas, and infiltrating our economy."[49] Russian environmentalists see China's huge market demand and the asymmetrical economic development of the two countries in last two decades as stimulating illegal and ecologically detrimental activities in Russia:

> People are poor in our villages. They just want cars, prostitutes [*sic*], and houses like in China. And so, here you have a normal village where people finally figured out how to get easy and quick money. So, they go to the forest and they cut trees. There is little work in the villages and so people live there on growing marijuana, on illegal logging, and on golden root [an endangered Red Data Book plant with, medicinal roots]. Their money is illegal. In our region, it is impossible to get money legally and fast. Because of the Chinese market demand you cut a tree and get 100 dollars right away.[50]

Russian environmentalists expressed their abhorrence and fear about the rapid expansion of China's market economy. A member of WWF called China's commerce "a bottomless market that swallows everything."[51] Another informant compared China with a vacuum-cleaner, devouring everything. Aversion towards the environmentally insensitive and exploitative behavior of Chinese sojourners who are investing and working in the RFE is not unusual among the interviewees. One interviewee burst out:

> Chinese citizens under the name of Russian helpers are taking territory for agriculture! According to the law Chinese citizen cannot do this, but Russians can, so they pay Russians and then use the land. In order to build greenhouses they cut trees along the streams. In Russia it is prohibited to do so along the waterways. They do not even think about Russian legislation! . .. They catch whatever moves and eat, soon there will be no frogs in the area![52]

Given the gravity of the situation, almost every Russian NGO that we visited in Vladivostok and its precincts expressed genuine desire and a sense of urgency to partner with Chinese NGOs in combating cross-boundary environmental problems. In spite of this, very few have taken the initiative to forge cooperative ties with Chinese NGOs. Attempts that have been made by Russian NGOs can best be said to be sporadic. For instance, BROC participated in a few international conferences on border environmental issues but no further actions were made to forge ties with the Chinese side. The Amur-SEU sent a student who speaks Chinese to look for Chinese NGOs in Harbin but she returned without results. To date, the most promising attempt that is likely to foster ties between NGOs of two sides is the Amur/Heilongjiang Ambassador Campaign that started in June 2005.

When asked why they have not formed cooperative ties with their Chinese counterparts, some Russian NGOs in the RFE pointed to the language barrier as the primary reason. It is true that few Russian NGO members can speak Chinese well enough and few Chinese activists speak Russian well enough to enable smooth communication. Furthermore, the Russian NGOs complained about the poor English communicative ability of the Chinese NGOs. Due to financial dependence on the West, the working language of big environmental NGOs in the RFE is English. Many local small NGOs, however, do not interact at the international level, and cannot speak English. Language can truly create problems when small local environmental groups of the two sides try to communicate, but this is less of a problem for student groups. For established NGOs, language should not be an insurmountable barrier if they earnestly want to reach the Chinese counterparts.

While small Russian NGOs in the RFE had little knowledge of Chinese NGOs, big Russian NGOs often complained that they could not find Chinese counterparts with which to cooperate. A few NGO representatives we interviewed in April 2004 did not even think that there was an environmental movement in China. In their opinion, since the Chinese environmental NGOs dared not confront their own government, they should not be counted as a vital force. They also did not value the student environmental groups very highly. One NGO representative, when asked about his opinion of the NGOs in China, said that there were only a number of student groups on the other side of the border, whose role was trivial and insignificant. The Russian NGOs doubt what these "handful of tree-planting, swamp cleaning" student groups on the other side of the Amur River could do to help solve the border environmental problems. Such an assessment and attitude partly explains why the more established Russian NGOs whle crying for the help of Chinese NGOs were not particularly enthusiastic about taking practical steps to facilitate partnership.

Unfortunately, the Russian NGOs' indignant view can potentially stymie cross-border cooperation. Especially as some of the criticisms turned "the Chinese" into an abstract object, and prevented the Russians from actively seeking Chinese partnership, and it would very likely turn away the Chinese NGOs. A historical and structural analysis of the border situations, in which their position and views are reflected and contextualized, is much needed for the Russian NGOs. It must be pointed out that the way in which Russian NGOs framed "Chinese guilt" reflects the fear and abhorrence of the population at large in the RFE towards the Chinese "demographic expansion." This fear has historical roots and it has emerged again, this time against the background of a severe post-Soviet economic crisis, a decrease in production across the board, a significant population flight from the RFE, and an internationally weakened Russia.[53] Human history has proven once and again that during difficult times, a scapegoat, usually an outsider, is found to account for abysmal conditions and to strengthen in-group solidarity.

Concluding observations: perspectives for cooperation?

In this contribution, we have reviewed the situation of NGOs in the RFE and Northeast China. It seems that there is a significant asymmetry between the two civil societies in the region. We want to point out that this is one of the most important reasons for the lack of Sino-Russian partnership. The NGOs in the RFE border region are substantially greater in number than those in neighboring Northeast China. Furthermore, in the RFE, there exist a number of environmental NGOs that are professional, well connected to Western foundations, and skillful in securing grants. Chinese NGOs of this level are concentrated mainly in other places, such as Beijing, Shanghai, or Yunnan. Russian NGOs and branches of international NGOs are, to a large extent, independent of state control. They are therefore ready and willing to adopt a critical attitude toward the government and to confront it over environmental issues. They command substantial resources and mobilization power. These environmental organizations are the most active and powerful players in the RFE, and are also the ones most likely to approach the Chinese NGOs.

On the Chinese side of the border, this type of established NGOs simply does not exist. College student environmental groups are the principal players of the NGO community in Heilongjiang. Although different student groups exhibit different degrees of independence, all student groups are under the control of the university CYLs, and thus are controlled by the state. GONGOs in this region are largely government-led entities. The few existing non-student local NGOs are not truly independent, and they generally describe themselves as "government helpers."[54] Most of these Chinese NGOs are not versed in grant application procedures and lack organizational skills. Their locus is local and their capacity weak. They are poorly connected with the international community.

This asymmetry has prevented Russian NGOs from viewing Chinese NGOs in the border region as equal and worthy partners. They have reservations about the capacity of the NGOs on the other side of the Amur River and wonder what they actually can do for the solution of the cross-border problems. As forging transnational ties with Chinese NGOs is a costly undertaking, the Russian NGOs are concerned that cost might greatly overweigh benefits. Moreover, if isomorphic organizations are more likely to form close ties than organizations with disparate structures and cultures,[55] the asymmetry in terms of organizational structure, capacity, size, and degree of professionalization certainly does not encourage the formation of partnership between the parties on the two sides of the Sino-Russian border. Even among student NGOs within the city of Harbin, the one that is more independent from the CYL's control will tend to think that the other organization, with less degree of organizational independence, is its inferior.

A question arises: Where did this asymmetry in organizational form and culture come from? This asymmetry has primarily been shaped by the differential participation and investment by Western foundations and NGOs in the RFE and China's Northeast. In deciding where to carry out operations and invest resources, Western foundations and environmental NGOs often employ the "hot-spot strategy," i.e. to highly concentrate their resources globally in a few sites which possess great ecological value. For example, Yunnan province in Southwest China, where the rich biodiversity has attracted a large number of international organizations and the local NGO community has experienced empowerment, is comparable with the capital Beijing in terms of the number and capacity of NGOs. Likewise, the ecological value of the ecosystems makes the RFE a high priority and a "hot spot" for Western funding. Therefore, a great number of Western foundations channeled money to the RFE and various international NGOs set up their offices there. International NGOs through small grant programs create and maintain a large number of grassroots NGOs in the RFE. By contrast, the China side of the China–Russia border, especially parts of Heilongjiang province, is primarily agricultural and industrial. As a result of industrialization and the state-organized large-scale land reclamation campaigns from the late 1950s to the 1970s, few places retain their pristine state, and so are not considered as biodiversity hot spots by international foundations. As a result, much less international funding or effort has come into China's Northeast to foster the growth of a grassroots NGO community compared with that flowing into the RFE. In other words, the hot-spot strategy of Western foundations and international organizations, and the high dependency on Western funding of Russian and Chinese NGOs, as well as the different ecological compositions of the Russian and Chinese border areas, make the growth of local NGOs on the two sides of the border diverge. While the RFE has a vibrant NGO community in which many are well established, the Chinese NGOs in Heilongjiang are much fewer and less developed. Most of them are almost invisible at the international level.

Despite the positive results of the hot-spot strategy on the development of civic initiatives, there are also various negative impacts. First, the artificially accelerated development of local groups does not guarantee their drive to resolve the environmental problems. In addition, artificially created groups are usually disconnected with their communities. They primarily cater for the interests of their donor. They are also financially unstable and ephemeral. In response to the hot-spot strategy, there is a high density of NGOs in the RFE and a number of NGOs are located in the same niche. This intensifies their competition for funding. Instead of cooperating to resolve environmental issues, NGOs are engaged in petty conflicts and waste energy in promoting their image and logo. Lastly, meaningful conservation work can only be done in accurately delineated natural ecosystems. In the Amur Basin, all major natural ecosystem types go across the border. So if the hot spot is confined to just the RFE, it is defined improperly and less meaningfully. In addition, since the main root causes to various environmental problems in the RFE are transnational in nature, and China's increasing economic influence on the RFE implies a rising negative environmental impact on the RFE, any solution to these problems that fails to address the Chinese side is bound to be partial and ineffective.[56]

It is not our intention to leave readers with an impression that the picture is all bleak. On the contrary, we want to show below there is potentially great room for both sides to come together. Chinese NGOs are not unwilling to develop cross-boundary ties. NGOs in developing countries welcome the opportunity to work with foreign NGOs, especially those that possess great resources. Not only will this enable the immediate influx of funding and other resources that they badly need, but it will bring in international attention and new opportunities. Furthermore, NGOs view cooperation with foreign NGOs as a learning experience to improve their organizational skill and member competence. It will also become a highlight in the NGOs' achievements, which is good for gaining social recognition and publicity. GONGOs show interest in cooperating with foreign NGOs, specifically from Hong Kong, Japan, and America. They have less interest in forging ties with Russia NGOs perhaps, but if cooperative projects can be demonstrated to be mutually beneficial, they could hardly resist.

It is a mistake on the side of the Russian NGOs to underestimate the capacity and potential of the student groups in Northeast China. After all, a few prominent ones have been there for almost a decade and keep on growing. They are backed up by a supportive network which is constituted by the CYLs and school administrations, faculty advisers, alumni, student green groups nationwide, local media, and international NGOs like GGF. Some of these student groups, like Green Union of HIT, have built on their good legacy and entered a benign circle of development. They continue to attract fresh blood, expand their programs, and undergo institutionalization. Moreover, student groups, compared with local small groups and GONGOs, tend to be more globally conscious and would care more for the ecosystem of another country. Also, cooperation with student groups would perhaps be more manageable because language would be less of a barrier for the Chinese students who study Russian or English and are able to handle basic communication. The Amur/Heilong Ambassador Campaign, sponsored and organized by WWF-China and WWF-Russia in the summer of 2005, is a case that shows the great potential of the Chinese and Russian student groups to come together and for their interest to be directed to cross-border environmental issues. It is too early to assess the success of the project. Nonetheless it serves as a good start that hopefully more similar initiatives will follow.

The local small group's focus is narrower. However, for the groups working close to the border, if there is an external agency or project to match them up with Russian NGOs on issues of similar interest, formation of cooperative ties is not impossible. There is a value to incorporating them since there is a possibility that poachers and illegal harvesters may come from these villages. On the Russian side, there is already a strong desire to cooperate with the Chinese NGOs. With the clear understanding that the solution of environmental problems in the RFE cannot be possibly achieved without cooperation from the Chinese side, Russian NGOs need to have more initiative, persistence and long-time commitment in their endeavors to form ties with Chinese counterparts and be prepared for all kinds of difficulties and frustrations that accompany this. The cooperative projects should involve different entities of the NGO community in the RFE. For example, *zapovedniks*, which have a long tradition of hosting Russian student groups, can consider hosting Chinese student groups and other Chinese groups as well. Through appreciating the wilderness and rich biodiversity of the RFE, the Chinese groups will be motivated to contribute to the solution of cross-border environmental problems. Funding should not be too much of a problem, if Western funding agencies start to adopt a more holistic conception about the RFE, i.e. to put the RFE within a more broadly defined and more dynamic system. Fortunately, a few international NGOs are becoming increasingly aware of the necessity to facilitate the cooperation between the civil societies of the two sides.[57]

Notes

1 Research for this paper was supported by Starr Collaborative Grants Program 2003–4, International Research and Exchange Board (IREX). The authors express their gratitude to Svetlana Pchelkina, researcher of the Centre for Independent Social Research in St Petersburg, who provided her interviews conducted in the RFE in 2003 for use in this paper. We are especially grateful to Richard Louis Edmonds, Peter Ho, Eugene Simonov, Dingxin Zhao, and anonymous reviewers for their advice and comments on earlier drafts. Our sincere thanks also go to our informants both in China and Russia. Without their generous help, this paper could not possibly have been done.

2 R. Johnson, "Laws, States, and Superstates: International Laws and the Environment," *Applied Geography*, Vol. 12 (1992), pp. 211–28; T. Princen and M. Finger, *Environmental NGOs in World Politics: Linking the Local and Global*, London: Routledge, 1994, pp. 11, 30, 62–3; T. Princen, M. Finger and J. Manno, "Nongovernmental Organizations in World Environmental Politics," *International Environmental Affairs*, Vol. 7 (1995), p. 50; O. Young., G. Demko, and K. Ramakrishna, "Global Environmental Change and International Governance," summary and recommendations of a conference held at Dartmouth College, Hanover, NH, 1991, p. 6; G. Porter and J. Brown, *Global Environmental Politics*, Boulder, CO: Westview Press, 1991, pp. 35–46.

3 E. Meidinger, "Forest Certification as a Global Civil Society Regulatory Institution," in E. Meidinger, C. Elliot and G. Oesten (eds) *Social and Political Dimensions of Forest Certification*, at www.fortsbuch.de, Freiburg, 2003, pp. 265–89; E. Meidinger, "Forest Certification as Environmental Law Making by Global Civil Society," in Meidinger *et al.*, pp. 293–329; D. Fisher, "Global and Domestic Actors within the Global Climate Change Regime: Toward a Theory of Global Environmental Systems," *International Journal of Sociology and Social Policy*, Vol. 23 (2003), pp. 5–30; C. Elliot and S. Rodolphe, "Global Governance and Forest Certification: A Fast Track Process for Global Policy Change," in Meidinger *et al.*, pp. 199–217; V. Haufler, "New Forms of Governance: Certification Regimes as Social Regulations of the Global Market," in Meidinger *et al.*, pp. 237–47.

4 P. Wapner, "Politics beyond the State: Environmental Activism and World Civic Politics," *World Politics*, Vol. 47 (1995), pp. 311–40; P. Wapner, "Horizontal Politics: Transnational Environmental Activism and Global Cultural Change," *Global Environmental Politics*, Vol. 2 (2002), pp. 37–62; T. Lewis, "Transnational Conservation Movement Organizations: Shaping the Protected Area Systems of Less Developed Countries," *Mobilization*, Vol. 5 (2000), pp. 105–23; J. McCormick, 'The Role of Environmental NGOs in International Regimes" in N. Vig and R. Axelrod (eds) *The Global Environment*, Washington, DC: CQ Press, 1999, pp. 52–71; T. Tvedt, "Development NGOs: Actors in a Global Civil Society or in a New International Social System?" *Voluntas*, Vol. 13 (2002), pp. 363–75. P. Nelson, "New Agendas and New Patterns of International NGO Political Action," *Voluntas*, Vol. 13 (2002), pp. 377–92. M. Tysiachniouk and J. Reisman, "Transnational Environmental Organizations and the Russian Forest Sector," in J. Kortelainen and J. Kotilainen (eds) *Environmental Transformations in the Russian Forest Industry*, Joensuu: University of Joensuu, 2002, pp. 56–72.

5 J. Smith, C. Chatfield, and R. Pugnucco (eds), *Transnational Social Movements and Global Politics: Solidarity Beyond the State*, Syracuse, NY: Syracuse University Press, 1997, pp. 59–80. M. Keck and K. Sikkink, *Activists beyond Borders: Advocacy Networks in International Politics*, Ithaca, NY: Cornell University Press, 1998.

6 For example, M. Rodrigues, "Advocating for the Environment: Local Dimensions of Transnational Networks," *Environment*, Vol. 46 (2004), pp. 14–25.

7 Primary fieldwork was conducted during the spring and summer of 2004. We updated some of the information in October 2005.

8 Following two border treaties signed respectively in 1991 and 1994, China and Russia signed the Supplementary Agreement on the Eastern Section of the China–Russia Boundary Line in Beijing on 14 October, 2004, thus completing the delimitation of the 4,300 km boundary line between the two countries.

9 Although the environmental problems are by no means limited to the three, these are the most frequently mentioned by our interviewees. For a comprehensive review of the environmental problems existing on the Sino-Russian border, please refer to E. Simonov, *Amur/Heilong River Basin Reader* (book manuscript).

10 M. Yamane and W. Lu, "The Recent Russia–China Timber Trade – an Analytical Overview," an interim report of IGES (Institute for Global Environment Strategies) Forest Conservation Project, 2000; W. Lu, "The Recent Changes of Forest Policy in China and its Influences on the Forest Sector," IGES interim report, Forest Conservation Project, 1999; X. Sun, E. Katsigris and A. White, "Meeting China's Demand for Forest Products: An Overview of Import Trend, Ports of Entry, and Supplying Countries, with Emphasis on the Asia-Pacific Region," a report of Forest Trend, 2004. X. Sun, N. Cheng, A. White, R. West, and E. Katsigris, "China's Forest Product Import Trends 1997–2002: Analysis of Customs Data with Emphasis on Asia-Pacific Supplying Countries," a report of Forest Trend, 2004.

11 Sun, Cheng, White, West, and Kafsigris *et al.*, 2004.

12 Yamane and Lu, op. cit.

13 P. Vandergert and J. Newell, "An Analysis of Illegal Logging and Trade in the Russian Far East and Siberia," *International Forestry Review*, online, available at www.atypon-link.com/CFA/loi/ifor>, 2003, Vol. 5: 303–6.

14 The Red Data Book of the Russian Federation, also known as Red Book or Russian Red Data Book, is a state document established for documenting rare and endangered species of animals, plants and fungi, as well as some local subspecies that exist within the territory of the Russian Federation and its continental shelf and marine economic zone. The book has been adopted by Russia and all CIS states to enact a common agreement on rare and endangered species protection.

15 CITES' proposal "Inclusion in Appendix II of Roots of Panax Ginseng," last updated 7 December 2000.

16 *Vladivostok News*, "Chinese Frog-Hunters Said to Poison Rivers," 26 March 2003.

17 Reports of such confiscations can be found at: www.phoenix.vl.ru/zoom/bust.htm and www.phoenix.vl.ru/zoom/shipment.htm (accessed 27 July 2005).

18 *Tichookeanskaya Zvezda* [Pacific Ocean Star] "*Ochota za Amur Phenolom*" [Hunting for amur-phenol], 5 July 2000; *Tichookeanskaya Zvezda*, "*Phenol v Vode, Pomoika na Beregu*" [Phenol in water, dumping ground on the bank], 8 October 2002.

19 *Molodoi Dalnevostochnik* [Young Far Easter] "*Zona Spetsialnogo Vnimania*" [Zone of special attention], 11–18 December 2002.

20 See also E. Simonov, op. cit., Chapter 4.

21 TRAFFIC is a program operating under WWF with the aim to monitor the Convention on International Trade in Endangered Species of Wild Fauna and Flora (CITES).

22 This paper focuses on NGOs operative in the border regions. National NGOs both in Russia and China are rarely mentioned because they have not notably been engaged in combating cross-border environmental problems.

23 In our examination of Chinese NGOs along the border, we have focused our fieldwork in Heilongjiang province. However, we have also traveled to the Hunchun Tiger–Leopard Reserve in Jilin province, which is separated only by mountains from the protected habitat in Borisovkoe and Barsovy protected areas on the Russian side of the border.

24 For example, many of the activities organized by Green Union and Green Angel were beyond the confines of their campuses and had social impact. Green Union had hosted a Sunday radio program on the environment with the Harbin Art and Culture Radio Station for three consecutive years. Its yearly spring tree-planting attracted enthusiastic participation of students from other universities and Harbin citizens alike, with the number of participants ranging from 700 to 5,000. Green Angel published its own journal "Dandelion" (*Pugongying*) and organized its first summer camp to several nature reserves near the Chinese–Russian border area in July–August, 2004. The camp admitted members from other student green groups as well. Moreover, facilitated by the WWF–Russia consultant studying at NFU, Green Angel had an exchange program with the Russian student summer camps in the RFE.

25 Among other things, both received grants from GGF. GGF in recent years has played a pivotal role, through the work of its China coordinator, in promoting the growth of grassroots environmental organizations, especially student environmental organizations, by distributing small grants to them.

26 For example, Green Union was honored as the Nation's Exemplary Student Groups of Excellence (*quanguo youxiu xuesheng shetuan biaobing*) by CYL of China, Ministry of Education and All-China Students' Federation in March 2005. Five months later, Green Angel was awarded as one of the Nation's Best Ten (Student) Environmental Groups of Excellence (*quanguo shijia youxiu huanbao shetuan*) by the Chinese Association for Promoting Environmental Culture (*Zhongguo huanjing wenhua cujin xiehui*).

27 H. Lu, "Bamboo Sprouts after the Rain: The History of University Student Environmental Associations in China," *China Environment Series*, Vol. 6, (2003), pp. 55–67.

28 In addition to disunion and attrition, Green Longjiang had to surmount yet another difficulty, i.e. to register with the Bureau of Civil Affairs in Heilongjiang in order to obtain legal status. The current governmental regulations permit only one NGO in a particular sphere of activity to be registered at each administrative level. Given that the Environmental Volunteers' Association of Heilongjiang Province (EVAHP) was already registered as an environmental NGO at the provincial level, Green Longjiang was left with not much choice but to register as a secondary social organization affiliated with the EVAHP, which was a first-level social organization.

29 The roots of the Russian environmental movement can be traced back to as early as the 1920s. See D. Weiner, *A Little Corner of Freedom*, Berkeley, CA: University of California, 1999, pp. 312–29.

30 Known as *Druzhina* in Russian.

31 O. Yanitsky, *Rossia: Ekologicheskii vizov* [Russia: Environmental Challenge], Novosibirsk: Sibirski Chronograph, 2002, pp. 98–107.

32 *Zapovedniks* were created in the beginning of the twentieth century according to the climatic zones with the purpose predominantly for research. Human access for recreation as well as any commercial activity was prohibited.

33 The relationship between Nature Protection Corps and VOOP varied in universities, some more conflictive, others more cooperative. CYL and VOOP often competed over the ability to control and influence Nature Protection Corps. Both CYL and VOOP were interested in claiming credits for involving the youth. In certain cases this competition positively affected the operation of Nature Protection Corps as both entities helped Nature Protection Corps operations and Nature Protection Corps leaders were involved in both CYL and VOOP, which allowed them through mediation and manipulation to receive additional resources. The relationship with univer-

sity leadership, faculty advisors and deans mattered greatly for Nature Protection Corps. The Russian sociologist Yanitsky calls universities the "engendering milieu" for Nature Protection Corps as universities protected the groups from Party-state threats, shaped their activities, provided knowledge on the environment, funding, access to libraries and other support. See O. Yanitsky, *Russian Greens in a Risk Society*, Helsinki: Kikimora Publications, 2000, p. 60.

34 For example, in Tomsk University one of the CYL leaders tried to engage the Nature Protection Corps in monitoring university forest lands, while the Nature Protection Corps wanted to be involved in catching illegal fisherman in wilderness areas. The conflict ended with the dissolution of the Tomsk University Nature Protection Corps. In 1973 the Central Committee of Communist Youth League tried to co-opt Nature Protection Corps by converting Nature Protection Corps into mass action across all 900 universities and demand Nature Protection Corps to recruit all kinds of youths into their organization. Nature Protection Corps, however, were not prepared for such a massive mobilization effort, and they were interested in maintaining themselves with dedicated environmentalists who shared the same values. Therefore, they resisted the Central Committee of the CYL demand and the CYL finally acknowledged that Nature Protection Corps, although on a small scale, were anyway an effective mechanism for involving youth in nature protection activities. See S. Mukhachev, and S. Zabelin, *30 let dvizeniya. Neformalnoe prirodoochrannoe dvizenie v SSSR: Fakti I dokumenti* [30 Years of the Movement. Informal Nature Protection Movement in the USSR: Facts and Documents], Kazan: Sotsialno-Ekologicheskii Soiuz.

35 *Glasnost* was one of Mikhail Gorbachev's policies introduced to the Soviet Union in 1985. The term in Russian means "publicity," "openness."

36 It is the Russian word (which passed into English) for the economic reforms introduced in June 1987 by the Soviet leader Mikhail Gorbachev. Its literal meaning is "restructuring," which refers to the restructuring of the Soviet economy.

37 Yanitsky, op. cit., 1996, p. 50

38 Weiner, op. cit., pp. 431–3.

39 Interview with the public outreach officer of Ussuriiski Zapovednik in Ussuriisk and participant observation in the buffer zone in 2004.

40 See NRC correspondent, "*Rusland stelt NGOs onder staatscontrole*" [Russia places NGOs under state control], *NRC Handelsblad*, 22 December 2005, p. 4.

41 Interview with BROC activist, 2003.

42 A project of the Institute for Sustainable Communities with a central office in Vermont and a branch in Moscow.

43 It means "cedar" in Russian. Kedr was a group of four men in a jeep, equipped with communication technology, computer databases, and guns, while checking logging trucks for the wood's legal documentation. Kedr existed as an individual brigade within Tiger, but dealt specifically with illegal logging.

44 Interview with BROC, 2003.

45 Interview with WWF, 2003.

46 Interview with ISAR, 2003.

47 Interview with BROC representative in 2003.

48 *Molodoi Dalnevostochnik* [Young Far Easter], "Cholera rmesto ribi" [Cholera instead of Fish], 4–11 July 2003.

49 Interview with WWF representative in 2003.

50 Interview with BROC, 2002.

51 Interview with head of WWF–RFE's Forest Program in 2002.

52 Interview with Amur-SEU representative in 2003.

53 A. Lukin, "The Image of China in Russia Border Regions," *Asian Survey*, Vol. 38 (1998), pp. 821–35; A. Lukin, *The Bear Watches the Dragon: Russia's Perceptions of China and the Evolution of Russian-Chinese Relations Since the Eighteenth Century*, Armonk, NY: M.E. Sharpe, 2003.

54 From the interview with the founder of Green Mother Earth in 2004. Indeed, such phrases as "government helper" or "to assist the government" are most commonly heard from Chinese grassroots NGOs when they define their own role.
55 M. Diani, *Green Networks: A Structural Analysis of the Italian Environmental Movement*, Edinburgh: Edinburgh University Press, 1995.
56 We would like to thank Eugene Simonov for his insights on this point.
57 Interviews with Phoenix, ROLL, and GGF-USA in 2003.

9 Transnational advocacy at the grassroots

Benefits and risks of international cooperation[1]

Katherine Morton

Over the past few years an increasing number of Chinese non-governmental organizations have established extensive international networks.[2] They now participate in international conferences; promote Chinese concerns in international campaigns; and work with international partners to bring about change on the ground. Perhaps even more significant is the fact that many of China's more autonomous NGOs are totally dependent upon international funding. Yet, few attempts have been made to examine the impact of international support for NGOs in China.[3] This is a large and complex terrain that cannot be adequately mapped out without extensive research. It involves examining the roles and behavior of multiple agencies including multilateral and bilateral donors, NGOs and increasingly corporations.[4] The purpose of this contribution is not to attempt such a challenge but rather to provide some insights into the evolving relationship between local and international NGOs (INGOs).

From an environmental perspective, it is important to examine this relationship for at least two reasons: first, the increasing density of local–international NGO linkages is a critical indicator of the extent to which Chinese environmental groups are able to move beyond purely local concerns to address the trans-boundary nature of China's environmental degradation. Second, the overlap between transnational and local social spaces has implications for the development of greater openness and accountability within Chinese society, which in turn is likely to affect China's capacity to deal with its environmental crisis, especially at the local level.

It is important to stress here that international support for environmental NGOs in China is highly contingent upon broader developments within Chinese civil society (*gongmin shehui*). As we have seen from the previous contribution, the potential for INGOs to act as a catalyst for environmental advocacy often exists but remains embedded in China's current semi-authoritarian context. It is likely to vary considerably across regions and issues. Sun and Tysiachniouk suggest that the nature of local state–society relations together with the priorities and funding prerogatives of INGOs are key factors behind the failure to develop NGO advocacy across the China–Russia border. It may well be the case that the local associational space on the Chinese side needs to develop further before cross-border environmental issues can be effectively addressed. Clearly, without the impetus from below, externally driven transnational cooperation is unlikely to make a significant impact. It is for this reason that in this contribution I take a bottom-up perspective that seeks to address three central questions: how are international linkages affecting the development of environmental advocacy at the grassroots in China? What are the key enabling factors and constraints? And how can we assess the broader implications for democratic change within Chinese society?

The contribution begins with a theoretical discussion of the broader parameters of transnational advocacy at the global level. Particular attention is given to the ways in which NGOs are able to mobilize support and advocate for reforms within authoritarian settings. This discussion is then linked to developments within Chinese civil society. A focus on advocacy is increasingly relevant in the Chinese context. This is because many new environmental NGOs are beginning to shift their attention towards policy influence in addition to public education campaigns and more service-oriented activities. After providing an overview of the nature and scope of environmental advocacy in China, I then focus on two internationally oriented campaigns: the campaign to protect the Tibetan antelope and grasslands in Qinghai, and the campaign to reduce pesticide use in Yunnan.

The selected cases are by no means representative, but they do illustrate some of the potential benefits that can be gained from international support. My focus upon China's poorer and more remote regions is intentional. To date, much attention has been given to the growing number of relatively independent environmental NGOs located in urban areas, especially in Beijing. Far less attention has been given to the small number of independent NGOs working in poorer regions of China where arguably the environmental challenge is greatest, state capacity is weakest, and the demand for international support is particularly high.

At a broader level, my intention in this contribution is not simply to highlight the positive. Nor is it my intention to suggest that the real value of international support for NGOs in China lies in the potential to bring about a transition to democracy at the national level. It is often assumed by academics and policymakers alike that facilitating civil society in authoritarian or transitional states (through either state or non-state means) will, in the longer term, lead to improved democratic governance.[5] Yet the reality appears to be more sobering. A number of comparative studies have revealed that this development trajectory is by no means inevitable.[6] It is also by no means clear that democracy in the procedural sense of multiparty elections will lead to better environmental outcomes.[7]

In the case of China, I would argue that what is required is greater public participation and government accountability at the local level. These democratic norms are needed to enhance the quality of environmental management, to restrain corruption, and to act as a check against the concentration of power in the hands of a few with a strong interest in pursuing economic interests at the expense of environmental concerns. Strengthening participation and accountability can provide ordinary citizens with a political voice and, therefore, a means of exercising oversight over local public affairs. As other scholars have noted, these norms can be promoted from the bottom up; they do not require a huge transformation of the political system.[8]

My observations to date have confirmed that cooperation between local and INGOs can contribute to a more vibrant local democratic culture in two significant ways. First, it can reinforce the importance of public participation, and second, it can strengthen the capacity of local NGOs to advocate for policy reforms within a restricted political environment. It is too early to systematically assess whether the benefits of international support for Chinese NGOs are likely to outweigh any possible costs. What is clear, however, is that transnational advocacy is now a significant driving force behind the construction of a more autonomous and inclusive social space in China. It is, therefore, a feature of civil society development that can no longer be ignored.

Bringing in the transnational dimension

The practice of transnational advocacy is not new. Local and regional civil societies have always been historically interconnected.[9] For example, there exists a long history of religious communities working transnationally – Catholic missionaries, Sufi orders and Buddhist monks – that pre-dates the modern state. What is new is the dramatic increase in the number of NGOs working beyond territorial boundaries together with the growing density of transnational advocacy networks.[10] Over the past decade, these networks have attracted considerable attention. For many scholars of International Relations, NGOs are seen as a powerful stimulus for global reform, especially in relation to human rights, gender equity and the environment.[11] The assumption is that NGOs working across borders constitute a progressive force for change by mobilizing support for state reforms and greater democratic freedoms.

In looking at the bigger picture it is important to acknowledge that the rise of NGOs over the past decade has not been confined to the West.[12] As is clear from the case of China, even authoritarian states have witnessed a proliferation of NGOs working at the margins of the state. An important test case for assessing the progressive potential of NGOs is the degree to which they can operate and, indeed, be effective in authoritarian settings. In their study of transnational advocacy networks, Keck and Sikkink identified a "boomerang pattern" within which domestic social groups were able to link up to international organizations, which, in turn, put pressure on a repressive state to reform. In the field of human rights and the environment, a wealth of studies suggest that some authoritarian states, if sensitively targeted, can be persuaded to change their polices to align with international norms.[13]

The problem with this outsider perspective is that it underestimates the extent to which state–society relations constrain the behavior of NGOs working at the grassroots. Studies on NGOs working in developing and authoritarian contexts reveal that they are just as likely to reinforce the status quo as they are to change it.[14] The potential for these organizations to bring about change is greatly hindered by cultural and historic impediments, state obstacles, and internal constraints. Consequently, the tendency is to maintain a low profile. Before the lifting of martial law in Taiwan, for example, environmental NGOs mobilized around relatively safe issues such as welfare and consumer protection.[15] And during the Soeharto regime in Indonesia, NGOs focused on trying to increase public participation but within the prevailing mechanisms of government.[16]

In actuality, when the political space to influence the state is limited, as in the case of China, advocacy groups are more likely to take a pragmatic approach in attempting to influence government officials when it is deemed likely to be effective. By practicing advocacy that features "self-imposed censorship" and a "de-politicization of politics" (see the introduction)[17] they also stand to benefit from the support of government connections.[18] Arguably an "engaged civil society" is more likely to reshape the state and in time provide an impetus for greater democratic freedoms. But this approach could take years (if not decades). It also underestimates the importance of mobilizing support for democratic freedoms from within society. This requires collective action to bring about changes in social conditions and behavior as well as state practices.

In the two cases detailed below we shall see that the potential of transnational advocacy to bring about progressive change in China certainly exists, but that a number of core challenges remain. To provide a sense of perspective, the following section looks at some of the more recent developments within Chinese civil society and attempts to define the broader parameters of environmental advocacy.

The nature and scope of environmental advocacy in China

A major issue for scholars conducting research on environmental organizations in China is that they are notoriously difficult to define.[19] According to the Not-for-Profit Organization (NPO) network in Beijing there are currently over 50 environmental organizations operating in China.[20] A large number remain under government control. These government-organized NGOs (GONGOs) (*ban guanfang zuzhi*) include the China Environmental Science Foundation, the China Wildlife Conservation Association, and the China Environment Protection Fund.[21] Many Chinese environmental NGOs are also academic in nature. They are probably best defined as people-based non-profit organizations (*minban fei qiye danwei*) rather than social advocacy organizations (*minjian zuzhi*), although the boundaries between academic research and social advocacy are often blurred. The Center for Biodiversity and Indigenous Knowledge (CBIK) in Yunnan is a good example of a research organization that is relatively independent from government, primarily focused on research and consultancy, and also engaged in promoting the interests of poor ethnic communities.[22]

The total number of independent and legal (i.e. formally registered) environmental NGOs in China is difficult to obtain. Most attention to date has been focused upon a small number of high profile NGOs in Beijing such as Friends of Nature (FON) and Global Village (GVB) that were set up in the mid-1990s to coincide with the central government's increasing commitment towards environmental protection. Over the past four years, an increasing number of independent environmental NGOs have also been set up in the regions, especially in Yunnan province that has benefited from a concentration of donor funding.

Many commentators have stressed the pioneering role of Chinese environmental NGOs.[23] It is widely assumed that they are positioned at the vanguard of China's emerging civil society and enjoy a higher level of autonomy relative to other NGOs by virtue of the fact that their agendas mesh comfortably with that of the central government. From the government perspective, the advantage of environmental organizations seems to lie in their service delivery role. As in many other developing countries, NGOs in China are stepping into the void left by the state to provide basic social and environmental services such as water and sanitation, low cost energy supply, and environmental education. This bias towards the functional rather than the democratic potential of NGOs lends a somewhat different meaning to the notion of environmental advocacy.

In China, the advocacy role of NGOs, in conjunction with the media, has typically been limited to the public supervision of environmental laws and regulations. The charter of Friends of Nature, for example, states that it "gives support to as well as conducts supervision over the government."[24] But, in practice, this has meant acting as a check against the excesses of local governments rather than petitioning for national reform.

More recently, it would appear that a second generation of environmental NGOs has started to engage in a new form of environmental advocacy that is more aligned with international experience.[25] More specifically, it is aimed at the twin imperatives of public participation and policy reform, and it draws its strength from representing the interests of the people rather than the government. Self-censorship remains a unique characteristic of this kind of advocacy as reflected in the use of language. "Advocacy" translates into Chinese as *chang dao* or *chang yi*, yet the term *shoufu* is more commonly used to refer to policy influence or persuasion because it is considered to be less politically sensitive.[26]

The emphasis upon public participation has always been central to the work of independent environmental NGOs in China. Since the late 1990s, this has become easier with increased government support in recognition of the need to involve the public in environmental protection efforts. It was officially sanctioned as a guiding principle in a White Paper on Poverty Reduction in 2001.[27] Although government support for participation in practice is still weak and unevenly spread across regions, at least the concept is becoming more familiar. For example, a new environmental NGO, Community Action, registered in Beijing in 2003, dedicates its whole mission to improving public participation in community development.

The shift towards policy advocacy, however, is new. It appears to have been motivated by three factors. First, there exists a growing recognition amongst environmental NGOs that the distinction that is often made, between operational NGOs, providing essential services, and advocacy NGOs, primarily focused on lobbying governments, cannot be maintained in practice. In short, they are increasingly convinced that their efforts on the ground to improve basic environmental services also require broader policy reforms in order to be sustainable.

Second, support from the academic community in China has also helped to create a new space for NGOs to meet with government officials and articulate their concerns in a non-confrontational manner. In particular, international conferences, seminars and workshops organized by the Center for NGO Research at Tsinghua University have helped to raise the highly controversial issue of China's restrictive regulatory environment.[28] As noted by Zhao Liqing, to be effective NGOs cannot simply assume a service delivery role; "they also need to mobilize the necessary resources to influence the organizational structures within which they operate."[29]

Third, in the absence of concrete evidence, it is reasonable to assume that increasing levels of international funding have had some impact on the direction of Chinese NGOs. In particular, higher levels of cooperation between local and international NGOs have created a unique opportunity to exchange ideas and learn from international experience. INGOs, like their local counterparts, are also under increasing pressure to explore alternative means of influencing policy reform to ensure the sustainability of their projects on the ground.

New developments within Chinese civil society are, in turn, providing opportunities for INGOs and the development of transnational relations. Although a comprehensive analysis of transnational advocacy in China is beyond the scope of this contribution, we can at least distinguish three levels. The first level involves INGOs working in China from the outside. Given their access to financial resources, these external players enjoy the benefit of sharing social space relatively free from state control on condition that their activities do not conflict with the interests of the state. It is for this reason that the majority of INGOs work on poverty alleviation, the environment, health or women's issues rather than human rights or security. They tend to channel their funds through the Chinese Association for NGO Cooperation (CANGO), which was specifically set up for this purpose. The State Environmental Protection Administration has also set up a special unit to work with NGOs.

Problems relating to implementation have encouraged many INGOs to establish a presence in China, thus creating a second level of transnational advocacy. At a rough estimate, there are now approximately 20 INGOs established in China (including eight in Hong Kong) with a central focus on the environment.[30] What is interesting is that these organizations are currently operating in an administrative vacuum, although the recent new rules on the registration and management of international foundations approved by the State Council in February 2004 suggest that this situation may soon change.[31]

In theory, INGOS are expected to operate in a similar fashion to their domestic counterparts according to the Chinese law on organizations promulgated in 1998 (*zhu zhi fa*). But, in practice, they tend to negotiate their own terms with government agencies and local partners. Although clarity over tax status is urgently needed, the current ambiguities in the system allow for greater flexibility and freedoms. For some INGOs, as they become more localized into the Chinese environment, the issue of international versus domestic status may become irrelevant. The WWF, for example, is a Chinese NGO in all but name and chief representative. Many other INGOs appear to be following suit.

The third level of transnational advocacy in China is taking place at the grassroots. Just as an increasing number of INGOs are localizing their operations in China, local NGOs are moving in the opposite direction. The internet has allowed grassroots NGOs, even in remote parts of China, to establish extensive international networks. An important advantage of the internet is that it provides a direct link to potential donors. Many NGOs have been able to overcome significant financial constraints by taking advantage of small grants schemes offered by international agencies for development projects and capacity building. In addition to informal networks established over the internet, local NGOs are also building *ad hoc* international partnerships and participating in regional and international fora. For example, in 2002, for the first time in Chinese history, a delegation of 18 environmental NGOs participated in the NGO forum at the World Summit on Sustainable Development held in Johannesburg.[32]

The full range and density of transnational advocacy at the local level is difficult to grasp. It is clearly much easier to focus on a few well-known INGOs in Beijing and track their activities across China than it is to attempt to capture the more complex processes and interactions that are taking place at the grassroots. Nevertheless, despite the difficulties involved, I would argue that this kind of research is needed to provide an insider perspective on the emergence of transnational advocacy in China.

In the cases that follow, I have singled out two environmental campaigns in the provinces of Qinghai and Yunnan (see Figure 9.1). In keeping with the assumption that the value of international support is dependent upon how it is realized by the beneficiaries, I will focus primarily on the perspective of local NGOs. This has the added advantage of increasing our understanding of how individual NGOs actually operate in China on a day-to-day basis. As mentioned earlier, these are not intended to be representative cases. Instead, they merely provide illustrative examples of the ways in which international support can broaden the associational space within which local NGOs operate.[33]

Advocating action at the grassroots: the international campaign to protect the Tibetan antelope and grasslands of Qinghai

The high alpine grasslands of southwest Qinghai province support a unique ecosystem as well as a rich cultural heritage. At the center of the Qinghai–Tibetan plateau this region, known locally as the Sanjiangyuan, lies at the source of China's three main rivers – Changjiang (Yangzte), Huang (Yellow) and Lancang (Mekong). Largely inhabited by semi-nomadic herdsmen, it is also home to a wide variety of endangered species including the Tibetan antelope, wild yak, snow leopard, and black-necked crane. But over the past three decades unsustainable development practices, overgrazing, and illegal

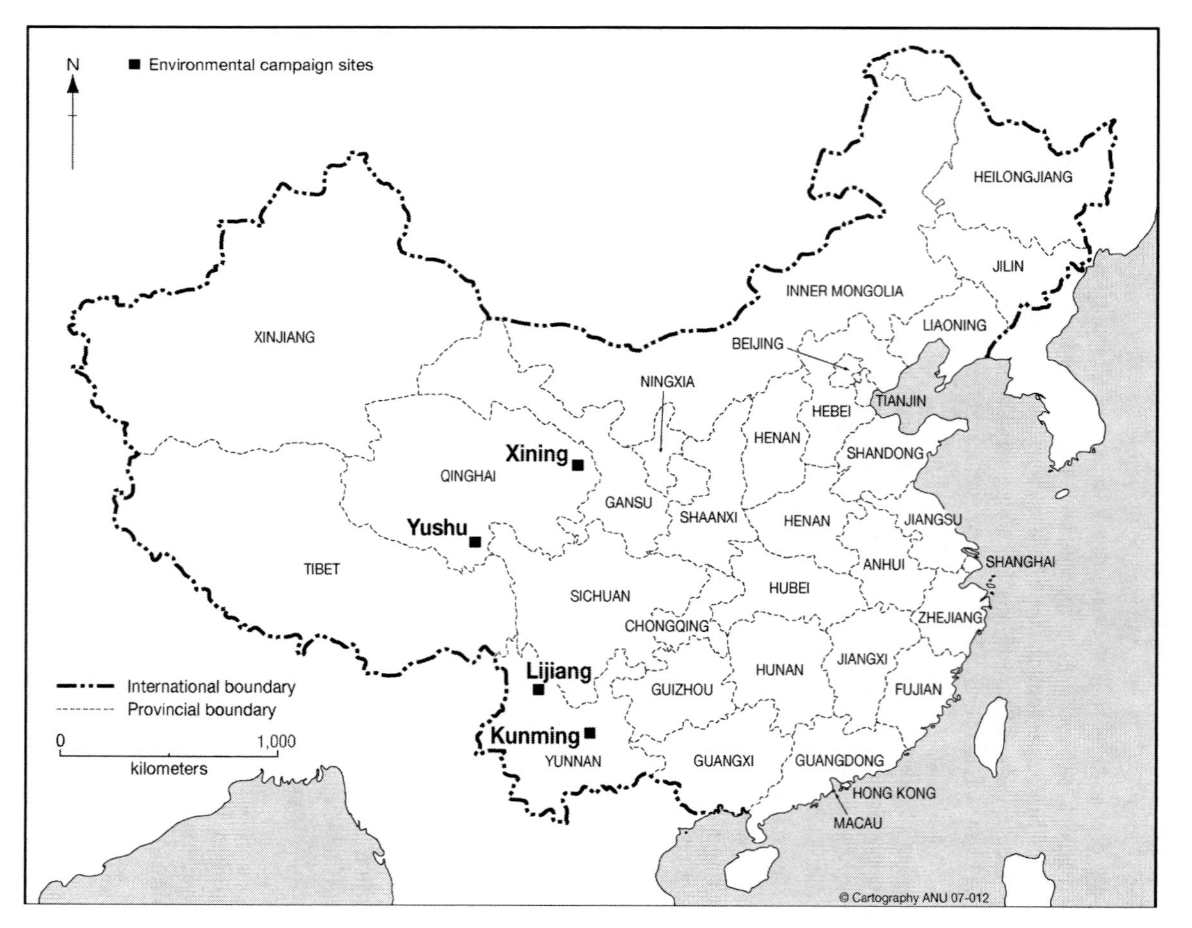

Figure 9.1 Map of China showing environmental campaign sites.

poaching together with global climate shifts have led to a marked decline in native wildlife and the productivity of traditional Tibetan pastoralism.

International attention has largely focused on the plight of the Tibetan antelope. Classified under Appendix 1 of the Convention of International Trade in Endangered Species (the highest legal protection for any species) the population of the Tibetan antelope or chiru declined to under 75,000 in the 1990s – equal to just 10 percent of the population 100 years ago.[34] Field studies conducted by international teams of biologists revealed that as many as 20,000 antelopes were being slaughtered annually to supply the trade in luxury shawls woven from the ultra-fine under-fleece called *shahtoosh* (Persian for "king of wools").[35] Since 1981, the Chinese government has banned the poaching and trading of this species but enforcement has remained sporadic and largely unsuccessful.

The local struggle to conserve the Tibetan antelope has been no less than heroic, winning both international and domestic support. At the height of the slaughtering in the mid-1990s, a small group of local people took authority into their own hands. Jiesang Sonan Dorjay,[36] an ethnic Tibetan, gathered together some police officers from Zhiduo county to hunt down the poachers.[37] The Wild Yak Brigade, as it became known, organized patrols of the Kekexili area that covers parts of Qinghai, Xinjiang and Tibet. In January 1994, the arrest of 15 poachers with over 1,300 skins led to the death of Sonan Dorjay. Strengthened in their resolve, the anti-poaching team (now totaling 30 people) continued under the leadership of Sonan Dorjay's brother-in-law, Zhaba Dorjay.

A turning point came in 1997 when the squad discovered a birthing ground and sought external support to set up a permanent base camp to protect the thousands of endangered female antelopes. Friends of Nature in Beijing donated two Jeeps; Green River, an NGO in Sichuan province run by the well-known environmental photographer and journalist Yang Xin, raised funds for the establishment of the Sounandajie Natural Protection Station;[38] and the International Fund for Animal Welfare (IFAW) became the first international organization to fund the patrol with a donation of US $10,000. A year later, the patrol witnessed a large-scale massacre of the chiru that further galvanized international support. An international conference in the provincial capital of Qinghai in October 1999 led to the Xining Declaration that called for all countries to strengthen law enforcement against the trade in *shahtoosh*. It also appealed to governments to recognize the influence of NGOs.[39]

Tensions, however, were running high at the local level. At the end of the year, the Wild Yak Brigade suffered a second tragedy. Zhaba Dorjay was shot dead at his home. His death remains a mystery, although it was officially proclaimed to be suicide. The patrols persisted until the Brigade was finally disbanded by the provincial government at the end of 2000.[40] By that time, the authorities had taken over control by establishing Kekexili as a nature reserve with official police stations under the direction of the State Forestry Administration.

Politics at the local level does not appear to have stalled the international campaign. IFAW continued to sponsor training workshops, reserve management and *in-situ* conservation research. In 2001, it produced a report on the scale of the illegal trade in *shahtoosh* in collaboration with the Wildlife Trust of India.[41] The report revealed that raw *shahtoosh* was being transported from Qinghai to the Indian state of Jammu and Kashmir where it was being woven into shawls and then sold for up to $17,000 a piece on the European market. On the basis of the findings, the governments of China, the United Kingdom, and India pledged to cooperate on cracking down on the trade.

In the same year, a report in the *Renmin Ribao* claimed that "with international cooperation, illegal poaching declined" in the first half of 2001.[42] In the absence of reliable scientific data, such claims cannot be verified. What is clear, however, is that the international campaign has succeeded in dampening the demand for *shahtoosh*, not least because of the combined efforts of an international NGO and an extraordinary group of local people.

In the longer term, the fate of the Tibetan antelope depends upon a high level of sustained cooperation on multiple levels – between the governments of China, India and the United Kingdom, between the regions of Qinghai, Tibet and Xinjiang, and between the local communities of the Tibet–Qinghai plateau. With such ambitious goals in sight, maintaining the momentum of transnational advocacy at the local level will be critical. No matter how much international attention is generated, the campaign is unlikely to work unless it is grounded in local realities.

In 1998, in recognition that the conservation and sustainable management of the alpine grasslands requires the active participation of the people, Hashi Drashi Dorjay (then a member of the Wild Yak Brigade) set up the Upper Yangtze Conservation and Development Organization (*Qingzang gaoyuan huan Changjiang yuan shengtai jingji cujin hui*), or UYO for short. This grassroots NGO, registered under the Zhiduo county government, was described by the Xinhua news agency as China's first Tibetan people's organization. It has three broad aims: to promote the sustainability of animal husbandry, to protect biodiversity, and to maintain Tibetan cultural values.

Transnational advocacy began with a joint collaborative project with the Earth Island Institute in San Francisco.[43] Some of the project activities included setting up four community-based protected areas (for the Tibetan antelope, snow leopard, black-necked cranes and Tibetan wild ass), community consultation, habitat surveys, and local training for farmers on biodiversity management.[44] More recently, UYO has been working with Plateau Perspectives, a Canadian NGO, to investigate the health status and risk factors of nomadic pastoralists in Zhiduo and Qumalai counties.

In 2001, Hashi Drashi Dorjay was democratically elected to become the Vice-Secretary General of Snowland Great Rivers Environmental Protection Association (*xueyu dahe Sanjiangyuan shengtai huanjing baohu xiehui*). With five full-time staff and funds provided by a local businessman, this new NGO is formally registered with the Yushu prefectural government.[45] Like UYO it is essentially a Tibetan NGO. The chairman is Lajiari Jiangbasuonanjian (Jiangge), the living Buddha of Luxiuloujaxi lamasery in Yushu. But whereas UYO is limited to Zhiduo county, Snowland covers the whole region of Sanjiangyuan.

The majority of its 80 members, who are essentially volunteers, are local herdsmen. The central aim is to disseminate new ideas on sustainable development while at the same time re-emphasizing the importance of traditional ecological ethics. Recent initiatives include a wild yak car to promote environmental education "from tent to tent," and an educational volunteer scheme, in collaboration with a green web organization in Beijing, to place college students amongst local herdsmen for a period of two weeks. Volunteers outside of the organization are as important as those within it. What is remarkable is that young Chinese volunteers have been prepared to put their lives at risk for the cause.[46]

According to Snowland's leaders, the Buddhist belief in the sanctity of wildlife and the environment has meant that local participation has proven to be less of a challenge than policy advocacy. In regard to the latter, it is currently building on the previous efforts of UYO to bring pastoralists into dialogue with the provincial authorities as well as creating model villages for demonstrating the importance of locally managed conservation. This message will be further reinforced with the establishment of a comprehensive scientific research and training base to monitor the complex relationships between traditional and modern approaches to environmental protection. In particular, Snowland is currently conducting evaluation studies on the government's approach to nature reserve management as well as its policy on permanent settlement for nomads under the *si pei tao* scheme.[47]

The intention is to promote policy advocacy on the basis of local and international support. In the words of Hashi Drashi Dorjay, "our NGO provides a platform for policy advocacy. It serves the interests of the local people and seeks the support and attention of the outside world in order to realize the goal of sustainable conservation and environmental protection in Sanjiangyuan."[48] Drashi Dorjay has visited over 30 NGOs in China to learn from their experiences and exchange ideas. He and his colleagues are also working with Flora and Fauna International based in Beijing to develop participatory village action plans and stakeholder analysis. This collaboration with the world's oldest international conservation NGO to convince the State Forestry Administration of the need to embrace alternative models of natural resource management is ongoing.[49]

The fundamental challenge for transnational advocacy is to convince the Chinese leadership of the need to integrate local concerns into development planning. At the provincial level, there are some signs of progress. Joint efforts between local and INGOs culminated in a conference held in Xining in January 2003 to bring together pastoralists, government officials, and NGOs to develop a biodiversity community management plan. But with the encroachment of large-scale development projects such as the Golmud–Lhasa railway as part of the central government's "Develop the West" campaign, the question remains whether locally based transnational advocacy will be enough to bring about policy reform.[50] At stake here is not only the protection of the Tibetan antelope and the alpine grasslands, but also the hope for China's democratic future.

Breaking the silence: the campaign to reduce pesticide use in China

China's place as the world's second largest producer and consumer of synthetic pesticides, producing more than 70,000 tons per annum,[51] singles it out for international attention. According to the government's classification system, an estimated 70 percent of pesticides used in China are highly toxic.[52] Yet, reliable data on the health and environmental impact is difficult to find. A study in 2000 estimated that up to 120,000 Chinese were affected by pesticide poisoning annually.[53] As in other developing countries, a lack of protective clothing, leaking knapsack sprayers, and repeated exposure exacerbates the risks of pesticide use. In addition to the negative health effects, pesticide overuse has also led to the severe contamination of rivers and lakes, and in some regions has dramatically reduced fish stocks.[54] The social and environmental effects are particularly acute in Yunnan province where the shift towards large-scale vegetable production has led to a dramatic rise in pesticide abuse.

The pesticide problem in China has not, as yet, captured the imagination of the international NGO community to the same extent as the threatened extinction of the Tibetan antelope. In part, this is because the slaughter of an endangered species is easier to spotlight in the media than the widespread pollution of rivers and lakes and the slow degeneration in health of hundreds of Chinese farmers. It is also important to consider that whereas the trade in endangered species in China is illegal, the production and use of pesticides is not. Hence, fewer avenues are open at the international level to seek redress. Given these limitations, it is probably best to view the pesticide campaign in China as a work in progress. In other words, it is a case of transnational advocacy in the making.

Low-level media coverage, together with government concerns over food security and the virtual silence within the scientific research community have also reduced the scope for citizen action at the local level. Very little is known about the human and ecological risks of pesticide use in China. Public knowledge is limited for two reasons. First, as noted by Jessica Hamburger, government agricultural extension centers that are responsible for the regulation and control of pesticides also receive their funding from pesticide sales.[55] Hence there exists little incentive to promote informed choice. Second, Chinese scientists working on pesticide issues tend to be isolated from the rest of society and they are often disengaged from social issues.

In recognition of the urgent need for independent research and public advocacy, China's first NGO dedicated to reducing harmful pesticides was recently established in Kunming in Yunnan province by a group of concerned scientists. As an offshoot of the Yunnan Entomological Society, the NGO was originally set up in 2000 as a government research institute responsible for advancing pesticide alternatives within the province. Frustrated by a lack of autonomy, especially in relation to appointing personnel, the institute initiated a dialogue with the North American and Asia Pacific branches of the Pesticide Action Network (PAN) – a global coalition of citizen groups working to reduce pesticides in over

50 countries. Spurred on by international support, the Pesticide Eco-Alternative Center Yunnan Thoughtful Action (*Yunnan sili nongyao tidai jishu zhongxin*), or PEAC, was formed in February 2002 and later registered with the Civil Affairs Administration as an independent social organization.

PEAC has a five-member Board of Directors headed by an eminent scientist, Professor Xiang Rong. They provide guidance to seven full-time staff, largely made up of young female graduates. The overall approach of this new NGO is to establish strong local roots on the basis of an international vision. In the words of the Director, Professor Kuang Rongping, "while PEAC's primary focus is on China's pesticide issues, its approach is very much an international one. Working with international partners allows the lessons to be learned from elsewhere to be applied to the Chinese situation."[56]

A unique characteristic of the organization is its focus on the plight of rural women. This is partly a legacy of its decision to merge with Green Mountain Women, an NGO that was set up two years earlier to promote green agriculture in remote mountain villages. The pro-women focus of PEAC is also in keeping with a central priority within the PAN Asia Pacific Network. The two organizations have worked closely to investigate ways of improving the health of rural women. This need is particularly acute given the fact that drinking pesticide is now the most common cause of suicide in China.[57] A joint report by the World Bank, World Health Organization, and Harvard University in 1999 revealed that 56 percent of female suicides globally occur in China, mostly in the countryside.[58] The ready availability of highly toxic pesticides on the Chinese market means that tackling this problem will require the enactment of legislation as well as the establishment of counseling and support networks. For the former to succeed, a coordinated campaign is needed to de-link pesticides from the broader issue of food security in the minds of government officials.

In general, breaking the pesticide habit in the Chinese countryside will take many years of sustained efforts. Achievements to date have been modest but nevertheless significant. In the field, PEAC's activities have focused on gathering and documenting information on traditional pest control techniques and introducing new eco-agricultural practices. To help raise awareness, Zuowei village in Jiangchuan county located 50km southeast of Kunming has been set up as a demonstration site for the promotion of eco-agriculture. With a local economy heavily dependent upon the sale of vegetables, the village is slowly being weaned off its pesticide dependency to the benefit of the environment and the community's health. Convincing village leaders and farmers of the need to develop sustainable agricultural practices has been helped by their exposure to outside opinions. Considerable efforts have been made to bring local farmers into a dialogue with scientists and government officials from Kunming. PAN Asia Pacific Network has also funded a series of projects in Lijiang aimed at raising awareness amongst farmers of the need to take responsibility both for their own health and for that of the local environment.

In Kunming, attention has largely focused on consumer advocacy. PEAC's approach has been to build a consumer advocacy network made up of individual consumers, scientists, consumer associations, the Yunnan People's Congress, and the Yunnan People's Political Consultative Conference. In 2003, it initiated a number of consumer awareness campaigns in Kunming that included a participatory workshop, an exhibition of posters relating to pesticide risks in the front entrance of the Carrefour supermarket, and a public forum at Nodika teahouse in cooperation with another environmental NGO, Green Watershed.

To complement locally driven initiatives, PEAC has cooperated with Greenpeace China to facilitate a dialogue between city officials and large retailers operating in Southwest China such as Pricemart, Walmart and Carrefour. PEAC's local knowledge combined with its international partner's long-time experience in advocating for corporate responsibility has proven to be a winning formula. On 1 December 2003, the Kunming government announced that it was going to conduct regular inspections for pesticide residues on vegetables sold at large supermarkets. The results will be published in the local newspapers and a hotline has been set up for consumers to contact inspection agencies should they detect a high level of pesticide residue.[59]

Beyond direct action on the ground, the benefits of transnational advocacy have perhaps been greatest in relation to knowledge dissemination and capacity-building. With regard to the former, PEAC has recently published the first comprehensive report on the social and ecological risks of pesticide use in China dating back to the 1950s. The report is the first of its kind to be funded by an international NGO and therefore independently from the interests of Chinese pesticide companies. It has been distributed across China and internationally via the PAN network.

As the China link in the chain of NGOs operating globally, PEAC has also benefited directly from international support. Greenpeace, PAN, the Rockefeller Brothers Fund, and the Global Greengrants Fund (GGF) have all provided advice on organizational development as well as financial support for PEAC members to attend international conferences. Internal capacity-building remains an important issue. In the words of Professor Kuang, "the development of PEAC is a work in progress but it is well on the way to becoming a model of an independent and democratic Chinese NGO."[60]

The real challenge for the future lies in addressing the complex issue of the trans-border transfer of pesticide pollution. With regard to the production of pesticides, China is suffering a double whammy. On the one hand, local factories are illegally dumping cheap pesticide products on the market, while on the other hand foreign chemical companies in China are taking advantage of lax regulations for pesticide production.[61] In 2002, for example, the world's largest agrochemical corporation, Syngenta (formally Zeneca, ICI), set up a new factory in China to produce paraquat (this highly controversial herbicide has been banned by 13 countries, including Malaysia).

On 3 December 2003, PEAC joined a coordinated international campaign in calling for a ban on the production and use of paraquat. To mark global "No Pesticides Day," in commemoration of the Bhopal pesticide factory accident that killed thousands of people in India in 1984, it hosted a public forum for students, activists and scientists. Clearly, fighting this battle is well beyond the scope of one small NGO in Kunming. It will require transnational cooperation on a large scale. Consequently, Kuang Rongping believes that the future priority for PEAC lies in building a fully integrated transnational network.[62] If the current momentum continues then this aspiration may actually become a reality.

Opportunities, risks, and future expectations

The above cases provide grounds for optimism both in relation to the emergence of independent NGOs and transnational advocacy. As in the case of NGOs in general, charismatic leadership is central to the drive and ambition of China's environmental organizations (although this may also pose a threat to NGOs, as noted by Ho in this volume). Also noteworthy is the emphasis upon democratic representation. Participation is not something that only relates to activities outside of the organizations; it is also being promoted from within.

There are, of course, obvious differences between the NGOs in Qinghai and Yunnan. UYO and Snowland evolved from spontaneous grassroots action. It was essentially the Wild Yak Brigade, a local social movement, that acted as the catalyst for a more formal social organization under the guidance of prominent and well-respected leaders within the community. In comparison, PEAC is an offspring of a government institution. The shift towards autonomous social organization has been facilitated by scientists. PEAC's leadership style of "scholar activism" is now a common characteristic of environmental organizations across China. To put it differently, whereas UYO and Snowland run on the basis of a self-help philosophy that keeps them rooted within the community, PEAC plays more of a mediating role between the local and the global. But all three organizations are similar in one important respect: they rely upon international support to widen the opportunities for facilitating change.

The fact that transnational advocacy is actually taking place in China is in itself a major breakthrough. But how can we assess its future potential? Are Chinese citizens on the way to becoming active members of a broader transnational or global civil society?[63] Or is the current push towards international cooperation likely to backfire? The optimistic scenario is that transnational advocacy will continue to expand the associational space within which democratic processes (both formal and informal) can take place.

From the perspective of local NGOs, the two cases highlight a number of benefits that can be gained from international cooperation. Most obvious is the ability to survive without financial dependency upon the state. UYO, Snowland and PEAC are currently able to carry out their activities on the basis of funding from international donor agencies and NGOs. Financial independence has the added benefit of creating a greater distance between NGOs and the bureaucracy. From my discussions with NGOs in Qinghai in particular, it was clear that cutting free from the endless red tape usually associated with government-funded development projects was allowing them to implement their own projects at half the cost. Furthermore, the NGOs believed that their ability to attract international funding raised their credibility in the eyes of the local government.

As recognized by the Chairman of PEAC, equally important is the moral support provided by INGOs. In his words, "with the support of INGOs we feel that our work here at PEAC is very valuable."[64] This aspect should not be underestimated. The majority of staff working for independent NGOs in China do so on a voluntary basis. Those who are on salaries typically earn around RMB 400 per month (approximately US$48). The young women working for PEAC are all graduates, fluent in English, and capable of attracting relatively high salaries working for private companies. Their ability to interact with like-minded people across the globe provides encouragement and helps to reinforce commitment to the cause. This is particularly important when recognition amongst peers and the broader public is relatively low.

INGOs also play a critical role in capacity-building. This was less evident in the case of the Qinghai NGOs because of the fundamentally grassroots nature of these organizations. Nevertheless, like their counterparts in Yunnan, they have benefited greatly from better access to scientific knowledge as well as from support for financial management. As in other parts of the world, the issue of accountability has attracted considerable attention within China. And it is likely that this will become an important measure of NGO success in China in years to come.

On a more sober note, it is important to recognize that international support can also have its downside. The Sun and Tysiachniouk contribution provides a clear example of the distorting effects of international funding on NGO environmental activity. At a broader level, comparative studies of North–South NGO partnerships have also identified a range of potential risks.[65] Of paramount concern is the risk that transnational cooperation could undermine domestic political agency by stifling the potential of civil society actors to genuinely represent the interests of the people. International dependency not only distracts attention away from local concerns, but also affects public perceptions of local NGOs.[66]

Conclusion

This contribution has explored the transnational dimension of China's emerging civil society by looking at the growing linkages between local and international NGOs. The rise of transnational advocacy in China is being driven both by a shift towards advocacy amongst Chinese NGOs and by the global development trend towards building civil society. An examination of the two environmental campaigns in Qinghai and Yunnan has demonstrated the benefits that can be gained from transnational advocacy, especially in the poorer regions of China where international support for local NGOs is a matter of survival.

Local environmental NGOs in China are working under severe economic and political constraints. They also face formidable challenges with respect to the implementation of their environmental agendas, which are inextricably linked to China's modernization drive. Regardless of whether the focus is upon protecting endangered species, conserving grasslands, or preserving human health and sustainable agriculture, significant progress is unlikely without a fundamental shift in thinking away from development at all costs and towards environmentally sustainable development that takes into account the needs of the Chinese people. Despite the complexities involved, the NGOs presented in this contribution are commendably clear about their objectives: they are seeking to improve conditions at the local level by strengthening public participation, and they are also concerned with advocating for policy reform. Working with INGOs is helping to strengthen their financial and organizational capacity as well as their ability to operate within a restrictive political environment. In essence, international support is providing a wedge between the state and civil society in China that is, in turn, increasing the opportunities for citizen action.

In the final assessment, a sense of perspective is important. Transnational advocacy is still new to China and its effects are likely to be mixed and uneven. But at least in the cases presented in this contribution, the benefits appear to be outweighing the costs. In some ways, this is not surprising. It is in the interests of both parties to work towards equitable partnerships that bring together the resources of INGOs with the valuable experience and knowledge of local NGOs. The latter are far better placed to exploit opportunities and avoid confrontation with the government.

From a broader perspective, the political implications are more difficult to assess. I have argued that if we are to understand the democratic potential of transnational advocacy we need to consider the likely impact on public participation and government accountability. The underlying assumption here is that transnational advocacy at the grassroots can help to lay the groundwork for the eventual reform of state practices by creating the necessary conditions for a more just and democratic Chinese society. But optimism for the future must be tempered by our existing knowledge of the potential risks involved in substituting state dependency with international support. In working at the transnational level, local NGOs in China face two key challenges. First, to find a way of engaging with INGOs that avoids the trap of dependency. Second, to maintain the contract with the people whom they claim to represent. Chinese NGOs are already under considerable pressure to divert scarce resources away from participatory practices and towards fund-raising activities. A competitive funding culture may improve efficiency and accountability within the NGO community, but this may also be at odds with the imperative of building an environmental consensus – the risk is that the voices of those who have most at stake will be silenced. Ultimately, what distinguishes NGOs from profit-driven organizations is their claim to represent the people in the pursuit of a particular cause. In future research, an important measure of the democratic potential of transnational advocacy will be the extent to which it is able to maintain this focus.

Notes

1 I would like to thank the Department of International Relations at the Australian National University for providing the financial assistance for this research. My thanks also to Richard Louis Edmonds, Peter Ho, Tak-Wing Ngo, and other anonymous reviewers for their helpful comments. For my fieldwork investigation, I am particularly indebted to Hashi Drashi Dorjay, Kuang Rongping, Xiang Rong-Jiong, Yan Mei, and William Bleisch. I do of course take full responsibility for any factual errors in the contribution.

2 I acknowledge the multiplicity of meanings that can be attached to the label NGO, especially in the China context. NGOs can be defined as non-governmental, non-profit, voluntary, and formal organizations. I will briefly discuss the various forms of Chinese NGOs later in the contribution.

3 For a more general survey of international support for NGOs in China see J.G. Bentley, "The Role of International Support for Civil Society Organizations in China," *Harvard Asia Quarterly*, Vol. 7, No. 1 (2003), pp. 11–19.

4 The role of corporations in promoting environmental advocacy in China is still at an early stage. For an introductory survey see J. Turner, "Cultivating Environmental NGO–Business Partnerships," *The China Business Review*, November–December 2003.

5 See, for example, G. Rodan, "The Prospects for Civil Society and Political Space" in A. Acharya, B.M. Frolic and R. Stubbs (eds) *Democracy, Human Rights and Civil Society in South East Asia*, Toronto: Centre for Asia Pacific Studies, 2001; and C. Kumar, "Transnational Networks and Campaigns for Democracy" in A. Florini (ed.) *The Third Force: The Rise of Transnational Civil Society*, Tokyo: Japan Center for International Exchange, 2000.

6 See, for example, E. Weinthal and P.J. Luong, "Environmental NGOs in Kazakhstan: Democratic Goals and Nondemocratic Outcomes," in S.E. Mendelson and J.K. Glenn (eds) *The Power and Limits of NGOs: A Critical Look at Building Democracy in Eastern Europe and Eurasia*, New York: Columbia University Press, 2002; and S. Dicklitch, "NGOs and Democratization in Transitional Societies: Lessons from Uganda" in D.N. Nelson and L. Neak (eds) *Global Society in Transition: An International Politics Reader*, The Hague: Kluwer Law International, 2002.

7 Cross-national studies carried out by the Research Unit for Environmental Policy at the Free University of Berlin confirm that democratic institutions on their own seem to play a minor role in the development of sound environmental polices. See M. Jänicke, "Conditions for Environmental Policy Success," *The Environmentalist*, Vol. 12 (1992), pp. 47–68.

8 G. White, *Riding the Tiger: The Politics of Economic Reform in Post-Mao China*, London: Macmillan, 1993; and J. Dreze and A. Sen, *India: Development and Participation*, New Delhi: Oxford University Press, 2002.

9 J. Keane, *Global Civil Society?* Cambridge: Cambridge University Press, 2003.

10 M. Keck and K. Sikkink, *Activists Beyond Borders: Advocacy Networks in International Politics*, Ithaca, NY: Cornell University Press, 1998.

11 P. Wapner, "Politics Beyond the State: Environmental Activism and World Civic Politics," *World Politics*, Vol. 47, No. 3 (1995), pp. 311–40; J. Matthews, "Power Shift," *Foreign Affairs*, Vol. 76, No. 1 (1997), pp. 50–66; and S. Khagram, J. Riker and K. Sikkink (eds) *Restructuring World Politics: Transnational Social Movements, Networks, and Norms*, Minneapolis: University of Minnesota Press, 2002.

12 There now exists a large literature on NGOs in the developing world. For general overviews see S. Khotari, "Rising from the Margins: The Awakening of Civil Society in the Third World," *Development*, Vol. 3 (1996), pp. 11–19; and J. Haynes, *Democracy and Civil Society in the Third World: Politics and Social Movements*, Cambridge: Polity Press, 1997.

13 Keck and Sikkink, *Activists Beyond Borders.* See also T. Princen, M. Finger and J. Manno, "Non-governmental Organizations in World Environmental Politics," *International Environmental Affairs*, Vol. 7, No. 1 (1995); A.M. Clark, "A Calendar of Abuses: Amnesty International's Campaign on Guatemala," in C. Welch (ed.) *NGOs and Human Rights: Promise and Performance*, Philadelphia: University of Pennsylvania Press, 2001; A. Klotz, "Transnational Activism and Global Transformations: The Anti-Apartheid and Abolitionist Experiences," *European Journal of International Relations*, Vol. 8, No. 1 (2002), pp. 49–76.

14 A. Fowler, "The Role of NGOs in Changing State–Society Relations: Perspectives from Eastern and Southern Africa," *Development Policy Review*, Vol. 9, No. 1 (1991); Haynes, *Democracy and Civil Society in the Third World*; and S. Cleary, *The Role of NGOs under Authoritarian Political Systems*, Basingstoke: Macmillan, 1997.

15 E.X. Gu, "'Non-Establishment' Intellectuals, Public Space, and the Creation of Non-Governmental Organizations in China: The Chen Ziming–Wang Juntao Saga," *The China Journal*, No. 39 (1998), pp. 39–58.

16 M. Ganie-Rochman, *An Uphill Struggle: Advocacy NGOs under Soeharto's New Order*, Jakarta: LabSosio, 2002.

17 This resonates with the literature on the role and influence of advocacy groups in Eastern Europe and the former Soviet Union before the revolutions in 1989. See A. Michnik, *Letters from Prison and Other Essays*, trans. M. Latynski, Berkley: University of California Press, 1985.

18 A.J. Saich, "Negotiating the State: The Development of Social Organizations in China," *The China Quarterly*, No. 161 (2000), pp. 125–41.

19 P. Ho, "Greening without Conflict? Environmentalism, NGOs and Civil Society in China," *Development and Change*, Vol. 32, No. 5 (2001), pp. 893–921, at 907.

20 Online, at www.fon.org.cn/index.php?id=2684> (in Chinese).

21 For a comprehensive survey of environmental GONGOs in China see Fengshi Wu, "New Partners or Old Brothers? GONGOs in Transnational Environmental Advocacy in China," *China Environment Series*, issue 5 (2002), pp. 45–58.

22 According to the Deputy Director of CBIK, a decision was made to shift towards advocacy because its researchers were finding it difficult to work in poor ethnic areas without giving something back to the community. Interview with Quan Jie, 23 April 2004.

23 Ho, "Greening without Conflict?"; Zhang Ye, "China's Emerging Civil Society," CNAPS Working Paper, Washington, DC: Centre for Northeast Asian Policy Studies, Brookings Institute, August 2003; and J. Turner, "Clearing the Air: Human Rights and the Legal Dimensions of China's Environmental Dilemma," China Environment Forum, Woodrow Wilson International Center for Scholars, Princeton, 27 January 2003.

24 C. Liang, "The Building Up of Environmental NGOs in China: The Road of Friends of Nature," in Z. Liqing and C.I. Irving (eds) *The Non-Profit Sector and Development*, Hong Kong: Hong Kong Press for Social Science, 2001, p. 301.

25 Based upon a comparison of my discussions with environmental NGOS in Beijing and Yunnan in May 1999 and October 2003.

26 I am grateful to Wang Yunxian from Oxfam Hong Kong's office in Beijing for drawing my attention to this point.

27 See State Council, *White Paper on Poverty Reduction in Rural Areas* [*Zhongguo nongcun fupin kaifa gangyao*], Beijing: State Council, October 2001. Note that the term "participation" is translated into Chinese as *zhengfu zhudao, gongtong canyu* meaning joint participation under the guidance of the government.

28 The first international conference on "The Non-profit Sector and Development," hosted by Tsinghua University in 1999, had a major focus on regulatory reform. It was also a central concern of a more recent conference on "Professionalism and Accountability in the Third Sector," hosted by Beijing University in October 2003.

29 Z. Liqing and C.I. Irving (eds) *The Non-profit Sector and Development: The Proceedings of the International Conference in Beijing in July 1999*, Hong Kong: Hong Kong Press for Social Sciences, 2001, p. 33. On this point, see also W. Ming, L.G. Han and H.J. Yu, *Zhongguo shetuan gaige – congzhen fuxuan zhedao shetuan xuanzhe* [Chinese Society Reform: From Government Choice to Social Choice], Beijing: Shehui kexue wenxian chubanshe, October 2001.
30 Based on the Directory of INGOs China Development Brief, at www.chinadevelopmentbrief.com, and individual NGO websites.
31 *Renmin Ribao*, 12 March 2004. The new regulations came into effect on 1 June 2004.
32 For an account of the experience see www.gvbchina.org/English/newsletter/2002012.htm.
33 The case studies are based on interviews conducted in Xining and Kunming in September and October 2003 and in Jiegu township, Yushu prefecture, Qinghai in April 2004.
34 *Renmin Ribao*, 19 November 2002.
35 This estimate is based on an assessment conducted by the veteran wildlife biologist George Schaller for the Wildlife Conservation Society. For further details see www.earthisland.org/tpp/antelope.htm.
36 All references to names in this section are in Tibetan rather than Chinese.
37 Zhiduo county in Yushu prefecture has six townships covering 80,000km in the area of the Upper Yangzte river basin.
38 Given that NGOs in China are forbidden from public fund-raising activities, Yang Xin was able to raise private funds from the sale of his book, *The Spirit of the Yangzte* published in 1997.
39 For the full text of the Xining Declaration on the Conservation of and Control of Trade in Tibetan Antelope, see www.cites.org/common/notifs/1999/98-a.pdf.
40 Liang Congjie from Friends of Nature together with 17 Chinese journalists wrote a letter to then Vice-Premier Wen Jiabao in support of the anti-poaching squad. He apparently ordered the local government to reconsider the issue. But this was not sufficient to influence the final decision. Interview Liang Congjie, Friends of Nature, Beijing, 27 October 2003.
41 International Fund for Animal Welfare, "Wrap Up the Trade: An International Campaign to Save the Endangered Tibetan Antelope," USA: IFAW, 2001, p. 69.
42 *Renmin Ribao*, 28 June 2001.
43 The Earth Island Institute runs a Tibetan Plateau Project that seeks to promote conservation and the sustainable development of mountain communities in India, Bhutan, Nepal, Pakistan and Tibet. For more details see www.earthisland.org/tpp/about.htm.
44 See J. Lowe, "Defending Qinghai," *Earth Island Journal*, Vol. 14, No. 3 (1999).
45 Initial funding of $50,000 came from a so-called green company producing bottled mineral water in Jiegu township. Interview Snowland Great Rivers Environmental Association, Jiegu, 18 April 2004.
46 See "Young Environmental Volunteer Dies on High Plateau," at www.usembassy-china.org.cn/sandt/ptr/estnews121302-prt.htm.
47 *Si pei tao* refers to the four stages of nomadic settlement. It involves enclosing pastureland, using fodder in place of free-range grazing, building houses, and building animal shelters. Anecdotal evidence suggests that fence-building doesn't work well in practice because animals often break down the fences, which can cause conflicts between communities, and eventual social breakdown. According to Snowland members, this policy is more about poverty alleviation than environmental protection. Interview Jiegu township, Yushu prefecture, Qinghai, 18 April 2004.
48 Interview with Hashi Drashi Dorjay and other members of Snowland, Beijing, 29 September 2003.
49 Interview Fauna and Flora International, Beijing, 28 September 2003.
50 For an account of the environmental implications of China's "Go West" campaign see

E. Economy, "China's Go West Campaign: Ecological Construction or Ecological Exploitation," *China Environment Series*, Issue 5 (2002), pp. 1–11.

51 *PEAC Newsletter*, Vol. 1, No. 1 (July 2002).

52 *PEAC Annual Report*, Kunming: PEAC, 20 December 2003.

53 J.K. Huang, F.B. Qiao, L.X. Zhang, and S. Rozelle, "Farm Pesticide, Rice Production, and Human Health," CCAP's Project Report, Beijing: Center for Chinese Agricultural Policy, Chinese Academy of Agricultural Sciences, 2000.

54 J. Becker, "Putrid Lake Proof Environmental Polices Have Failed to Hold Water," *South China Morning Post*, 15 October 2001.

55 J. Hamburger, "Pesticides in China: A Growing Threat to Food Safety, Public Health, and the Environment," *China Environment Series*, Issue 5 (2002), pp. 29–44.

56 Cited in *PEAC Newsletter*, Vol. 1, No. 1 (July 2002).

57 See "Suicide by Pesticide," at www.usembassy-china.org.cn/sandt/estnews0330.htm.

58 E. Rosenthal, "Suicides Reveal Bitter Roots of China's Rural Life," *New York Times*, 24 January 1999.

59 *PEAC Annual Report*.

60 Interview PEAC in Kunming, 10 October 2003.

61 Hamburger, "Pesticides in China," p. 32.

62 Interview PEAC in Kunming, 10 October 2003.

63 It is interesting that Chinese scholars are beginning to discuss the idea of a global civil society. See J.X. Jin and W.Y. Deng, *Quanqiu gongmin shehui – fei yingli bumen shijie* [Global Civil Society – Vision of the Non-profit Sector], Beijing: Shehui kexue wenxian chubanshe, July 2002.

64 Interview PEAC in Kunming, 10 October 2003.

65 L.D. Brown, "Social Change through Collective Reflection with Asia Nongovernmental Development Organisations," *Human Relations*, Vol. 46, No. 2 (1993), pp. 249–73; B. Eccleston, "Does North–South Collaboration Enhance NGO Influence on Deforestation Policies in Malaysia and Indonesia?" in D. Potter (ed.) *NGOs and Environmental Policies in Asia and Africa*, Portland, OR: F. Cass, 1996; M. Edwards and D. Hulme, *NGOs, States and Donors: Too Close for Comfort?* New York: St Martin's Press in association with Save the Children, 1997.

66 Bentley, "The Role of International Support," p. 17.

10 Perspectives of time and change

Rethinking green environmental activism in China

Peter Ho and Richard Louis Edmonds

The sociological and political science literatures hold that social movements emerge and evolve as a function of several critical conditions: the level of cleavages and conflicts in society; the movement's norms and values ("cultural frames") that bind together its participants; its organization and capacity to mobilize resources; and the movement's specific political opportunities.[1] In other words, a social movement's emergence, dynamics, and its development trajectory are measures of these parameters. Furthermore, social movements are generally seen as capable of effecting structural political and social changes – be they for wider civil rights or a better environment. To study the dynamics and potential of social movements in China, this volume zoomed in on one of the country's earliest and most active sectors of civil society since the beginning of the Chinese reforms in 1979: the environmental realm.

Environmental activism in China critically diverges from the popular image of social movements as "masses of people taking to the streets and erecting barricades" in opposition to an established order. Chinese environmentalism differs from what the world has witnessed in some of the ex-socialist countries in East and Central Europe where green movements acted as a catalyst for a wider democratization in society. It also differs from what could be seen in Ukraine, Georgia, and Kyrgyzstan where social movements transformed into "color revolutions" that toppled autocratic regimes. In fact, since its emergence in the early 1990s, Chinese environmentalism has been mostly fragmentary, highly localized, and nonconfrontational. To explain these features some scholars have portrayed, and at times criticized, China as a country under a totalitarian regime where any attempt at autonomous association is either suppressed or incorporated into the structures of the "Leninist Party-state."[2] Others have talked about a "state-led civil society" or "state corporatism" whereby social activists and their organizations are co-opted into the state agenda.[3]

However, the concepts of a state-led civil society or state corporatism fail to capture the current dynamics of Chinese social activism, as they cannot account for two of its critical features. For one thing, in China today the state cannot dictate what society needs to do. Current state–society relations can be better likened to an ongoing negotiation or a "negotiated symbiosis," although not necessarily a negotiation between equals. Moreover, as a result of this semi-authoritarian setting, Chinese social activists have to rely on an informal and diffuse, rather than a formal, network of relations. Due to this informal web of relations the divides between the Party-state and society can be successfully bridged, gaining Chinese green activism an unmistakable social legitimacy, as well as political influence.

Chinese environmentalism is different from what is popularly regarded as a "green social movement." However, at the same time it is evident that the parameters that could propel a broadly supported social movement are present in contemporary Chinese society. In this light, we tried to gauge the effect of the limited political space for the development and dynamics of a social movement in China. We also examined whether the possibility for, or the actual occurrence of, a social movement is a *conditio sine qua non* for political change and the development of civil society. Lastly, we explored the prospects for the emergence of a social movement in China, and how it would relate to transnational environmentalism.

Implications of a limited political space

In order to answer the first question posed in the preceding – the consequence of China's semi-authoritarian context for the development and dynamics of a social movement – it is necessary to revert to the conceptualization of a social and environmental movement. Some insisted that a shared collective identity and a sense of mission are critical preconditions to label social groups as a "movement."[4] Discussions have also focused on the constituent elements of social and environmental movements – loosely organized grassroots cells or also well-institutionalized, specialized organizations.[5] However, considering the specific features of Chinese environmentalism, it might be useful to remind ourselves of what Doyle and McEachern rightly remarked about social movements: "There is no unifying teleological purpose that drives them. There is no single causal reality that made them. Interpretations of their origins and significance are as contested as the movements themselves."[6]

The people actively involved in environmental activism in China include individuals, groups, loose networks, and organizations with varying degrees of institutionalization and formalization, such as NGOs, green salons, student associations, but also the "government-organized NGOs." However, equating Chinese green activism with, for example, the May Fourth Movement that followed the Versailles negotiations in 1919,[7] or with the students' movement that swept over Tian'anmen Square and the rest of China in 1989 is a comparison that falls short because of its fragmentary, highly localized and nonconfrontational nature.[8] For this reason, we have chosen to dub Chinese green activism "embedded environmentalism." It is the resultant of a limited, semiauthoritarian political space for civil society.

Since the revitalization of Polanyi's notion of "embeddedness" formulated by Granovetter, the concept has been applied in various disciplines of the social sciences.[9] In economic sociology it has come to denote the idea that economic action is more than merely a function of prices and markets; it also critically depends on the wider sociopolitical, cultural, and historical context.[10] In political science it has been used to explain economic success in relation to the relative autonomy of the state from the claims of interest groups in society.[11] Vice versa, in organizational studies "embeddedness" is employed as a measure of autonomy of civic organizations from the state. However, in the proliferating, and at times contradictory, writings about embeddedness, we can distil two critical features that are relevant to the Chinese experience: contextualization and networks.[12]

First, embeddedness refers to the contextualization of human action. In a different wording, although China's semiauthoritarian environment is still heavily influenced by the Party and state, which set out seemingly invisible limits that paradoxically seem to be known by all, social action is not a matter of state versus society, or state dictating society. Different from a fully authoritarian context,[13] social action between state and society crucially is an interaction and ongoing negotiation. In this sense, embedded environmentalism in China is a situation of a *negotiated symbiosis* with the Party and state (see Zhu and Ho, this volume). Second, interaction and negotiation are effected through social networks and ties. This, of course, holds true for any society. However, the crux of the matter is to what extent personal, informal, and "weak" ties are instrumental in relation to the level of formality of society. It is generally posited that in less formal environments, the evolvement of "thick" personalized relations is critical to secure sufficient trust for effective human interaction.[14] In the Chinese case, where the central state leaves little maneuvering space for formal associations independent from the state, it is vital for NGOs and individual activists to rely on informal, personal ties to achieve their ends. Moreover, the nature of these ties – or *guanxi* as it is termed in Chinese – is rooted in the kind of interpersonal Confucian-style connections that have long existed in China. For these reasons, we argue that environmental activism in China is not an activity that can operate fully autonomous from the Party and state – as some foreign observers and scholars silently hope. However, this by no means implies that environmental activism in China is "state-led" or suppressed. What sets Chinese embedded activism aside from a state-led activism is that it has evolved within a semiauthoritarian environment, which limits activism while enabling it. This specific setting has caused green activists to be enmeshed in a diffuse web of informal relations, unwritten rules, and shared missions with the Party-state. It is this informal network that can effectively bridge the divides between the Party-state and society, as well as between China and the outside world. Yet, perhaps as important, Chinese green activists engage in collective action in the name of, and thus, not necessarily out of a concern for, environmental conservation and protection. This brings us to the next question, namely whether the occurrence of a social movement is a precondition for political change and the development of civil society in China.

The contradiction of embeddedness: restricted while liberated

Classical studies by de Tocqueville and Tilly offer the view that social movements are critically linked to revolution and regime change.[15] They are seen as vehicles for a societal transition towards more democratic, transparent, and accountable governance structures. In the former socialist economies of Bulgaria and Hungary, environmentalism was propelled by a demand for political change, rather than by environmental protection per se. However, as overt political opposition was a sheer impossibility within the repressive, authoritarian environment at the time, citizens looked for alternative, and less politically sensitive, channels to expand civil and political rights. They found it in part in environmentalism. In fact, green activism proved to be a very effective means to mobilize people in protest against the communist administrations, with the ultimate result of regime change. Environmentalism was thus inextricably linked to wider issues of human rights and democracy.[16] When one of China's first environmental NGOs – Friends of Nature – was established in the early 1990s, some foreign observers and the international media hoped for a Hungarian or Bulgarian scenario unfolding in China as well. Yet, after a decade such scenarios seem far away.

The Chinese authorities have shown on several occasions that any nationwide movement in opposition of the central state will not be tolerated. Even more so, the gradually expanding political and civil freedoms that led to the emergence of voluntary green groups in the 1990s have been increasingly rescinded in the aftermath of the color revolutions. The end of 2005 saw a sudden stepping up of control over civil organizations, particularly environmental NGOs. By demanding renewed registration, creating new "nongovernmental" institutions,[17] and controlling activists' channels for communication such as the internet and the media,[18] the Chinese state sought to keep the environmentalist spirit in the bottle.[19] Moreover, in order to expose potential links with "foreign hostile and destabilizing forces" NGOs were asked to open up their financial accounts for government inspection. This action culminated in a vicious attack on Global Village Beijing, a renowned green NGO, which was heavily criticized for having received funding from the Heinrich Boell Stiftung, the development foundation of the German Green Party.[20] The fact that the Heinrich Boell Stiftung is one of the foreign foundations that openly support more controversial projects related to human rights, press freedom, and the environment rendered Global Village Beijing quite vulnerable at the time.

In China's semiauthoritarian context today, it is no surprise that the Chinese green movement lacks what characterizes environmentalism in Europe and North America: the capacity and opportunity to mobilize a nationwide, popular movement against national policies. By contrast, Chinese environmentalism is fragmentary, small-scaled, and organized around local rather than national political issues. Chinese environmentalism consciously steers away from possible confrontations with the national government through "depoliticized politics" and "self-imposed censorship." The organizational spectrum of Chinese activism features a substantial number of NGOs that fail to apply for formal registration.[21] On the other hand, the organizations that do succeed in registering often have strong ties with the government, or result directly from the privatization and disbanding of former state institutions. Some critics question whether a docile and depoliticized green activism will be able to achieve any "real" sociopolitical and environmental changes. In fact, the lack of autonomy and the fear to openly confront the national authorities have caused some observers to wonder whether one can talk about a Chinese green movement at all.[22]

However, is an embedded Chinese environmentalism an entirely negative matter? Not necessarily, as embeddedness might be seen as a specific social response to the current legal-political conditions, while allowing for organizational survival over the long run. Furthermore, it is a grave misconception to portray Chinese environmentalism as strictly confined within state-defined and controlled spaces. Embeddedness has earned Chinese environmental activism a certain legitimacy in the central government's eyes, a firm place within society, as well as links with international NGOs. Chinese environmentalism has amply demonstrated an ability to engage in successful and effective pressure politics.

An excellent illustration of the political leverage of embedded activism is the case study of the Nu River anti-dam campaign described by Guobin Yang and Craig Calhoun in this volume. This movement grew out of an informal salon, subsequently developed into a loose network of activists, academics, journalists, and sympathetic officials, and eventually became an institutionalized "green partnership" of eight environmental NGOs, with transnational connections as well. Although the struggle over the dam is still ongoing at the time of writing, this environmental partnership succeeded in attracting high political attention, and the temporary halting of the dam construction. One of the main reasons for the movement's success lies in its clever use of the political opportunities provided by the state. Rather than pushing against political limits or demanding greater freedoms and rights, the movement chose to work within the political spaces opened up by what Ho and Vermeer termed the "greening of the state."[23] As Yang and Calhoun also note, the movement carefully avoided framing the campaign in political terms, and instead explicitly used legal terms. In this sense, the proclamation of the new Environmental Impact Assessment Law shortly before the environmental movement gained full momentum proved crucial. By referring to this law, the movement succeeded in exerting such strong political pressure on the central government that Premier Wen Jiabao in person called for a renewed environmental impact assessment of the dam.

What we can learn from this case is that embeddedness is most certainly not a matter of subjecting oneself to the authoritarian restrictions of the state, or being silenced for voicing dissent, as some in the international media might want us to believe. Rather, embedded environmentalism is a resourceful and negotiated strategy employed by activists to gain maximum political and social influence, at least in name, by professing to uphold the principles of the Chinese Communist Party and state. This is the contradictory essence of the embeddedness of Chinese activism: limiting while enabling.

The lapse of time

Some observers have voiced the opinion that Chinese environmentalism can only then be considered a "full-fledged movement" if it can – similar to its counterparts in the West – engage in mass mobilization and radical strategies.[24] In this view, the current dynamics of Chinese green activism is a result of state-imposed restrictions rather than choice. The founder and director of Global Village Beijing, Liao Xiaoyi, once said, "Chinese environmentalism cannot always rely on nonconfrontational tactics to achieve its aims."[25] With this remark she touched on the final two questions of this special issue – the future prospects for the emergence of a social movement in China, and its potential relations with a transnational or global environmentalism.

In this volume, we believe that the answer to the former question is of significantly less relevance than the observation that Chinese environmentalism has found a fairly effective response to a semiauthoritarian context, which enables it to play an increasingly critical role in the greening of industries, the government, and consumer lifestyles. Part of the criticism of Chinese activism as docile and a mere supporter of the state also derives from the fact that these sociopolitical analyses are a snapshot in time rather than a series of pictures. In some transitional countries, environmentalism started out as a broadly supported movement that managed to mobilize huge masses of people, and subsided in later years, only to return as professionalized, formal organizations. For instance, Jancar-Webster wrote that environmentalism in East and Central Europe has changed "from being a mobilising agent for populist protest against the *totalitia* of the Communist regime … and in its place have emerged pragmatic, goal-oriented professional organisations."[26] Her characterization appeals to the wider assumption that the environmentalist development trajectory generally starts from a popular, radical movement, and over time changes into an institutionalized, "social movement organization" oriented to incremental environmental reforms by lobbying and tripartite negotiations (i.e. between NGOs, government, and business).[27]

In China we seem to witness a reverse development. Rather than starting out as a broad protest movement, activists are forced by the semiauthoritarian context to abandon any radical, confrontational, and mass mobilization tactics to achieve political objectives. Conversely, different from a fully authoritarian context, there are ample avenues left to gain political leverage through an embedded form of activism. More important, what all the contributions in this book demonstrate without exception is that embeddedness buys time for civic organizations to consolidate and institutionalize. Although many NGOs fail to register as an official "social organization," their nonregistered status does not prevent them from building up expertise in specific areas as their activities develop. We see this happening for widely divergent organizations ranging from well-established NGOs that focus on something as technical as combating agricultural pesticide use, to groups of student volunteers that gather together for bird-watching or tree-planting. Moreover, the type of organization also differs with its specialization, and as such, different NGOs need to gain different managerial and political experience in running their organizations.

The element of time is critical for a balanced assessment of the opportunities and constraints of Chinese green activism. Apart from being a "full-fledged movement," Chinese embedded environmentalism in fact shows virtually all organized forms of green activism we can see elsewhere in the world: grassroots networks of volunteers, lobby and pressure groups, and NGOs. Moreover, despite the highly localized nature of today's embedded environmentalism, some campaigns have clear consequences for national politics. For example, as Yang and Calhoun show, in the beginning the Nu River anti-dam campaign was primarily an issue between local activists and the Yunnan provincial government. However, as the movement grew into a partnership with the involvement of Beijing-based NGOs, it soon took on national dimensions. The intervention by the central Party leadership to temporarily stop the construction is proof of this.

The hallmark and, in fact, the success of China's reforms lie in their strategy of incremental change. In this light, we might view embedded environmentalism as a transient phase which is itself changing through time, a transitional feature of a burgeoning civil society in a semiauthoritarian context.[28] Since its initial emergence in the early 1990s, environmental activism has undeniably gained in political leverage and international linkages because it resourcefully adapted to rather than opposed the political conditions of its era. Anti-dam protests were unthinkable at the time when the National People's Congress voted on the construction of the Three Gorges (Sanxia) Dam in April 1992. At the time of writing, anti-dam protests are a political reality in China. In industrialized nations, the occurrence of social movements is also seen as an indicator of the degree of trust that the state has in its political, social, and economic institutions. In other words, the question is whether institutions are sufficiently robust and mature to deal with the force of social movements without risking the disintegration of the state. The question that we might ask ourselves is not so much whether China will open up the space for embedded environmentalism to employ confrontational, radical, and mass mobilization tactics, but rather *when* and *under what conditions*. In determining these conditions, allowing for the lapse of time is vital.

Notes

1 See, for instance, S. Tarrow, *Power in Movement: Social Movements and Contentious Politics*, Cambridge: Cambridge University Press, 1998, pp. 10–25.

2 R. Baum and A. Shevchenko, "The 'State of the State," in M.Goldman and R. MacFarquhar (eds) *The Paradox of China's Post Mao Reforms*, Cambridge, MA: Harvard University Press, 1999, p. 348.

3 M. Frolic, "State-led Civil Society" in T. Brook and M. Frolic (eds) *Civil Society in China*, Armonk, NY: M.E. Sharpe, 1997, pp. 15–38; J. Unger and A. Chan, "Corporatism in China: A Developmental State in an East Asian Context," in B.L. McCormick and J. Unger (eds) *China after Socialism: In the Footsteps of Eastern Europe or East Asia*, Armonk, NY: M.E. Sharpe, 1996, pp. 95–129.

4 M. Diani, "The Concept of a Social Movement," *Sociological Review*, No. 40 (1992), pp. 1–15.

5 For an overview of the various definitional discussions of environmental and social movements, see also C.A. Rootes, "Environmental Movements and Green Parties in Western and Eastern Europe" in M. Redclift and G. Woodgate (eds) *The International Handbook of Environmental Sociology*, Cheltenham: Edward Elgar, 1997, pp. 325–6. An interesting analogy is the likening of a social movement to a "palimpsest" – a piece of papyrus used in ancient times to write and rewrite, as a result of which several layers of text were created. A social movement is thus seen as consisting of five layers: the individual; the network; the group; the organization; and the movement itself. See T. Doyle and D. McEachern, *Environment and Politics*, London: Routledge, 1998, p. 62.

6 Doyle and McEachern, *Environment and Politics*.

7 After the defeat of Germany, the Allied powers decided to pass on the German concessions in Shandong province to Japan. The Chinese frustrations about this were great, and on 4 May 1919 around 5,000 students of Peking University demonstrated against the Versailles Treaty, which eventually culminated in a nationwide movement and boycott against Japan and Japanese goods.

8 The same also counts for more recent (religious) movements, such as the Falun Gong, which managed to attract thousands of followers in opposition to the Chinese government. For more information on the chain of events that led to the emergence of the Falun Gong in China, see "Wang Zhaoguo on Fight against Falungong," World News Connection, 23 July 1999, from World Reporter (TM) cited in Human Rights Watch, "Dangerous Mediation: China's Campaign against Falungong," 2002, online, available at http://hrw.org/reports/2002/china/China0102–02.htm#P331_49488 (accessed 1 February 2002); and H. Seiwert, "China's Repression of Folk Religions: The Battle against Falun Gong in Historical Context," *Neue Zurcher Zeitung* (Zurich), 13 July 2001, online, available at www.nzz.ch/english/background/2001/07/13_china.html (accessed 14 July 2001). Note that Chinese religious mass movements date back to at least 200 BC.

9 K. Polanyi, *The Great Transformation*, New York: Holt, Rinehart, and Winston, 1944.

10 For a discussion on embeddedness in relation to economic action, see M. Granovetter, "Economic Action and Social Structure: The Problem of Embeddedness," *American Journal of Sociology*, Vol. 91, No. 3 (1984), pp. 481–510; and A. Martinelli and N.J. Smelser (eds) *Economy and Society: Overview in Economic Sociology*, London: Sage, 1990.

11 P. Evans, *Embedded Autonomy. States and Industrial Transformation*, Princeton, NJ: Princeton University Press, 1995; P. Evans, "Government Action, Social Capital and Development: Reviewing the Evidence on Synergy," *World Development*, Vol. 24, No. 6 (1995), pp. 1119–32; and B. Trezzini, "Embedded State Autonomy and Legitimacy: Piecing Together the Malaysian Development Puzzle," *Economy and Society*, Vol. 30, No. 3 (2001), pp. 324–53.

12 See, for instance, R.P. Weller, *Alternate Civilities: Democracy and Culture in China and Taiwan*, Boulder, CO: Westview Press, 2001, pp. 66–70. This idea has been specifically employed for environmental agencies by D.P. Angel and M.T. Rock, "Engaging Economic Development Agencies in Environmental Protection: The Case for Embedded Autonomy," *Local Environment*, Vol. 8, No. 1 (2003), pp. 45–59.

13 This is not to say that the role of agency in a fully authoritarian society is reduced to zero. However, in relative terms human action is more strongly contextualized in a semiauthoritarian than in a full-fledged authoritarian setting.

14 Granovetter showed that "weak," informal and personal ties can be of great significance in human action. See M. Granovetter, "The Strength of Weak Ties," *American Journal of Sociology*, Vol. 78, No. 6 (1973), pp. 1360–80; R. Putnam, *Making*

Democracy Work: Civic Traditions in Modern Italy, *Princeton*, NJ: Princeton University Press, 1993. For a work on trust in a postsocialist context, see W. Mishler and R. Rose, "Trust, Distrust and Skepticism: Popular Evaluations of Civil and Political Institutions in Post-Communist Societies," *Journal of Politics*, Vol. 59 (1997), pp. 418–51.

15 C. Tilly, *From Mobilization to Revolution*, Reading, MA: Addison-Wesley, 1978; A. de Tocqueville, *The Old Regime and the French Revolution*, trans. S. Gilbert, Garden City, NY: Doubleday Anchor, 1954.

16 B. Jancar-Webster, "Environmental Movement and Social Change in the Transition Countries," *Environmental Politics*, Vol. 7, No. 1 (1998), pp. 69–90. Some authors argue that the same may be said about the Falun Gong Sect, which is not driven by a desire for free religious praxis alone but has clear political aims as well. See J. Tong, "An Organizational Analysis of the *Falun Gong*: Structure, Communications, Financing," *The China Quarterly*, Vol. 171 (2002), pp. 636–60.

17 Such a newly created institution included the All China Federation of Environmental Protection. See also the contribution by Ho in this volume, which describes the developments for environmental NGOs shortly after the color revolutions took place.

18 J. Ma, "Green Groups Fall under Microscope in Beijing," *South China Morning Post*, 18 August 2005, p. 12. According to Lin, the Chinese government has recently also stepped up control over international NGOs, fearing that they might destabilize Chinese society. See Y. Lin, "China Worries that Foreign NGOs Are Importing a 'Color Revolution'," Sun/Central News Agency, 2005, online, available at www.chinalaborwatch.org (accessed 6 August 2005).

19 The concept of "building a harmonious society" was launched by the Chinese leadership in the face of challenges posed to the government by the "independent thinking of the general public, their newly developed penchant for independent choices and thus the widening gap of ideas among different social strata." According to President Hu Jintao, a harmonious society should "feature democracy, the rule of law, equity, justice, sincerity, amity, and vitality." Foreign observers have interpreted the new concept as a way to pre-empt the occurrence of social movements in China. See also Xinhua News Agency, "Building Harmonious Society Crucial for China's Progress," *People's Daily* online, 27 June 2005, available at www.english.people.com.cn/200506/27/eng20050627_192495.htm (accessed 27 June 2005).

20 See also M. Busgen, "The Campaign against the Nujiang Dam," MA thesis, Institute of Social Studies, The Hague, 2005, p. 58.

21 P. Ho, "Greening without Conflict? Environmentalism, Green NGOs and Civil Society in China," *Development and Change*, Vol. 32, No. 5 (2001), pp. 893–921.

22 Y. Lu, *Environmental Civil Society and Governance in China*, London: Chatham House, 2005; F. Wu, "New Partners or Old Brothers? GONGOs in Transnational Environmental Advocacy in China," *China Environment Series*, No. 5 (2002), pp. 47, 53.

23 P. Ho and E.B. Vermeer (eds) *China's Limits to Growth: Prospects for Greening State and Society*, Oxford: Blackwell, 2006. See also P. Ho (ed.) *Greening Industries in Newly Industrializing Countries: Asian-style Leapfrogging?* London: Kegan Paul, 2007.

24 Exactly this issue, namely the use of radicalist strategies and the large-scale mobilization of people for environmental politics, was heatedly debated during a meeting among mainland Chinese, Hong Kong, and Taiwanese NGOs. See Y. Chiu, "Same War, Different Battles," *Taipei Times*, 21 April 2001, online, available at www.taipeitimes.com/news (accessed 15 November 2006).

25 This remark immediately sparked a heated discussion among Liao Xiaoyi, Liu Guozheng, deputy director general of the DG for Environmental Information of the State Environmental Protection Administration, and Wang Canfa, the director of the Centre for Legal Assistance to Pollution Victims. The discussion was part of a

delegation visit in the framework of the SENGO Project (Strengthening Environmental NGOs in China), which took place from 6 to 16 November 2003. The SENGO project (2001–5; budget _752,000) of which this article is the result, was funded by the Dutch Ministry of Foreign Affairs, and the Ministry of Economic Affairs.

26 B. Jancar-Webster, "Environmental Movement and Social Change".

27 See, for instance, D.A. Sonnenfeld, "Social Movements and Ecological Modernization: The Transformation of Pulp and Paper Manufacturing," *Development and Change*, Vol. 33 (2002), pp. 1–27.

28 For a discussion of the early developments of civil society in China, see G. White, *Riding the Tiger: The Politics of Economic Reform in Post-Mao China*, London: Macmillan, 1993.

Bibliography

Adams, W.M. (2001) *Green Development: Environment and Sustainability in the Third World*, 2nd edn, London and New York: Routledge.

Ahearne, J. (1995) *Michel de Certeau: Interpretation and Its Other*, Stanford: Stanford University Press.

Akhavan-Majid, R. (2004) "Mass Media Reform in China: Toward a New Analytical Framework," *Gazette*, 66–6: 553–65.

Alford, W. and Shen, Y. (1997) "Limits of the Law in Addressing China's Environmental Dilemma," *Stanford Environmental Law Journal*, 16: 125–48.

Angel, D.P. and Rock, M.T. (2003) "Engaging Economic Development Agencies in Environmental Protection: The Case for Embedded Autonomy," *Local Environment*, 8–1: 45–59.

Anheier, H., Glasius, M. and Kaldor, M. (2001) "Introducing Global Civil Society," in H. Anheier, M. Glasius and M. Kaldor (eds) *Global Civil Society Yearbook 2001*, Oxford: Oxford University Press.

Apter, D.E. and Sawa, N. (1984) *Against the State: Politics and Social Protest in Japan*, Cambridge, MA: Harvard University Press.

Atwood, C. (1992) "National Party and Local Politics in Ordos, Inner Mongolia (1926–1935)," *Journal of Asian History*, 26–1: 1–30.

Bao H. and Liu B. (2006) "*Zhongsheng xuanhua: 'jingwei ziran' da taolun*" [The Big Debate Regarding "Respect or Fear Nature"], in C. Liang (ed.) *Huanjing lü pishu 2005 nian: Zhongguo de huanjing weiju yu tuwei* [Green Book of Environment 2005: Crisis and Breakthrough of China's Environment], Beijing: Shehui Kexue Wenxian Chubanshe.

Bao, Qingwu (2001) personal communication with Hong Jiang.

Baoriledai (1965) "*Baoriledai jianghua*" [Baoriledai speech], unpublished.

—— (1968) "*Zhongyu maozhuxi shi zuida zuidai de gong*" [Loyalty to Chairman Mao is the Utmost Highest Collective] in Collection 3, unpublished material.

Barfield, T.J. (1989) *The Perilous Frontier. Nomadic Empires and China, 221 BC to AD 1757*, Cambridge, MA: Blackwell.

Barmé, G.R. and Davies, G, (2004) "Have We Been Noticed Yet? Intellectual Contestation and the Chinese Web" in E. Gu and M. Goldman (eds) *Chinese Intellectuals Between State and Market*, London: Routledge-Curzon.

Baum, R. and Shevchenko, A. (1999) "The 'State of the State'" in M. Goldman and R. MacFarquhar (eds) *The Paradox of China's Post Mao Reforms*, Cambridge, MA: Harvard University Press.

Baumol, W.J. and Oats, W.E. (1979) *Economics, Environmental Policy, and the Quality of Life*, Englewood Cliffs, NJ: Prentice-Hall.

Bayin (1998) *Menggu youmu shehui de bianqian* [Change in Mongolian Nomadic Society], Hohhot: Neimenggu Renmin Chubanshe.

Beccalli, B. (1994) "The Modern Women's Movement in Italy," *New Left Review*, 204: 86–112.

Becker, J. (1996) *Hungry Ghosts: Mao's Secret Famine*, New York: The Free Press.

—— (2001) "Putrid Lake Proof Environmental Polices have Failed to Hold Water," *South China Morning Post*, 15 October.

Bentley, J.G. (2003) "The Role of International Support for Civil Society Organizations in China," *Harvard Asia Quarterly*, 7–1: 11–20.

Bernauer, T. and Koubi, V. (2004) "Political Determinants of Environmental Quality," paper presented at the Annual Convention of the American Political Science Association, Chicago, 2–5 September.

Bernstein, T. and Lü, X. (2003) *Taxation Without Representation in Contemporary Rural China*, Cambridge: Cambridge University Press.

Bo, W. (1998) "Greening the Chinese Media" (Woodrow Wilson Center), *China Environment Series*, 2: 39–45.

Bolotova, A., Tysiachniouk, M. and Vorobiov, D. (1999) "*Analiz I klassifikatsia ekologicheskich nepravitelstvennich organizatsii Sankt-Peterburga*" [Analysis and Classification of Non-governmental Organizations in St Petersburg,], in E. Zdravomislova and M. Tysiachniouk (eds) *Ekologicheskoe dvigenie v Rossii* [Ecological Movements in Russia], St Petersburg: Tcentr nezavisimich sotsiologicheskich issledovaniy [Center for Independent Social Research].

Brechin, S.R. and Kempton, W. (1994) "Global Environmentalism: A Challenge to the Postmaterialism Thesis?" *Social Science Quarterly*, 75–2: 245–69.

Brettell, A. (1997) "Environmental Nongovernmental Organizations in the People's Republic of China: Civil Society or a Co-opted Environmental Movement?" paper presented at the Pacific Affairs Study Society Conference, Washington DC, 25 April.

—— (2000) "Environmental Non-governmental Organizations in the People's Republic of China: Innocents in a Co-opted Environmental Movement?" *Journal of Pacific Asia*, 6: 27–56.

—— (2003) "The Politics of Public Participation and the Emergence of Environmental Proto-movements in China," PhD dissertation, University of Maryland.

—— (2007) "Environmental Disputes: Contentious Politics and Citizen Access to Justice" in M. Scheures and J. Turner (eds) *Environmental Justice in China*, Washington, DC: Woodrow Wilson Center Press (forthcoming).

—— (2007) "China's Pollution Challenge: The Impact of Economic Growth and Environmental Complaints on Environmental and Social Outcomes" in S. Guo (ed.) *Challenges Facing Chinese Political Development*, Lanham, MD: Lexington-Rowman and Littlefield (forthcoming).

Broadbent, J. (1999) *Environmental Politics in Japan*, Cambridge: Cambridge University Press.

Brook, T. and Frolic, B.M. (1997) "The Ambiguous Challenge of Civil Society" in T. Brook and B.M. Frolic (eds) *Civil Society in China*, Armonk, NY: M.E. Sharpe.

Brown, L.D. (1993) "Social Change through Collective Reflection with Asia Nongovernmental Development Organisations," *Human Relations*, 46–2: 249–73.

Bryant, R. (2002) "False Prophets? Mutant NGOs and Philippine Environmentalism," *Society and Natural Resources*, 15: 629–39.

Bulag, U.E. (2000) "From Inequality to Difference: Colonial Contradictions of Class and Ethnicity in 'Socialist' China," *Cultural Studies*, 14–3/4: 531–61.

—— (2002) *The Mongols at China's Edge: History and Politics of National Unity*, Lanham, MD: Rowman and Littlefield.

—— (2004) personal communication with Hong Jiang.

Bullock, R.J. and Tubbs, M.E. (1987) "The Case Meta-Analysis Method for OD," *Research in Organizational Change and Development*, 1: 171–228.

Busgen, M. (2005) "The Campaign against the Nujiang Dam," unpublished MA thesis, Institute of Social Studies, The Hague.

Cai, Y. (2004) "Managed Political Participation in China," *Political Science Quarterly*, 119–3: 425–51.

Calhoun, C. (1989) "Tiananmen, Television and the Public Sphere: Internationalization of Culture and the Beijing Spring of 1989," *Public Culture*. 2–1: 54–71.

—— (1995) "'New Social Movements' of the Early Nineteenth Century" in M. Traugott (ed.) *Repertoires and Cycles of Collective Action*, Durham: Duke University Press.

—— (2004) "Information Technology and the International Public Sphere" in D. Schuler and P. Day (eds) *Shaping the Network Society: The New Role of Civil Society in Cyberspace*, Cambridge, MA: MIT Press.

Carothers, T. (1999/2000) "Civil Society," *Foreign Policy*, 117.

Carson, R. (1962) *Silent Spring*, Boston: Houghton Mifflin.

Castells, M. (1977) *The Urban Question*, London: Edward Arnold.

—— (1983) *The City and the Grassroots*, Berkeley: University of California Press.

Cellarius, B.A. and Staddon, C. (2002) "Environmental Nongovernmental Organizations, Civil Society and Democratization in Bulgaria," *East European Politics and Societies*, 16–1: 182–222.

Certeau, M. de (1984) *The Practice of Everyday Life*, trans. S. Rendall, Berkeley, CA: University of California Press.

Chamberlain, H.B. (1993) "On the Search for Civil Society in China," *Modern China*, April: 199–215.

Chan, A. (1993) "Revolution or Corporatism? Workers and Trade Unions in Post-Mao China," *Australia Journal of Chinese Affairs*, 29: 31–61.

Chan, A., Madsen, R. and Unger, J, (1992) *Chen Village under Mao and Deng*, revised edn, Berkeley, CA: University of California Press.

Chan, C.S. (2004) "The *Falun Gong* in China: A Sociological Perspective," *The China Quarterly*, 179: 665–83.

Chang, G.H. and Wen, G.J. (1997) "Communal Dining and the Chinese Famine of 1958–1961," *Economic Development and Cultural Change*, 46–1: 1–34.

Chatterjee, P. and Finger, M. (1994) *The Earth Brokers: Power, Politics and World Development*, London: Routledge.

Cheng, T. and Selden, M. (1994) "The Origins and Social Consequences of China's Hukou System," *The China Quarterly*, 139: 644–69.

China Development Brief (2001) "Civil Society in the Making: 250 Chinese NGOs," Beijing: China Development Brief.

—— (2003) "Imminent Foundation Rules will Create Two-track System and Embrace International Organizations," 6–2, Beijing: China Development Brief.

China's Agenda 21: White Paper on China's Population, Environment, and Development in the 21st Century (1994) (Adopted at the 16th Executive Meeting of the State Council of the People's Republic of China), Beijing: Zhongguo Huanjing Kexue Chubanshe.

Chinese Communist Party Politburo (1998) "*Zhonggong Zhongyang Bangongting Guowuyuan Bangongting guanyu dangzheng jiguan lingdao ganbu bu jianren shehui tuanti lingdao zhiwu de tongzhi*" [Notice by the CCP Politburo and the State Council Regarding the Fact that Leading Cadres in Party and Government Organizations Should Not Hold Concurrent Leading Posts in Social Organizations], 17: 1–2.

Chiu, Y. (2001) "Same War, Different Battles," *Taipei Times*, 21 April. Online, available at www.taipeitimes.com/news (accessed 15 November 2006).

Clark, A.M. (2001) "A Calendar of Abuses: Amnesty International's Campaign on Guatemala" in C. Welch (ed.) *NGOs and Human Rights: Promise and Performance*, Philadelphia: University of Pennsylvania Press.

Cooley, A. and Ron, J. (2002) "The NGO Scramble: Organisational Insecurity and the Political Economy of Transnational Action," *International Security*, 27–1: 5–39.

Croll, E. (1994) *From Heaven to Earth: Images and Experiences of Development in China*, London: Routledge.

Crook, D. and Crook, I. (1959) *Revolution in a Chinese Village: Ten Mile Inn*, London: Routledge and Paul Kegan.

Dacin, M.T., Ventresca, M.J. and Beal, B.D (1999) "The Embeddedness of Organizations: Dialogue and Directions," *Journal of Management*, 25–3: 317–56.

Dasgupta, S. and Wheeler, D. (1997) *Citizen Complaints as Environmental Indicators: Evidence from China*, Washington, DC: World Bank, Policy Research Department.

Dasgupta, S., Huq, M. and Wheeler, D. (1997) "Bending the Rules: Discretionary Pollution Control in China," Working Paper, Environment, Infrastructure and Agriculture Division, Policy Research Department, World Bank, February. Online, available at www.worldbank.org/html/dec/Publications/Workpapers/WPS1700series/wps1761/wps 1761.pdf.

Dawson, J. (1996) *Eco-Nationalism: Anti-Nuclear Activism and National Identity in Russia, Lithuania, and Ukraine*, Durham, NC: Duke University Press.

Deng, G. (2000) "*Beijing NGO wenjuan diaocha fenxi*" [Analysis of the Questionnaires on NGOs in Beijing] in Wang M. (ed.) *Zhongguo NGO yanjiu* [Research on NGOs in China], Beijing: UNCRD Research Report Series No. 38.

Deng, G. and Gu, L. (eds) (2001) *Fei yingli zhuzhi pinggu* [Non-profit Organization Evaluation], Beijing: Shehui Kexue Wenxian Chubanshe.

Diamond, L. and Myers, R.H. (2000) "Elections and Democracy in China," *The China Quarterly*, 162: 365–86.

Diani, M. (1992) "The Concept of a Social Movement," *Sociological Review*, 40: 1–15.

—— (1995) *Green Networks: A Structural Analysis of the Italian Environmental Movement*, Edinburgh: Edinburgh University Press.

Dicklitch, S. (2002) "NGOs and Democratization in Transitional Societies: Lessons from Uganda" in D.N. Nelson and L. Neak (eds) *Global Society in Transition: An International Politics Reader*, The Hague: Kluwer Law International.

Doyle, T. and McEachern, D. (1998) *Environment and Politics*, London: Routledge.

Dunlap, R.E. (1989) "Public Opinion and Environmental Policy" in J.P. Lester (ed.) *Environmental Politics and Policy*, Durham, NC: Duke University Press.

Dunlap, R.E., Gallup, G.H., Jr and Gallup, A.M. (1993) "Of Global Concern: Results of the Health of the Planet Survey," *Environment*, 35–9: 7–39.

Duo J. (ed.) (1996) *Shehui tuanti guanli gongzuo* [Work on the Administration of Social Organizations], Beijing: Zhongguo Shehui Chubanshe.

Durkheim, E. (1951) *Suicide: A Study in Sociological Interpretation*, Glencoe, IL: Free Press.

Dwivedi, R. (2001) "Environmental Movements in the Global South. Issues of Livelihood and Beyond," *International Sociology*, 16–1: 11–31.

Eade, D. (1997) *Capacity Building. An Approach to People Centered Development*, Oxford: Oxfam.

Eccleston, B. (1996) "Does North–South Collaboration Enhance NGO Influence on Deforestation Policies in Malaysia and Indonesia?" in D. Potter (ed.) *NGOs and Environmental Policies in Asia and Africa*, Portland, OR: F. Cass.

Economy, E.C. (2002) "China's Go West Campaign: Ecological Construction or Ecological Exploitation?" *China Environment Series*, 5: 1–11.

—— (2004) *The River Runs Black: Environmental Challenges to China's Future*, Ithaca: Cornell University Press.

—— (2006) "Environmental Governance: The Emerging Economic Dimension," *Environmental Politics*, 15–2: 171–89.

Edelman, M. (1999) *Peasants against Globalization*, Stanford, CA: Stanford University Press.

Edmonds, R.L. (1994) *Patterns of China's Lost Harmony: A Survey of the Country's Environmental Degradation and Protection*, London: Routledge.

—— (ed.) (2000) *Managing the Chinese Environment*, Oxford: Oxford University Press.

—— (2003) "China's Environmental Problems," in R.E. Gamer (ed.) *Understanding Contemporary China*, 2nd edition, Boulder: Lynne Rienner.

Edwards, M. and Hulme, D. (1997) *NGOs, States and Donors: Too Close for Comfort?* New York: St Martin's Press in association with Save the Children.

Eimer, D. (2005) "China Clamps Down on Environmental Monitoring Group," *Independent*, 28 October, pp. 1–2.

Elliot, C. and Schlaepfer, R. (2003) "Global Governance and Forest Certification: A Fast Track Process for Global Policy Change" in E. Meidinger, C. Elliot and G. Oesten (eds) *Social and Political Dimensions of Forest Certification*, Remagen-Oberwinter: Forstbuch.

Elvin, M. (1998) "The Environmental Legacy of Imperial China" in R.L. Edmonds (ed.) *Managing the Chinese Environment*, New York: Oxford University Press.

Elvin, M. and Liu, T. (eds) (1998) *Sediments of Time: Environment and Society in Chinese History*, Cambridge, UK: Cambridge University Press.

Evangelista, M. (1999) *Unarmed Forces: The Transnational Movement to End the Cold War*, Ithaca: Cornell University Press.

Evans, P. (1995) *Embedded Autonomy. States and Industrial Transformation*, Princeton: Princeton University Press.

—— (1996) "Government Action, Social Capital and Development: Reviewing the Evidence on Synergy," *World Development*, 24–6: 1119–32.

Fagan, A. and Jehlicka, P. (2003) "Contours of the Czech Environmental Movement: A Comparative Analysis of Hnuti Duha (Rainbow Movement) and Jihoceske Matky (South Bohemian Mothers)," *Environmental Politics*, 12–2: 49–70.

Falk, R. (1998) "Global Civil Society: Perspectives, Initiatives, Movements," *Oxford Development Studies*, 26–1: 99–111.

Farzin, Y.H. and Bond, C.A. (2006) "Democracy and Environmental Quality," *Journal of Development Economics*, 81–1: 213–35.

Feng, C. (2006) "Privatization and its Discontents in Chinese Factories," *The China Quarterly*, 185: 42–60.

Feng, L. (2000) "Cultural Center of the Global Village of Beijing," in Wang M. (ed.) *Zhongguo NGO yanjiu* [Research on NGOs in China], Beijing: UNCRD Research Report Series no. 38.

Fisher, Dana R. (2003) "Global and Domestic Actors within the Global Climate Change

Regime: Toward a Theory of Global Environmental Systems," *International Journal of Sociology and Social Policy*, 23–10: 5–30.

Fisher, Duncan (1993) "The Emergence of the Environmental Movement in Eastern Europe and its Role in the Revolutions of 1989," in B. Jancar-Webster (ed.), *Environmental Action in Eastern Europe: Responses to Crisis*, Armonk, NY: M.E. Sharpe.

Fisher, J. (1993) *The Road from Rio: Sustainable Development and the Nongovernmental Movement in the Third World*, Westport: Praeger.

Fowler, A (1991) "The Role of NGOs in Changing State–Society Relations: Perspectives from Eastern and Southern Africa," *Development Policy Review*, 9–1: 53–84.

Fraser, N. (1992) "Rethinking the Public Sphere – A Contribution to the Critique of Actually Existing Democracy" in C. Calhoun (ed.) *Habermas and the Public Sphere*, Cambridge, MA: MIT Press.

Friedman, E. (1978) "The Politics of Local Models, Social Transformation and State Power Struggles in the People's Republic of China: Tachai and Teng Hsiao-p"ing," *The China Quarterly*, 76: 873–90.

Friedman, E., Pickowicz, P.G., Selden, M. and Johnson, K.J. (1991) *Chinese Village, Socialist State*, New Haven: Yale University Press.

Frolic, M. (1997) "State-led Civil Society" in T. Brook and M. Frolic (eds) *Civil Society in China*, Armonk, NY: M.E. Sharpe.

Gamson, W. (1992) "The Social Psychology of Collective Action" in A. Morris and C. Mueller (eds) *Frontiers in Social Movement Theory*, New Haven, CT: Yale University Press.

Ganie-Rochman, M (2002) *An Uphill Struggle: Advocacy NGOs under Soeharto's New Order*, Jakarta: LabSosio.

Gao, Chao and Zhang, Y. (1997) *Kexie gongzuo jianming cidian* [Concise Dictionary for Associations of Science and Technology], Beijing: Zhongguo Keji Chubanshe.

Gao, Chao, Ying, G. and Tian, G, (1992) *Zhongguo kexie xue* [Study of Chinese Association of Science and Technology], Beijing: Zhongguo Keji Chubanshe.

Gao, Chunsuo (2003) "Embarrassed Trade Associations [*Tongye gonghui tai ganga*]" *China Industrial and Commercial Times*, 13 March.

Gerardo, N.J., Miller, J. and Liu, X. (eds) (2001) *Daoism and Ecology: Ways within a Cosmic Landscape*, Cambridge, MA: Harvard University Press.

Ghose, R. (1987) *Protest Movements in South and South-East Asia: Traditional and Modern Idioms of Expression*, Hong Kong: University of Hong Kong, Centre of Asian Studies.

Gold, T. (1990) "The Resurgence of Civil Society in China," *Journal of Democracy*, 1–1: 18–31.

Gold, T., Guthrie, D. and Wank, D. (eds) (2002) *Social Connections in China: Institutions, Culture and the Changing Nature of Guanxi*, Cambridge: Cambridge University Press.

Goldman, M.I. (1990) "Environmentalism and Nationalism: An Unlikely Twist in an Unlikely Direction" in J.M. Stewart (ed.) *The Soviet Environment: Problems, Policies and Politics*, Cambridge, UK: Cambridge University Press.

Goldman, M. and Perry, E.J. (eds) (2002) *Changing Meanings of Citizenship in Modern China*, Harvard: Harvard University Press.

Gongmeng (comp.) (2006) "*Zhongguo Xinfang Baogao*" [Report on Appeals in China], *Gongmeng*, Beijing, China. Online, available at www.gongmeng.cn/sub_list.php?zyj_mid=21.

Goodman, D.S.G. (1981) *Beijing Street Voices: The Poetry and Politics of China's Democracy Movement*, London: Marion Boyars.

Granovetter, M. (1973) "The Strength of Weak Ties," *American Journal of Sociology*, 78–6: 1360–80.

—— (1984) "Economic Action and Social Structure: The Problem of Embeddedness," *American Journal of Sociology*, 91–3: 481–510.

Greer, J. and Bruno, K. (1996) *Greenwash: The Reality Behind Corporate Environmentalism*, Penang: Third World Network.

Grossman, G.M. and Krueger, A.B. (1993) "Environmental Impacts of a North American Free Trade Agreement" in P.M. Garbier (ed.) *The Mexico–U.S. Free Trade Agreement*, Cambridge, MA: MIT Press.

Grossman, G.M. and Krueger, A.B. (1995) "Economic Growth and the Environment," *Quarterly Journal of Economics*, 112: 353–77.

Gu, E.X. (1998) "Non-Establishment Intellectuals, Public Space, and the Creation of Non-Governmental Organizations in China: The Chen Ziming–Wang Juntao Saga," *The China Journal*, 39: 39–58.

Guidry, J.A., Kennedy, M.D. and Zald, M.N. (eds) (2000) *Globalizations and Social Movements: Culture, Power and the Transnational Public Sphere*, Ann Arbor: University of Michigan Press.

Gujarati, D. (1995) *Basic Econometrics*, 3rd edn, New York: McGraw-Hill.

Guojia Huanjing Baohuju [National Environmental Protection Agency] (1987) *Baogao Huanjing Wuran yu Pohuai Shigu de Zhanxing Banfa* [Provisional Regulation Regarding Notification of Environmental Pollution and Accidents], Beijing; State Environmental Protection Agency, 10 September.

—— (1991) *Huanjing Baohu Xinfang Guanli Banfa*, Guojia Huanjing Baohuju (01 February, went into effect).

—— (1997) *Huanjing Xinfang Banfa*, Guojia Huanjing Baohu Ju (29 April, announced).

—— (2006) *Huanjing Xinfang Banfa*, Guojia Huanjing Baohushu. (24 June, announced).

Guowuyuan (1995) *Xinfang Tiaoli* [Regulation regarding complaints], Zhonghua Renmin Gongheguo Guowuyuan, promulgated 10 October.

—— (2005) *Xinfang Tiaoli* [Regulation regarding complaints revised], Zhonghua Renmin Gongheguo Guowuyuan, issued 10 January.

Guowuyuan Jigou Gaige Gailan [General Reader on the Restructuring of the State Council] (1998) Beijing: China News Publishing.

Guthrie, D, (1998) "The Declining Significance of *Guanxi* in China's Economic Transition," *The China Quarterly*, 154: 254–82.

Habermas, J. (1989) *The Structural Transformation of Public Sphere*, Cambridge, MA: MIT Press.

—— (1989) "The Public Sphere" in S. Seidman (ed.) *Jurgen Habermas on Society and Politics: A Reader*, Boston: Beacon Press.

—— (1992) "Further Reflections on the Public Sphere" in C. Calhoun (ed.) *Habermas and the Public Sphere*, Cambridge, MA: MIT Press.

Hamburger, J. (2002) "Pesticides in China: A Growing Threat to Food Safety, Public Health, and the Environment," *China Environment Series*, 5: 29–44.

Hamilton, G.G. (ed.) (1991, *Business Networks and Economic Development in East and Southeast Asia*, Hong Kong: Centre of Asian Studies, University of Hong Kong.

Hann, C. and Dunn, E. (1996) *Civil Society, Challenging Western Models*, London: Routledge.

Harre, R., Brockmeier, J. and Muhlhausler, P. (1999) *Greenspeak: A Study of Environmental Discourse*, Thousand Oaks: Sage.

Haufler, V. (2003) "New Forms of Governance: Certification Regimes as Social Regulations of the Global Market" in E. Meidinger, C. Elliot and G. Oesten (eds) *Social and Political Dimensions of Forest Certification*, Remagen-Oberwinter: Forstbuch.

Hawthorn, G. (2001) "The Promise of Civil Society in the South" in S. Kaviraj and S. Khilnani (eds) *Civil Society: History and Possibilities*, Cambridge: Cambridge University Press.

Haxi Zhaxiduojie (2002) "*Sanjiangyuan de huhuan*" [The call of the Three River Source], *Friends of Nature Newsletter*, 4. Online, available at www.fon.org.cn/index.php?id=3009 (accessed 2 April 2003).

Haynes, J. (1997) *Democracy and Civil Society in the Third World. Politics and New Political Movements*, Cambridge: Polity Press.

Heggelund, G. (2004) *Environment and Resettlement Politics in China: The Three Gorges Project*, London: Ashgate.

—— (2006) "Resettlement Programs and Environmental Capacity in the Three Gorges Dam Project" in P. Ho and E.B. Vermeer (eds) *China's Limits to Growth: Greening State and Society*, Oxford: Blackwell.

Hirsch, P. (1994) "Where are the Roots of Thai Environmentalism?" *TEI Quarterly Environment Journal*, 2: 5–15.

Ho, P. (2001) "Greening without Conflict? Environmentalism, Green NGOs and Civil Society in China," *Development and Change*, 32–5: 893–921.

—— (2001) "Who Owns China's Land? Policies, Property Rights and Deliberate Institutional Ambiguity," *The China Quarterly*, 166: 387–414.

—— (2005) *Institutions in Transition: Land Ownership, Property Rights and Social Conflict in China*, Oxford: Oxford University Press.

—— (ed.) (2005) "Greening Industries in Newly Industrializing Countries: Asian-style Leapfrogging?" *International Journal of Environment and Sustainable Development: A Special Issue*, 4–3: 209–26 (six articles).

—— (ed.) (2007) *Greening Industries in Newly Industrializing Countries: Asian-style Leapfrogging?* London: Kegan Paul.

Ho, P. and Vermeer, E.B. (eds) (2006) *China's Limits to Growth: Greening State and Society*, Oxford: Blackwell.

Ho, P., Vermeer, E.B. and Zhao, J.H, (2006) "Biotech and Food Safety in China: Consumers' Acceptance or Resistance?" in P. Ho and E.B. Vermeer (eds) "China's Limits to Growth: Prospects for Greening State and Society," *Development and Change*, 37–1: 227–53.

Hoffer, E. (1951) *The True Believer: Thoughts on the Nature of Mass Movements*, New York, NY: Harper and Row.

Howell, J. (ed.). (2004) *Governance in China*, Boulder, CO: Rowman and Littlefield.

Howell, J. and Pearce, J. (2001) *Civil Society and the State: A Critical Exploration*, Boulder, CO: Lynne Rienner.

Hsiao, M., Lai, O., Liu, H., Magno, F.A., Edles, L. and So, A.Y. (1999) "Culture and Asian Styles of Environmental Movements," in: Y.F. Lee and A.Y. So (eds) *Asia's Environmental Movements: Comparative Perspectives*, Armonk, NY: M.E. Sharpe.

Hu, W.A. (2000) "Friends of Nature" in Wang M. (ed.) *Zhongguo NGO yanjiu* [Research on NGOs in China], Beijing: UNCRD Research Report Series 38.

Huang, J., Qiao, F.B., Zhang, L.X. and Rozelle, S. (2000) "Farm Pesticide, Rice Production, and Human Health," Beijing: CCAP's Project Report, Center for Chinese Agricultural Policy, Chinese Academy of Agricultural Sciences.

Huang, W. (2000) "Individuals Changing the World: An Interview with Sophie Prize 2000 Winner Liao Xiaoyi," *Beijing Review*, 33 (14 August): 12–19.

Huang, Z. and Song, B. (1986) "*Neimenggu Yikezhao tudi shamohua jiqi fangzhi*" [Desertification and its control in Ih-Ju League, Inner Mongolia], *Zhongguo Kexueyuan Lanzhou Shamosuo*, Jikan, 3: 35–47.
Human Rights Watch (2002) "Dangerous Mediation: China's Campaign against Falungong," online, available at http://hrw.org/reports/2002/china/China0102–02.htm#P331_49488> (accessed 1 February 2002).
—— (2005) "The Petitioning System," online, available at http://hrw.org/reports/2005/china1205/4.htm.
Huntington, S.P. and Nelson, J.M. (1976) *No Easy Choice: Political Participation in Developing Countries*, Cambridge, MA: Harvard University Press.
International Fund for Animal Welfare (2001) "Wrap Up the Trade: An International Campaign to Save the Endangered Tibetan Antelope," USA: International Fund for Animal Welfare.
International Rivers Network (ed.) (2003) *Human Rights Dammed Off at Three Gorges: An Investigation of Resettlement and Human Rights Problems in the Three Gorges Dam Project*, Berkeley, CA: International Rivers Network.
Jahiel, A. (1994) "Policy Implementation through Organizational Learning: The Case of Water Pollution Control in China's Reforming Socialist System," PhD dissertation, University of Michigan.
Jamison, A., Eyerman, R., Cramer, J. with Laessøe, J. (1990) *The Making of the New Environmental Consciousness: A Comparative Study of the Environmental Movements in Sweden, Denmark and the Netherlands*, Edinburgh: Edinburgh University Press.
Jancar-Webster, B. (1998) "Environmental Movement and Social Change in the Transition Countries," *Environmental Politics*, 7–1: 69–90.
Jänicke, M. (1992) "Conditions for Environmental Policy Success," *The Environmentalist*, 12: 47–68.
Jankowiak, W.R. (1988) "The Last Hurrah? Political Protest in Inner Mongolia," *Australian Journal of Chinese Affairs*, 19/20: 369–88.
Jia, H. and Lin, Z. (1994) *Changing Central–Local Relations in China*, Boulder: Westview.
Jiang, H. (2004) "Cooperation, Land Use, and the Environment in Uxin Ju: A Changing Landscape of a Mongolian–Chinese Borderland in China," *Annals of Association of American Geographers*, 94–1: 117–39.
—— (forthcoming) "Poaching State Politics in Socialist China: Uxin Ju's Grassland Campaign during 1958–1966," *Geographical Review*.
Jin, C. (2003) "*Zhan chulai – shuohua*" [Step forward – and speak up!], *Cengjing shidi* [The Wetlands That Once Were], Green Camp: 44.
Jin, J. and Deng W. (2002) *Quanqiu gongmin shehui – fei yingli bumen shijie* [Global Civil Society – Vision of the Non-profit Sector], Beijing: Shehui Kexue Wenxian Chubanshe.
Jing, J. (1997) "Rural Resettlement: Past Lessons for the Three Gorges Project," *The China Journal*, 38: 65–94.
—— (2000) "Environmental Protest in Rural China" in E. Perry and M. Selden (eds) *Chinese Society; Change, Conflict and Resistance*, London: Routledge.
Johnson, I. (2004) *Wild Grass: Three Stories of Change in Modern China*, New York: Pantheon.
Johnson, R.J. (1992) "Laws, States, and Superstates: International Laws and the Environment," *Applied Geography*, 12: 211–28.
Kaimowitz, D. (1996) "Social Pressure for Environmental Reform" in H. Collinson (ed.) *Green Guerrillas*, New York: Monthly Review Press.

Kalland, A. and Persoon, G. (1998) *Environmental Movements in Asia*, Copenhagen: Nordic Institute of Asian Studies.

Keane, J. (2003).*Global Civil Society?* Cambridge: Cambridge University Press.

Keck, M.E. and Sikkink, K. (1998) *Activists Beyond Borders, Advocacy Networks in International Politics*, Ithaca, NY: Cornell University Press.

Keith, R.C. and Lin, Z. (2003) "The '*Falun Gong* Problem': Politics and the Struggle for the Rule of Law in China," *The China Quarterly*, 175: 623–42.

Khagram, S. (2000) "Toward Democratic Governance for Sustainable Development: Transnational Civil Society Organizing around Big Dams" in A.M. Florini (ed.) *The Third Force: The Rise of Transnational Civil Society*, Tokyo: Japan Center for International Exchange.

Khagram, S., Riker, J. and Sikkink, K. (eds) (2002) *Restructuring World Politics: Transnational Social Movements, Networks, and Norms*, Minneapolis: University of Minnesota Press.

Khotari, S. (1996) "Rising from the Margins: The Awakening of Civil Society in the Third World," *Development*, 3: 11–19.

Klandermans, B. (1991) "New Social Movements and Resource Mobilization: The European and the American Approach Revisited" in D. Rucht (ed.) *Research on Social Movements*, Boulder, CO: Westview Press.

Klotz, A. (2002) "Transnational Activism and Global Transformations: The Anti-Apartheid and Abolitionist Experiences," *European Journal of International Relations*, 8–1: 49–76.

Knup, E. (1997) "Environmental NGOs in China: An Overview," *China Environment Series*, 1: 9–15.

Kornhauser, W. (1959) *The Politics of Mass Society*, Glencoe, IL: Free Press.

Kumar, C. (2000) "Transnational Networks and Campaigns for Democracy" in A. Florini (ed.) *The Third Force: The Rise of Transnational Civil Society*, Tokyo: Japan Center for International Exchange.

Lafferriere, E. (1994) "Environmentalism and the Global Divide," *Environmental Politics*, 3–1: 91–113.

Lafferty, W.M. and Meadowcroft, J. (eds) (1996) *Democracy and the Environment: Problems and Prospects*, Cheltenham: Edward Elgar.

Lean, E. (2004) "The Making of a Public: Emotions and Media Sensation in 1930s China," *Twentieth-Century China*, 29–2: 39–62.

Lee, N. (2006) "The Development of Environmental NGOs in China," *China Brief*, 6–23: 7–10.

Lee, S. (1999) "Environmental Movements in South Korea" in Y.F. Lee and A.Y. So (eds) *Asia's Environmental Movements: Comparative Perspectives*, Armonk, NY: M.E. Sharpe.

Lewis, T.L. (2000) "Transnational Conservation Movement Organizations: Shaping the Protected Area Systems of Less Developed Countries," *Mobilization*, 5: 105–123.

Li, L. and O'Brien, K.J. (1996) "Villagers and Popular Resistance in Contemporary China," *Modern China*, 22–1: 28–61.

Li, X. (2002) "'Focus' (*Jiaodian Fangtan*) and the Changes in the Chinese Television Industry," *Journal of Contemporary China*, 11–30: 17–34.

Liang, B. (1991) *Yikezhao Meng de tudi kaiken* [Land Opening in Ih-Ju League], Hohhot: Neimenggu Daxue Press.

Liang, C. (2001) "The Building Up of Environmental NGOs in China: The Road of Friends of Nature" in Zhao, L. and C.I. Irving (eds) *The Non-Profit Sector and Development*, Hong Kong: Hong Kong Press for Social Science.

Lieberthal, K. (1995) *Governing China: From Revolution through Reform*, New York: W.W. Norton.

Lieberthal, K. and Lampton, D.M. (eds) (1992) *Bureaucracy, Politics and Decision Making in Post-Mao China*, Berkeley, CA: University of California Press.

Lieberthal, K. and Oksenberg, M. (1988) P*olicy Making in China: Leaders, Structures, and Process*, Princeton, NJ: Princeton University Press.

Lin, Yuguo, (2005), "China Worries that Foreign NGOs are Importing a 'Color Revolution'," Sun/Central News Agency, Hong Kong, online, available at www.chinalaborwatch.org (accessed 6 August 2005).

Lindblom, C. (1979) "Still Muddling, Not Yet Through," *Public Administration Review*, 39: 517–37.

—— (1992) "The Science of 'Muddling Through'" in J.M. Shafritz and A.C. Hyde (eds) *Classics of Public Administration*, 3rd edn, Belmont: Wadsworth.

Lipschutz, R. (1992) "Reconstructing World Politics: The Emergence of Global Civil Society," *Millennium: Journal of International Studies*, 21–3: 389–420.

Litzinger, R. (2004) "Damming the Angry River," *China Review Magazine*, 30: 1.

Lo, C.W.H. and Fryxell, G.E. (2003) "Enforcement Styles among Environmental Protection Officials in China," *Journal of Public Policy*, 23–1: 81–115.

—— (2005) "Governmental and Societal Support for Environmental Enforcement in China: An Empirical Study in Guangzhou," *Journal of Development Studies*, 41–4: 558–88.

—— (2006) "Effective Regulations with Little Effect? The Antecedents of the Perceptions of Environmental Officials on Enforcement Effectiveness in China," *Environmental Management*, 38–3: 388–410.

Lo, C.W.H. and Sai, W.L. (2000) "Environmental Agency and Public Opinion in Guangzhou: The Limits of a Popular Approach to Environmental Governance," *The China Quarterly*, 163: 677–704.

Lowe, J. (1999) "Defending Qinghai," *Earth Island Journal*, 14–3.

Lowe, S. (1986) *Urban Social Movement: After Castells*. London: Macmillan.

Lu, H. (2003) "Bamboo Sprouts after the Rain: The History of University Student Environmental Associations in China," *China Environment Series*, 6: 55–67.

Lu, W. (1999) "The Recent Changes of Forest Policy in China and Its Influences on the Forest Sector," interim report of the Institute for Global Environment Strategies Forest Conservation Project.

Lu, Y. (2005) *Environmental Civil Society and Governance in China*, London: Chatham House.

Luehrmann, L.M. (2003) "Facing Citizen Complaints in China, 1951–1996," *Asian Survey*, 43–5: 845–66.

Lukin, A. (1998) "The Image of China in Russia Border Regions," *Asian Survey*, 38–9: 821–35.

—— (2003) *The Bear Watches the Dragon: Russia's Perceptions of China and the Evolution of Russian–Chinese Relations Since the Eighteenth Century*, Armonk, NY: M.E. Sharpe.

Ma, J. (2005) "Green Groups Fall under Microscope in Beijing," *South China Morning Post*, 18 August: 12.

Ma, S.-Y. (1994) "The Chinese Discourse on Civil Society," *The China Quarterly*, 137: 180–93.

Ma, X. and Ortolano, L. (2000) *Environmental Regulation in China: Institutions, Enforcement, and Compliance*, New York: Rowman and Littlefield.

McAdam D., McCarthy, J.D. and Zald, M.N. (eds) (1996) *Comparative Perspectives on*

Social Movements: Political Opportunities, Mobilizing Structures, and Cultural Framings, Cambridge: Cambridge University Press.

McCormick, J. (1999) "The Role of Environmental NGOs in International Regimes," in J.N. Vig and R.S. Axelrod (eds) *The Global Environment*, Washington, DC: CQ Press.

McKean, M.A. (1981) *Environmental Protest and Citizen Politics in Japan*, Berkeley: University of California Press.

Manion, M. (2000) "Chinese Democratization in Perspective: Electorates and Selectorates at the Township Level," *The China Quarterly*, 163: 764–82.

Mao, W. and Tao, J. (2003) *Tuitujixia de Beiju* [The tragedy in front of the bulldozer], online, available at http://news.21cn.com/domestic/guoshi/2003/09/03/1254412.shtml.

Marks, R. (1998) *Tigers, Rice, Silk and Silt: Environment and Economy in Late Imperial South China*, New York: Cambridge University Press.

—— (2002) *The Origins of the Modern World: A Global and Ecological Narrative*, Lanham, MD: Rowman and Littlefield.

Martens, S. (2006) "Public Participation with Chinese Characteristics: Citizen Consumers in China's Environmental Management," *Environmental Politics*, 15–21: 211–30.

Martinelli, A. and Smelser, N.J. (eds) (1990) *Economy and Society: Overview in Economic Sociology*, London: Sage.

Matthews, J. (1997) "Power Shift," *Foreign Affairs*, 76–1: 50–66.

Mayer, M. (1991) "Social Movement Research and Social Movement Practice: The U.S. Pattern," in D. Rucht (ed.) *Research on Social Movements*, Boulder, CO: Westview Press.

Meidinger, E. (2003a) "Forest Certification as a Global Civil Society Regulatory Institution," in E. Meidinger, C. Elliot and G. Oesten (eds) *Social and Political Dimensions of Forest Certification*, Remagen-Oberwinter: Forstbuch.

—— (2003b) "Forest Certification as Environmental Law Making by Global Civil Society" in E. Meidinger, C. Elliot and G. Oesten (eds) *Social and Political Dimensions of Forest Certification*, Remagen-Oberwinter: Forstbuch.

Melucci, A. (1989) *Nomads of the Present: Social Movements and Individual Needs in Contemporary Society*, London: Hutchinson Radius.

Mertha, A. (2006) "Water Warriors: Political Pluralization in China's Hydropower Policy," unpublished manuscript; online, available at http://artsci.wustl.edu/~amertha/pdf/WaterWarriors.pdf.

Michelson, E. (2007) "Justice from Above or Justice from Below? Popular Strategies for Resolving Grievances in Rural China," *The China Quarterly*, forthcoming.

—— (2007) "Climbing the Dispute Pagoda: Grievances and Appeals to the Official Justice System in Rural China," *American Sociological Review*, forthcoming. Online: pre-copy edited version available at www.indiana.edu/~emsoc/Publications/Michelson_DisputePagoda.pdf.

Michnik, A. (1985) *Letters from Prison and Other Essays*, trans. Maya Latynski, Berkeley: University of California Press.

Milbrath, L.W. and Goel, M.L. (1977) *Political Participation: How and Why Do People Get Involved in Politics*? Chicago: Rand McNally College Publishing.

Ministry of Civil Affairs (1997) *Civil Affairs Statistics Yearbook of China: 1996*, Beijing: Ministry of Civil Affairs.

—— (ed.) (2000) *Minjian Zuzhi Guanli Zuixin Fagui Zhengce Huibian* [Compilation of the Newest Regulations and Policies for the Management of Civil Organizations], restricted material, speech by Deputy Minister Xu Ruixin, Beijing, 8 December 1999.

—— (2002) *Civil Affairs Statistics Yearbook of China: 2001*, Beijing: China Statistics Press.
Minzner, C.F. (2006) "Xinfang: An Alternative to Formal Chinese Legal Institutions," *Stanford Journal of International Law*, 42–2: 103–79.
Mishler, W. and Rose, R. (1997) "Trust, Distrust and Skepticism: Popular Evaluations of Civil and Political Institutions in Post-Communist Societies," *Journal of Politics*, 59: 418–51.
Mol, A. and Carter, N. (2006) "China's Environmental Governance in Transition," *Environmental Politics*, 15–2 (2006): 149–70.
Mollenkopf, J. (1983) *The Contested City*, Princeton, NJ: Princeton University Press.
Mooney, H.A. (1998) *The Globalization of Ecological Thought*, Oldendorf: Ecology Institute.
Morris, A. (1981) "Black Southern Student Sit-in Movement: An Analysis of Internal Organization," *American Sociological Review*, 46: 744–67.
—— (1984) *The Origins of the Civil Rights Movement*, New York: Free Press.
Morris, A. and Mueller, C.M. (eds) (1992) *Frontiers in Social Movement Theory*, New Haven: Yale University Press.
Morton, K. (2005) "The Emergence of NGOs in China and Their Transnational Linkages: Implications for Domestic Reform," *Australian Journal of International Affairs*, 59–4: 519–32.
Nash, K. (2000) *Contemporary Political Sociology: Globalization, Politics and Power*, Oxford: Blackwell.
Neimeng Xumuju [Inner Mongolia Department of Animal Husbandry] (1973) "*Wushenzhao Renmingongshe jiben caomuchang de jianshe qingkuang*" [Basic Situation of Uxin Ju Commune's Pastureland Construction] in Collection 4, unpublished material.
Neimenggu Ribao [Inner Mongolia Daily], 3 Dec. 1965; 7 Dec. 1965; 3 Feb. 1966; 14 Feb. 1966; 19 Aug. 1970; 4 Sept. 1970; 16 March 1973; 16 April 1973.
Nelson, P. (2002) "New Agendas and New Patterns of International NGO Political Action," *Voluntas*, 13: 377–92.
Norman, E. (1996) *Poisoned Prosperity: Development, Modernization, and the Environment in South Korea*, Armonk, NY: M.E. Sharpe.
O'Brien, K.J. (1996) "Rightful Resistance," *World Politics*, 49–1: 31–55
—— (2002) "Collective Action in the Chinese Countryside," *The China Journal*, 48: 139–54.
—— (2006) *Rightful Resistance: Contentious Politics in Rural China*, Cambridge: Cambridge University Press.
O'Brien, K.J. and Li, L. (1995) "The Politics of Lodging Complaints in Rural China," *The China Quarterly*, 143: 756–83.
Offenheiser, R. and Holcombe, S. (2001) "Challenges and Opportunities of Implementing a Rights-Based Approach to Development: An Oxfam Perspective," paper presented at "Northern Relief and Development NGOs: New Directions in Poverty Alleviation and Global Leadership Transitions" 2–4 July, Balliol College, Oxford, UK.
Oi, J. (1989) *State and Peasant in Contemporary China: The Political Economy of Village Government*, Berkeley: University of California Press.
Ostman, R.E. and Parker, J.L. (1987) "Impact of Education, Age, Newspapers, and Television on Environmental Knowledge, Concerns, and Behaviors," *Journal of Environmental Education*, 19: 3–9.
Pahl, R.E. (1969) "Urban Social Theory and Research," *Environment and Planning*, 1–2: 143–53.

Palmer, M. (1998) "Environmental Regulation in the People's Republic of China: The Face of Domestic Law," *The China Quarterly*, 156: 788–808.
Pepper, D. (1996) *Modern Environmentalism*, London: Routledge.
Perdue, P. (2006) *Exhausting the Earth: State and Peasant in Hunan, 1500–1850*, Cambridge, MA: Harvard University Press.
Perry, E.J. and Selden, M. (eds) (2003) *Chinese Society: Change, Conflict and Resistance*, 2nd edn, London: Routledge.
Pesticide Eco-Alternative Centre Yunnan Thoughtful Action (2002) *PEAC Newsletter*, 1–1, 1–2, and 2–1.
—— (2003) Annual Report, PEAC, Kunming, 20 December.
Pickvance, C.G. (ed.) (1975) "On the Study of Urban Social Movements," *Sociological Review*, 23–1: 29–49.
Polanyi, K. (1944) *The Great Transformation*, New York: Holt, Rinehart, and Winston.
Pomfret, J. and Laris, M. (1999) "China Confronts a Silent Threat," *Washington Post*, 30 October.
Porter, G. and Brown, J.W. (1996) *Global Environmental Politics*, Boulder, CO: Westview Press.
Potter, B.P. (2003) "Belief in Control: Regulation of Religion in China," *The China Quarterly*, 174: 317–37.
Power, J. (2005) "*In China komt de Democratie pas in 2035*" [Democracy will not come in China until 2035], translated from English, *NRC Handelsblad*, 4 June: 17.
Princen, T. and Finger, M. (1994) *Environmental NGOs in World Politics: Linking the Local and Global*, London: Routledge.
Princen, T., Finger, M. and Manno, J. (1995) "Nongovernmental Organizations in World Environmental Politics," *International Environmental Affairs*, 7–1: 42–58.
Putnam, R. (1993) *Making Democracy Work: Civic Traditions in Modern Italy*, Princeton: Princeton University Press.
Putten, J. van der (1998) "*Chinese dissidenten opgepakt*" [Chinese Dissidents Arrested], *Volkskrant*, 1 December: 4.
—— (1998) "*Bewind China vreest een naderende protestgolf*" [China's Regime Fears an Approaching Gulf of Protest], *Volkskrant*, 28 December: 5.
—— (1999) "*De mens is nog nooit zo diep gezonken*" [Man has Never Fallen so Low], *Volkskrant*, 27 April: 5.
Rankin, M.B. (1993) "Some Observations on a Chinese Public Sphere," *Modern China*, 19–2: 158–82.
Read, B.L. (2000) "Revitalizing the State's Urban 'Nerve Tips'," *The China Quarterly*, 163: 806–20.
—— (2003) "Democratizing the Neighborhood? New Private Housing and Homeowner Self-Organization in Urban China," *The China Journal*, 49: 31–59.
Renmin Daxue Social Survey Center (1996) *Survey of People's Environmental Awareness*, Beijing: China Environmental Protection Fund.
Renmin Ribao [*People's Daily*] (1965) 2 December.
Rex, J.A. (1968) "The Sociology of a Zone of Transition" in R.E. Pahl (ed.) *Readings in Urban Sociology*, Oxford: Pergamon Press.
Risse-Kappen, T. (1998) "Bringing Transnational Relations Back In" in T. Risse-Kappen (ed.) *Bring Transnational Relations Back In*, Cambridge: Cambridge University Press.
Rock, M.T. (2002) "Integrating Environmental and Economic Policymaking in China and Taiwan," *American Behavioural Scientist*, 45–9: 1435–55.

Rodan, G. (2001) "The Prospects for Civil Society and Political Space" in Amitav Acharya, B.M. Frolic, and R, Stubbs (eds) *Democracy, Human Rights and Civil Society in South East Asia*, Toronto: Centre for Asia Pacific Studies.
Rohrschneiger, R. and Dalton R.J. (2002) "A Global Network? Transnational Cooperation Among Environmental Groups," *Journal of Politics*, 64–2: 510–33.
Rootes, C.A. (1997) "Environmental Movements and Green Parties in Western and Eastern Europe" in M. Redclift and G. Woodgate (eds) *The International Handbook of Environmental Sociology*, Cheltenham, UK: Edward Elgar.
Rosemarin, A. (ed.) (2002) *China Human Development Report 2002: Making Green Development a Choice*, Oxford: Oxford University Press.
Rosenthal, E (1999) "Suicides Reveal Bitter Roots of China's Rural Life," *New York Times*, 24 January.
Ross, L. (1988) *Environmental Policy in China*, Bloomington, IN: Indiana University Press.
Ross, L. and Silk, M.A. (1987) *Environmental Law and Policy in the People's Republic of China*, New York: Quorum Books.
Rowe, W.T. (1993) "The Problem of 'Civil Society' in Late Imperial China," *Modern China*, 19–2: 143–48.
Ru, J. (2004) "Environmental NGOs in China: The Interplay of State Controls, Agency Interests and NGO Strategies," PhD dissertation, Stanford University.
Ruttan, V.W. (1971) "Technology and the Environment," *Journal of Agricultural Economics*, 53: 707–17.
Saich, T. (1994) "The Search for Civil Society and Democracy in China," *Current History*, September: 260–4.
—— (2000) "Negotiating the State: The Development of Social Organizations in China," *The China Quarterly*, 161: 124–41.
Salinger, D. (1999) *Peasant, Migrants, and Protest Contesting Citizenship in Urban China: Peasant, Migrants, the State, and the Logic of the Market*, Stanford, CA: Studies of the East Asian Institute.
Schaller, G.B. (1993) *The Last Panda*, Chicago: University of Chicago Press.
Schmitter, P.C. (1974) "Still a Century of Corporatism?" in F.B. Pike and T. Stritch (eds) *The New Corporatism: Social-Political Structures in the Iberian World*, Notre Dame: University of Notre Dame Press.
Schreurs, M. (2002) *Environmental Politics in Japan, Germany, and the United States*, Cambridge: Cambridge University Press.
Schroeder, L., Sjoquist, D. and Stephan, P. (1986) *Understanding Regression Analysis: An Introductory Guide*, Newbury Park, CA: Sage.
Schwartz, J. (2004) "Environmental NGOs in China: Roles and Limits," *Pacific Affairs*, 77–1: 28–49.
Scott, D. and Willits, F.K. (1994) "Environmental Attitudes and Behavior: A Pennsylvania Survey," *Environment and Behavior*, 26–2: 239–60.
Scott, J.C. (1985) *Weapons of the Weak: The Everyday Forms of Peasant Resistance*, New Haven: Yale University Press.
—— (1990) *Domination and the Arts of Resistance: Hidden Transcripts*, New Haven: Yale University Press.
Seiwert, H. (2001) "China's Repression of Folk Religions: The Battle against Falun Gong in Historical Context," *Neue Zurcher Zeitung* (Zurich), 13 July. Online, available at www.nzz.ch/english/background/2001/07/13_china.html (accessed 14 July 2001).
Selden, M. (1998) "Household, Cooperative, and State in the Remaking of China's Countryside" in E.B. Vermeer, F.N. Pieke and W.L. Chong (eds) *Cooperative and Collect-*

ive in China's Rural Development: Between State and Private Interests, Armonk, NY: M.E. Sharpe.

Shapiro, J. (2001) *Mao's War against Nature: Politics and the Environment in Revolutionary China*, Cambridge: Cambridge University Press.

Shi, J.T. (,2005) "Watchdog clears the way for power plant construction to restart," *South China Morning Post*, 17 February. Online, available at http://www.scmp.com.

Shi, T. (1997) *Political Participation in Beijing*, Cambridge, MA: Harvard College.

Shue, V. (1988) *The Reach of the State: Sketches of the Chinese Body Politics*, Stanford: Stanford University Press.

Simonov, E. "Amur/Heilong River Basin Reader," unpublished book manuscript.

Singh, K. (2001) "Handing Over the Stick: The Global Spread of Participatory Approaches to Development" in M. Edwards and J. Gaventa (eds) *Global Citizen Action*, London: Earthscan.

Siqing Gongzuozu [Four-cleanup Work Team] (1966) "*Wushenzhao Gongshe Siqing yundong zongjie*" [Uxin Ju Commune's Four-cleanup Summary], in "Collected material of Four-cleanup," unpublished material.

Skocpol, T. (1979) *States and Social Revolutions: A Comparative Analysis of France, Russia and China*, New York: Cambridge University Press.

Smelser, N.J. (1963) *Theory of Collective Behaviour*, New York: Free Press.

Smil, V. (1984) *The Bad Earth*, Armonk, NY: M. E. Sharpe.

—— (1993) *China's Environmental Crisis*, Armonk, NY: M. E. Sharpe.

Sneath, D. (1994) "The Impact of the Cultural Revolution in China on the Mongolians of Inner Mongolia," *Modern Asian Studies*, 28–2: 409–30.

Snow, D. and Benford R. (1988) "Ideology, Frame Resonance and Participant Mobilization," in B. Klanddermans, H. Kriesi and S. Tarrow (eds) *From Structure to Action: Comparing Social Movements Across Cultures*, Greenwich, CT: JAI Press.

—— (1992) "Master Frames and Cycles of Protest," in A. Morris and C. McClurg Mueller (eds) *Frontiers in Social Movement Theory*, New Haven, CT: Yale University Press.

Sonnenfeld, D.A. (2002) "Social Movements and Ecological Modernization: The Transformation of Pulp and Paper Manufacturing," *Development and Change*, 33: 1–27.

Stalley, P. and Yang, D. (2006) "An Emerging Environmental Movement in China?" *The China Quarterly*, 186: 333–56.

State Council (2001) *White Paper on Poverty Reduction in Rural Areas*, Beijing: State Council.

State Council Legal Affairs Office (1999) *Shehui Tuanti Dengji Guanli Tiaoli Shiyi* [Interpretation of the Administrative Regulations on the Registration of Social Organizations], Beijing: Zhongguo Shehui Chubanshe.

State Environmental Protection Agency (ed.) (1999) *Quanguo Gongzhong Huanjing Yishi Diaocha Baogao* [Research Report on Public Environmental Awareness in China], Beijing: Zhongguo Huanjing Kexue Chubanshe.

Sun, J. (1990) "*Eerduosi gaoyuan shengtai huanjing zhengzhi de zhanlue yanjiu*" [Strategic Research on the Improvement of Ecological Environment in the Ordos Plateau], *Ganhanqu Ziyuan yu Huanjing* [Journal of Arid Land Resources and Environment], 4–4: 45–51.

Sun, L., Jin, J. and He, J. (1999) "*Yi Shehui hua de Fangshi Chongzu Shehui ziyuan – dui xiwang gongcheng ziyuan dongyuan guocheng de Yanjiu*" [Re-organizing social resources through socialization – study on resource mobilization process of the China Project Hope], *International Conference on the Development of Nonprofit Organizations and the China Project Hope*, Beijing, 29 October – 1 November.

Sun, W. (1992) *Zhonghua Renmin Gongheguo Xingzheng Guanli Dacidian* [People's Republic of China Administrative Management Encyclopedia], Beijing: Renmin Ribao Chubanshe.

Sun, X. (2004) "Meeting Chinese Demand for Forest Products," a report of Forest Trend.

Sun, X, Cheng N., White A, West, R.A. and Katsigris, E. (2004) "China's Forest Product Import Trends 1997–2002: Analysis of Customs Data with Emphasis on Asia-Pacific Supplying Countries," a report of Forest Trend.

Sun, X., Katsigris, E. and White, A. (2004) "Meeting China's Demand for Forest Products: An Overview of Import Trend, Ports of Entry, and Supplying Countries, with Emphasis on the Asia-Pacific Region," a report of Forest Trend.

Taiyuan (1995) *Taiyuan Nianjian, 1994* [Taiyuan Yearbook, 1994], Taiyuan: Shanxi Renmin Chubanshe.

Tang, S. and Tang, C. (1999) "Democratization and the Environment: Entrepreneurial Politics and Interest Representation in Taiwan," *The China Quarterly*, 158: 350–66.

Tarrow, S. (1992) "Mentalities, Political Cultures and Collective Action Frames: Constructing Meanings through Action," in A. Morris and C. McClurg Mueller (eds) *Frontiers in Social Movement Theory*, New Haven, CT: Yale University Press.

—— (1996) "States and Opportunities: The Political Structuring of Social Movements" in D. McAdam, J.D. McCarthy and M.N. Zald (eds) *Comparative Perspectives on Social Movements. Political Opportunities, Mobilizing Structures and Cultural Framings*, Cambridge: Cambridge University Press.

—— (1998) *Power in Movement: Social Movements and Contentious Politics*, Cambridge: Cambridge University Pres).

—— (2001) "Silence and Voice in the Study of Contentious Politics: Introduction" in D. McAdam, S. Tarrow and C. Tilly (eds) *Dynamics of Contention*, New York: Cambridge University Press.

Taylor, B.R. (ed.) (1995) *Ecological Resistance Movements: The Global Emergence of Radical and Popular Environmentalism*, Albany: State University of New York Press.

Thompson, P. and Strohm, L.A. (1996) "Trade and Environmental Quality: A Review of the Evidence," *Journal of Environment & Development*, 5–4: 363–83.

Thornton, P.M. (2002) "Framing Dissent in Contemporary China: Irony, Ambiguity and Metonymy," *The China Quarterly*, 171: 661–81.

Tian, M. (2003) "Collapse of the Intermediate Stratum Under Unclear Rules" [*Qian guize xia de zhongjian ceng taxian*], *CEO and CIO in Information Time*, 11.

Tilly, C. (1978) *From Mobilization to Revolution*, Reading, MA: Addison-Wesley.

—— (1984) "Social Movements and National Policies" in C. Bright and S.Harding (eds) *State-Making and Social Movements*, Ann Arbor: University of Michigan Press.

Tocqueville, A. de (1954) *The Old Regime and the French Revolution*, trans. Stuart Gilbert, Garden City, NY: Doubleday Anchor.

Tomba, L. (2005) "Residential Space and Collective Interest Formation in Beijing's Housing Disputes," *The China Quarterly*, 184: 934–51.

Tong, J. (2002) "An Organizational Analysis of the *Falun Gong*: Structure, Communications, Financing," *The China Quarterly*, 171: 636–60.

Torgerson, D. (1994) "Strategy and Ideology in Environmentalism: A Decentered Approach to Sustainability," *Industrial and Environmental Crisis Quarterly*, 8: 295–321.

—— (2000) "Farewell to the Green Movement? Political Action and the Green Public Sphere," *Environmental Politics*, 9–4: 1–19.

Touraine, A. (1981) *The Voice and the Eye: An Analysis of Social Movements*, New York: Cambridge University Press.

—— (1984) "Social Movements: Special Area or Central Problem in Sociological Analysis?" *Thesis Eleven*, 9: 5–15.

Townsend, J.R. (1967) *Political Participation in Communist China*, Berkeley, CA: University of California Press.

Tremblay, J.-F. (2005) "Tempers Flare in China," *Chemical & Engineering News*, 83–39: 21–28; online, available at http://pubs.acs.org/cen/coverstory/83/8339china.html.

Trezzini, B. (2000) "Embedded State Autonomy and Legitimacy: Piecing together the Malaysian Development Puzzle," *Economy and Society*, 30–3: 324–53.

Tseren, P B. (1996) "Traditional Pastoral Practice of the Oirat Mongols and Their Relationship with the Environment," in C. Humphrey and D. Sneath (eds) *Culture and Environment in Inner Asia Vol. 2*, Cambridge, UK: White Horse Press.

Tumen and Zhu, D. (1995) *Kang Sheng yu "neirendang" Yuanan* [Kang Sheng and the "Inner Mongolia People's Party" Tragedy], Beijing: Zhonggong Zhongyang Dangxiao Press.

Turner, J. (1997) "Authority Flowing Downwards? Local Government Entrepreneurship in the Chinese Water Sector," PhD dissertation, Indiana University.

—— (2003) "Clearing the Air: Human Rights and the Legal Dimensions of China's Environmental Dilemma," China Environment Forum, Woodrow Wilson International Center for Scholars, 27 January.

—— (2003) "Cultivating Environmental NGO–Business Partnerships," *The China Business Review*, November–December.

Turner, J. and Wu, F. (2001) "Development of Environmental NGOs in Mainland China, Taiwan, and Hong Kong," *Green NGO and Environmental Journalist Forum*, Washington, DC: Woodrow Wilson Center.

Tvedt, T. (2002) "Development NGOs: Actors in a Global Civil Society or in a New International Social System?" *Voluntas*, 13–4: 363–75.

Tysiachniouk, M. and Reisman, J. (2002) "Transnational Environmental Organizations and the Russian Forest Sector" in J. Kortelainen (ed.) *Environmental Transformations in the Russian Forest Industry*, Joensuu: University of Joensuu, Publications of Karelian Institute. No. 136.

Ulanhu (1990) *Wulanfu lun muqu gongzuo* [Ulanhu's Talks on Pastoral Work]. compiled by Temuer, Hohhot: Neimenggu Renmin Chubanshe.

Unger, J. (1996) "'Bridge': Private Business, the Chinese Government and the Rise of New Associations," *The China Quarterly*, 47: 795–819.

—— (2000) "Power, Patronage, and Protest in Rural China," in T. White (ed.) *China Briefing 2000: The Continuing Transformation*, Armonk, NY and London: M.E. Sharpe.

Unger, J. and Chan, A. (1995) "China, Corporatism, and the East Asian Model," *Australian Journal of Chinese Affairs*, 33: 29–53.

—— (1996) "Corporatism in China: A Developmental State in an East Asian Context," in B.L. McCormick and J. Unger (eds) *China after Socialism: In the Footsteps of Eastern Europe or East Asia*, Armonk, NY: M.E. Sharpe.

Upham, F. (1987) *Law and Social Change in Postwar Japan*, Cambridge, MA: Harvard University Press.

US Congress, Congressional-Executive Commission on China (2005) *Environmental NGOs in China: Encouraging Action and Addressing Public Grievances*, roundtable before the Congressional-Executive Commission on China, 109th Congress, First

Session, 7 February; online, available at http://frwebgate.access.gpo.gov/cgi-bin/getdoc.cgi?dbname=109_house_hearings&docid=f:20182.pdf (accessed June 2005).
van Bergeijk, P.A.G. (1991) "International Trade and the Environmental Challenge," *Journal of World Trade*, 25–6: 105–15.
van Rooij, B. (2006) "Implementation of Chinese Environmental Law: Regular Enforcement and Political Campaigns," *Development and Change*, 37–1: 57–74.
Vandergert, P. and Newell J. (2003) "An Analysis of Illegal Logging and Trade in the Russian Far East and Siberia," *International Forestry Review*, 5–3: 303–6.
Verba, S. and Nie, N.H. (1972) *Participation in America: Social Equality and Political Democracy*, New York: Harper and Row.
Verba, S., Nie, N.H. and Kim, J. (1978) *Participation and Political Equality*, Cambridge: Cambridge University Press.
Wakeman, F., Jr (1993) "The Civil Society and Public Sphere Debate: Western Reflections on Chinese Political Culture," *Modern China*, 19–2: 108–38.
Walder, A.G. and Yang, S. (2003) "The Cultural Revolution in the Countryside: Scope, Timing and Human Impact," *The China Quarterly*, 173: 74–99.
Walter, I. (1982) "International Economic Repercussions of Environmental Policy: An Economist's Perspective" in S. Rubin (ed.) *Environment and Trade*, Totowa, NJ: Allanheld, Osmum.
Wang, C. (1997) *Huanjing Faxue Jiaocheng* [Lectures on Environmental Law], Beijing: Zhongguo Zhengfa Daxue Chubanshe.
Wang, C.F., Xu, X.K., Hu, J., Liu, M., Terao, T. and Otsuka, K. (eds) (2001) *Studies on Environmental Pollution Disputes in East Asia: Cases from Mainland China and Taiwan*, Tokyo: Institute of Developing Economies.
Wang, H., Mamingi, N., Laplante, B. and Dasgupta, S. (2002) *Incomplete Enforcement of Pollution Regulation: Bargaining Power of Chinese Factories*, Washington, DC: World Bank Policy Research Working Paper.
Wang, J.C.F. (1989) *Contemporary Chinese Politics: An Introduction*, 3rd edn, Englewood Cliffs, NJ: Prentice-Hall.
Wang, L. (2005) *Lü Meiti: Zhongguo Huanbao Chuanmei Yanjiu* [Green Media: Environmental Communication in China], Beijing: Tsinghua University Press.
Wang, M. (ed.) (1999) *Huanjing Jiufen Fangfan Yu Chuli Shiwu Quanshu* [Environmental Disputes: A Practical Guide for Prevention and Management], Beijing: Zhongguo Yanshi Chubanshe.
—— (2000) "Overview of China's NGOs," in Wang, M. (ed.) *Zhongguo NGO Yanjiu* [Research on NGOs in China], Beijing: UNCRD Research Report Series no. 38.
Wang, Ming, (2001) "NGO and its Cooperative Relationship with Government and Business in China," *Working Paper Series*, School of Public Policy and Management, Tsinghua University, NGO-019.
Wang, Ming, Liu, G. and He, J. (2001) *Zhongguo shetuan gaige – congzhen fuxuan zhedao shetuan xuanzhe* [Chinese Society Reform: From Government Choice to Social Choice], Beijing: Shehui Kexue Wenxian Chubanshe.
Wang, Shaoguang and He, J. (2004) "Associational Revolution in China: Mapping the Landscapes," *Korea Observer*, 35–3: 485–534.
Wang, Shuntong, Shen, Q. and Gao, Z. (1994) *Zhongguo Kexue Jishu Xiehui* [China Association of Science and Technology], Beijing: Xiandai Zhongguo Chubanshe.
Wang, Z. (1998) "Village Committees: The Basis for China's Democratization," in E.B. Vermeer, F.N. Pieke and W.L. Chong (eds) *Cooperative and Development in China's Rural Development: Between State and Private Interests*, Armonk, NY: M.E. Sharpe.

Wapner, P. (1995) "Politics Beyond the State: Environmental Activism and World Civic Politics," *World Politics*, 47–3: 311–40.

—— (2002) "Horizontal Politics: Transnational Environmental Activism and Global Cultural Change," *Global Environmental Politics*, 2: 37–62.

Warner, M. (2002) "Publics and Counterpublics," *Public Culture*, 14–1: 49–90.

Warwick, M.K. (2003) "Environmental Information Collection and Enforcement at Small-Scale Enterprises in Shanghai: The Role of the Bureaucracy, Legislatures and Citizens," PhD dissertation, Stanford University.

Wasserstrom, J. and Perry, E. (1994) *Popular Protest and Political Culture in Modern China*, Boulder, CO: Westview Press.

Weidner, H. (1989) "An Administrative Compensation System for Pollution-Related Health Damages" in S. Tsuru and H. Weidner (eds) *Environmental Policy in Japan*, Berlin, Germany: Edition Sigma Rainer Bohn.

Weiner, D.R. (1999) *A Little Corner of Freedom: Russian Nature Protection from Stalin to Gorbachev*, Berkeley: University of California Press.

Weinthal, E. and Jones Luong, P. (2002) "Environmental NGOs in Kazakhstan: Democratic Goals and Nondemocratic Outcomes" in S.E. Mendelson and J.K. Glenn (eds) *The Power and Limits of NGOs: A Critical Look at Building Democracy in Eastern Europe and Eurasia*, New York: Columbia University Press.

Weller, R.P. (1999) *Alternate Civilities: Democracy and Culture in China and Taiwan*, Boulder, CO: Westview Press.

Weller, R.P. and Hsiao, M. (1998) "Culture, Gender and Community in Taiwan's Environmental Movement" in A. Kalland and G. Persoon (eds) *Environmental Movements in Asia*, Richmond: Curzon Press.

White, G. (1993) "Perspective for Civil Society in China: A Case Study of Xiaoshan City," *Australian Journal of Chinese Affairs*, 29: 63–87.

—— (1993) *Riding the Tiger: The Politics of Economic Reform in Post-Mao China*, London: Macmillan.

White, G., Howell J. and Shang, X. (1996) *In Search of Civil Society: Market Reform and Social Change in Contemporary China*, Oxford: Clarendon Press.

Whyte, M.K. (1991) "Urban life in the People's Republic" in R. MacFarquhar and J.K. Fairbank (eds) *The Cambridge History of China: The People's Republic Part 2, Revolutions within the Chinese Revolution 1966–1982*, Cambridge: Cambridge University Press

—— (1992) "Urban China: A Civil Society in the Making?" in A.L. Rosenbaum (ed.) *State and Society in China: The Consequences of Reform*, Boulder: Westview Press.

Williams, J.F. and Chang, C. (1994) "Paying the Price of Economic Development in Taiwan: Environmental Degradation" in M.A. Rubinstein (ed.) *The Other Taiwan*, Armonk, NY: M.E. Sharpe.

Wong, K.-K. (2003) "The Environmental Awareness of University Students in Beijing, China," *Journal of Contemporary* China, 12–36: 519–36.

Woody, W. (1993) *The Cultural Revolution in Inner Mongolia*, Center for Pacific Asia Studies at Stockholm University Occasional Paper 20.

Wright, T. (1999) "State Repression and Student Protest in Contemporary China," *The China Quarterly*, 157: 142–72.

—— (2002) "The China Democracy Party and the Politics of Protest in the 1980s–1990s," *The China Quarterly*, 172: 906–26.

Wu, F. (2002) "New Partners or Old Brothers? GONGOs in Transnational Environmental Advocacy in China," *China Environment Series*, 5: 45–58.

Wu, G. (2000) "One Head, Many Mouths: Diversifying Press Structures in Reform China" in C.-C. Lee (ed.) *Power, Money, and Media: Communication Patterns and Bureaucratic Control in Cultural China*, Evanston, IL: Northwestern University Press.

Wushen Qi [Uxin Banner] (1998) "*Wushenqi zhi*" [Gazetteer of Uxin Banner], unpublished material.

Wushenzhao Dangwei [Uxin Ju Party Committee) (1972) (Title unknown) unpublished materials.

—— (1974) "*Women shi zenyang ba pi Lin pi Kong douzheng buduan yinxiang shenru de*" [How did we Keep Deepening the "Criticizing Lin and Confucius" Movement?], in Collection 7, unpublished material.

Wushenzhao Geweihui [Uxin Ju Revolutionary Committee] (1968) "*Women shi zenyang daban xuexiban, kaizhan geming de dapipan, zhuageming, cushengchan de*" [How have we Promoted Mao Study Meetings, Conducted Revolutionary Criticisms, and used Revolution to Lead Production?], unpublished material.

Wushenzhao Jiedaizhan [Uxin Ju Reception Station] (1974) "*Wushenzhao gaikuang jieshao cailiao*" [General Introduction of Uxin Ju], in Collection 20, unpublished material.

"*Wushenzhao tongji*" [Uxin Ju statistics] (2001) unpublished material.

"*Wushenzhao tuchu zhengzhi jiangshe caoyuan*" [Uxin Ju Emphasizes Politics in Grassland Construction] (1966) in Collection 4, unpublished material.

Xi, X. and Xu, Qi. (1998) *Zhongguo Gongzhong Huanjing Yishi Diaocha* [Surveys of Chinese Society's Environmental Awareness], Beijing: Zhongguo Huanjing Kexue Chubanshe.

Xinhua News Agency (2005) "Building Harmonious Society Crucial for China's Progress," *People's Daily*, 27 June. Online, available at http://english.people.com.cn/200506/27/eng20050627_192495.htm (accessed 27 June 2005).

Xinhuanet (2006) "*Huanbao minjian zuzhi shuliang he renshu jiang yi 10% zhi 15% sudu dizeng*" [Number and Staff of Environmental NGOs to Increase by 10 Percent to 15 Percent], 28 October. Online, available at http://news.xinhuanet.com/environment/2006–10/28/content_5261309.htm (accessed 29 October 2006).

Xiuzhi Bianshi Weiyuanhui [Gazetteer Compilation Committee, Inner Mongolia] (2000) *Neimenggu xumuye fazhangshi* [Development History of Animal Husbandry in Inner Mongolia], Hohhot: Neimenggu Renmin Chubanshe.

Yamane, M. and Lu, W. (2000) "The Recent Russia–China Timber Trade – An Analytical Overview," interim report of Institute for Global Environment Strategies Forest Conservation Project.

Yan, X. (2003) *Anhui qinyangxian nongming Zhu Zhengliang Tiananmen zifen zhengxiang* [The Fact of the Self-immolation in Tiananmen of Zhu Zhengliang, a Peasant from Qingyang County in Anhui Province]. Online, available at http://news.sina.com.cn/c/2003–09–18/1337775762s.shtml.

Yan, Y. (1996) *The Flow of Gifts: Reciprocity and Social Networks in a Chinese Village*, Stanford: Stanford University Press

Yang, G. (2002) "Civil Society in China: A Dynamic Field of Study," *China Review International*, 9–1: 1–16.

—— (2005) "Environmental NGOs and Institutional Dynamics in China," *The China Quarterly*, 181: 46–66.

—— (2006) "Between Control and Contention: China's New Internet Politics," *Washington Journal of Modern China*, 8–1: 30–47.

—— (2006) "Activists Beyond Virtual Borders: Internet–Mediated Networks and Informational Politics in China," *First Monday*, 11–9.

Yang, M.M.-H. (1994) *Gifts, Favors and Banquets: The Art of Social Relationships in China*, Ithaca: Cornell University Press.

—— (2004) "Spatial Struggles: Postcolonial Complex, State Disenchantment, and Popular Reappropriation of Space in Rural Southeast China," *Journal of Asian Studies*, 63–3: 719–55.

Yanitsky, O. (1996) *Ecological Movement in Russia*. Moscow: RAS.

—— (2000) *Russian Greens in a Risk Society*. Helsinki: Kikimora Publications.

—— (2002) *Rossia: Ekologicheskii vizov* [Russia: Environmental Challenge]. Novosibirsk: Sibirski Chronograph.

Yardley, J. (2004) "Beijing Suspends Plan for Large Dam," *International Herald Tribune*, 8 April. Online, available at www.irn.org/programs/nujiang/index.asp?id=041304_ihtnyt.html.

—— (2004) "Dam Building Threatens China's 'Grand Canyon'," *New York Times*, 10 March.

Yikezhao Meng Difangzhi Bianzuan Weiyuanhui [Ih-Ju Gazetteer Compilation Committee] (1994) *Yikezhao Mengzhi* [Ih-Ju Gazetteer], Vol. 2, Beijing: Xiandai Chubanshe.

Yin C. (2002) *Women and Energy: A Story in Baima Snow Mountain Reserve*, Beijing: South North Institute for Sustainable Development.

Young, N. (2001a) "250 Chinese NGOs; Civil Society in the Making: A Special Report for China Development Brief," *China Development Brief*, August, special edn.

—— (2001b) "Green Groups Explore the Boundaries of Advocacy," *China Development Brief*, 4–1.

—— (2003) "Constructive Engagement is the Only Option for Nature," *China Development Brief*, 4–2: 9–19.

Young, O.R., Demko, G.J. and Ramakrishna, K. (1991) "Global Environmental Change and International Governance," summary and recommendations of a conference held at Dartmouth College, Hanover, NH, June.

Yu, H. (2001) "*Zhuanxing qi hangye xiehui de fasheng zhuangkuang*" [Development of Industrial Association in Transitioning Period], *China Economic Times*, 16 March.

Zald, M.N. (1992) "Looking Backward to Look Forward: Reflections on the Past and Future of the Resource Mobilization Research Program" in A. Morris and C. Mueller (eds) *Frontiers in Social Movement Theory*, New Haven, CT: Yale University Press.

Zhang, H. and Ferris, R.J., Jr (1998) "Shaping an Environmental Protection Regime for the New Century: Environmental Law and Policy in the People's Republic of China," *Asian Journal of Environmental Management*, 6–1: 35–58.

Zhang Ye (2003) "China's Emerging Civil Society," CNAPS Working Paper, The Brookings Institution, August.

Zhao, D. (2001) *The Power of Tiananmen: State–Society Relations and the 1989 Beijing Student Movement*, Chicago: University of Chicago Press.

Zhao, J.H. and Ho, P. (2005) "A Developmental Risk Society? Genetically Modified Organisms (GMOs) in China," *International Journal for Environment and Sustainable Development*, 4–4: 370–94.

Zhao, L. and Irving, C. I. (eds) (2001) *The Non-profit Sector and Development: The Proceedings of the International Conference in Beijing in July 1999*, Hong Kong: Hong Kong Press for Social Sciences.

Zhao, X. (1999) *Fazhanzhong de Huanjing Baohu Shehui Tuanti* [Environmental Social Organizations in Development], Beijing: Tsinghua University NGO Research Center Book Series.

Zhao, Y. (2000) "From Commercialization to Conglomeration: The Transformation of

the Chinese Press within the Orbit of the Party State," *Journal of Communication*, 50–2: 3–26.
Zheng, S. (2005) "The New Era in Chinese Elite Politics," *Issues and Studies*, 41–1: 190–203.
Zhong, Xin (2000) "*Zhizhi xuanchuan fengjian mixin he wei kexue, qigong shetuan jiang bei qingli*" [Qigong Social Organizations will be Cleaned Up to Prevent the Public from Feudalistic Superstition and Pseudoscience], *Fuzhou Wanbao* [Fuzhou Evening News], 21 January.
Zhong, Xueping, Wang Zheng, and Bai Di (2001) *Some of Us: Chinese Women Growing Up in the Mao Era*, New Brunswick, NJ: Rutgers University Press.
Zhongguo Falü Nianjian Bianji Weiyuanhui (ed.) (1997) *Zhongguo Falü Nianjian 1997* [1997 China Law Yearbook], Beijing: Zhongguo Falü Nianjian Chubanshe.
Zhongguo Huanjing Nianjian Bianji Weiyuanhui (ed.) (1990) *Zhongguo Huanjing Nianjian, 1990* [China Environment Yearbook 1990], Beijing: Zhongguo Huanjing Kexue Chubanshe.
—— (1991) *Zhongguo Huanjing Nianjian, 1991* [China Environment Yearbook 1991], Beijing: Zhongguo Huanjing Kexue Chubanshe.
—— (1992) *Zhongguo Huanjing Nianjian, 1992* [China Environment Yearbook 1992], Beijing: Zhongguo Huanjing Kexue Chubanshe.
—— (1993) *Zhongguo Huanjing Nianjian, 1993* [China Environment Yearbook 1993], Beijing: Zhongguo Huanjing Kexue Chubanshe.
—— (1994) *Zhongguo Huanjing Nianjian, 1994* [China Environment Yearbook 1994], Beijing: Zhongguo Huanjing Kexue Chubanshe.
—— (1995) *Zhongguo Huanjing Nianjian, 1995* [China Environment Yearbook 1995], Beijing: Zhongguo Huanjing Kexue Chubanshe.
—— (1996) *Zhongguo Huanjing Nianjian, 1996* [China Environment Yearbook 1996], Beijing: Zhongguo Huanjing Kexue Chubanshe.
—— (1997) *Zhongguo Huanjing Nianjian, 1997* [China Environment Yearbook 1997], Beijing: Zhongguo Huanjing Kexue Chubanshe.
—— (1998) *Zhongguo Huanjing Nianjian, 1998* [China Environment Yearbook 1998], Beijing: Zhongguo Huanjing Kexue Chubanshe.
—— (1999) *Zhongguo Huanjing Nianjian, 1999* [China Environment Yearbook 1999], Beijing: Zhongguo Huanjing Kexue Chubanshe.
—— (2000) *Zhongguo Huanjing Nianjian, 2000* [China Environment Yearbook 2000], Beijing: Zhongguo Huanjing Kexue Chubanshe.
—— (2001) *Zhongguo Huanjing Nianjian, 2001* [China Environment Yearbook 2001], Beijing: Zhongguo Huanjing Kexue Chubanshe.
—— (2002) *Zhongguo Huanjing Nianjian, 2002* [China Environment Yearbook 2002], Beijing: Zhongguo Huanjing Kexue Chubanshe.
—— (2003) *Zhongguo Huanjing Nianjian, 2003* [China Environment Yearbook 2003], Beijing: Zhongguo Huanjing Kexue Chubanshe.
—— (2004) *Zhongguo Huanjing Nianjian, 2004* [China Environment Yearbook 2004], Beijing: Zhongguo Huanjing Kexue Chubanshe.
—— (2005) *Zhongguo Huanjing Nianjian, 2004* [China Environment Yearbook 2005], Beijing: Zhongguo Huanjing Kexue Chubanshe.
Zhou W. (2000) "*Zoujin xianshi, zaisi huanbao*" [Up Close to Reality, Rethinking Environmental Protection], in Green Camp (ed.) *Ba qian li lu yun he yue: 2000 Zhongguo daxuesheng lüseying wenji* [Eight Thousand Li of Road, Clouds, and Moonlight: Collected Works of Chinese University Students" green camp in 2000], Lüseying.

Zhou, Y. (2005) "Living on the Cyber Border," *Current Anthropology*, 46: 779–803.

Zhu, J. (2002) "*Guo yu jia zhijian: Shanghai linli de shimin tuanti yu shequ yundong de minzuzhi*" [Between the Family and the State: an Ethnography of the Civil Associations and Community Movements in a Shanghai *lilong* Neighborhood], PhD thesis, Department of Anthropology, Chinese University of Hong Kong.

Ziegler, C. (1991) "Environmental Politics and Policy under Perestroika" in J.B. Sedaitis and J. Butterfield (eds) *Perestroika from Below: Social Movements in the Soviet Union*, Boulder, CO: Westview Press.

Ziran zhi you [Friends of Nature] (2000) "*Zhongguo baozhi de huanjing yishi*" [Survey on Environmental Reporting in Chinese Newspapers], Ziran zhi you.

Zissis, C. (2006) "Media Censorship in China," New York: Council on Foreign Relations. Online, available at www.cfr.org/publication/11515/media_censorship_in_china.html?breadcrumb=%2Fissue%2F119%2Ftechnology_and_foreign_policy.

Index

Lightning Source UK Ltd.
Milton Keynes UK
23 March 2011

169723UK00001B/26/P